3222   com sci

Reading   chap 10.9

             chap 11   1. - 7

             chap 8
             chap 9. 1, 2, 3, 7 10, 110

Tue @ 2:00 1:30   4-13

STATISTICS

SECOND EDITION

NORMA GILBERT

Department of Mathematics
Drew University

SAUNDERS COLLEGE PUBLISHING

Philadelphia   New York   Chicago
San Francisco   Montreal   Toronto
London   Sydney   Tokyo   Mexico City
Rio de Janeiro   Madrid

Address orders to:
383 Madison Avenue
New York, NY 10017

Address editorial correspondence to:
West Washington Square
Philadelphia, PA 19105

This book was set in Times Roman by York Graphics Services, Inc.
The editors were Jim Porterfield and Carol Field.
The art director and cover designer was Nancy E.J. Grossman.
The text design was done by North 7 Atelier, Ltd.
The production manager was Tom O'Connor.
Von Hoffmann was the printer.

Front cover art: Watercolor by Richard Lutzke.

Library of Congress Catalog Card No.: 80-53939

STATISTICS

ISBN 0-03-058091-1

1234   032   98765432

CBS COLLEGE PUBLISHING
Saunders College Publishing
Holt, Rinehart and Winston
The Dryden Press

PREFACE

This book is intended for students who need to understand how statistical decisions are made but who have little mathematical background (a year or two of high school algebra is sufficient). So many statistics texts aimed at this group have appeared that some explanation—other than the pleasure I've had in writing it—is needed of why another appears. Here are several reasons:

First, many formulas appear in any statistics book. I have tried to explain where a formula comes from, why it should be used, and when it should not be used. Explanations are often intuitive and informal, but a student should get a grasp of why a decision is reached rather than just learn what is reached.

Second, in some texts the notation gets pretty involved. In this one it doesn't. (The exception is Chapter 15; by then a student is well accustomed to statistics. Some instructors will prefer to omit this chapter.)

Third, the book is suited to self-paced ("Keller Plan") study as well as to the usual lecture approach: An introductory section at the beginning of each chapter explains its scope and where emphasis should be put. About 250 examples are worked out completely. Answers are given in detail to about 600 of the exercises in the belief that it does no good to know that the probability of an event, for example, is .40 if the student doesn't know how .40 is derived.

Last, some texts avoid probability, but an understanding of statistics without it is impossible. Others dive into probability immediately. Many students have trouble with this topic, however, so I have begun with descriptive statistics and delayed probability until Chapter 5; it is then used in every succeeding chapter.

Chapters 1 through 8, and parts of 9 and 10 should ordinarily be taken in sequence; material from the remainder of Chapters 9 and 10 and Chapters 11 through 17 can be selected according to the interests and speed of the students and the length of the course.

The second edition differs from the first in that (1) hundreds of new exercises have been added; many of these are statistical problems from real life; (2) the concepts behind confidence intervals and hypothesis tests are emphasized again and again; (3) notation is simplified; and (4) the errors in arithmetic have been corrected.

I am especially indebted to Dr. Peter Nemenyi who suggested many of the new and interesting problems. Special thanks are due to others who read and criticized the first edition: Katharine J. Kharas, Karen Rappaport, Donald L. Evans, Stephen B. Vardeman, Neil W. Henry, Martin D. Fraser, Michael Jacobs, and Fred Morgan.

NORMA M. GILBERT

CONTENTS

CHAPTER 12

χ^2 (CHI SQUARE) 265

CHAPTER 13

ANALYSIS OF VARIANCE 288

1
WHAT MATHEMATICAL BACKGROUND IS NEEDED?

Most of this chapter should be familiar stuff. Summation notation (Section 1.3, pages 3–4) may be new to you. You will use the remaining parts of this chapter so frequently, however, that you will be foolish indeed if you let anything slip by without understanding it. Even students who are good at mathematics are sometimes confused about whether multiplication or addition is carried out first [$4 + 2(8) = 4 + 16$ or $6(8)$?] and how parentheses are used to change the usual order.

1.1 INTRODUCTION

Many students approach statistics with apprehension, having resolved after junior year in high school never to take another mathematics course. But be of good cheer for two reasons. The background in mathematics which you need is remarkably slight. And in your high school courses you may have learned how to solve quadratic equations, for example, without understanding **why** you were learning to do this. Now those of you who are concerned about using mathematics are studying statistics because a knowledge of this material is necessary background for your study of psychology, economics, zoology, or sociology. You will find an almost immediate application of the mathematics you learn to a subject that interests you, and this delightful situation—in contrast to your past experience—will give most of you a new and splendid ability to cope with mathematics.

Try always to gain some perception of the general arguments that produce results rather than accept them blindly. An idiot can be trained (or a computer can be programmed) to plug numbers into formulas; you must learn when and why a particular formula is used. Occasionally an explanation, in order to be mathematically proper, would require that you know calculus; in these cases the explanation is purely intuitive or is omitted. Fair warning will be given when such occasions arise.

A special notation will be used to warn you of traps that have been set for you: an unusual twist, a problem that cannot be solved, or some special reason to be wary.

1.2 THE ORDER OF OPERATIONS

The word "operations" here refers to squaring, multiplication or division, and addition or subtraction. The rules agreed on by mathematicians are that these are carried out in the order given below when more than one operation is to be performed:

> (a) First square.
> (b) Then multiply or divide.
> (c) Then add or subtract.

Example 1

$2(3) + 4 = ?$

$2^2(3) + 1 = ?$

$3 - \dfrac{12}{2^2} = ?$

$2(3) + 4 = 6 + 4 = 10.$

$2^2(3) + 1 = 4(3) + 1 = 12 + 1 = 13.$

$3 - \dfrac{12}{2^2} = 3 - \dfrac{12}{4} = 3 - 3 = 0.$

But suppose you don't want to do things in that order? What if, for example, 2 and 3 are first to be added and then the result squared? Parentheses ( ), square brackets [ ], or curly brackets { } are used, and a new rule is added:

> If there are parentheses or brackets, simplify what is inside them first before continuing.

Example 2

$6(4 + 1)^2 = ?$

$[2^2(3) - 2]^2(4 - 1) = ?$

$2(3^2 + 4 \div 2) - 3(2^2 - 4)^2 = ?$

$6(4 + 1)^2 = 6(5)^2 = 6(25) = 150.$

$[2^2(3) - 2]^2(4 - 1) = [4(3) - 2]^2(3) = [12 - 2]^2 3 = 10^2(3) = 100(3) = 300.$

$2(3^2 + 4 \div 2) - 3(2^2 - 4)^2 = 2(9 + 4 \div 2) - 3(4 - 4)^2$
$\qquad = 2(9 + 2) - 3(0)^2$
$\qquad = 2(11) - 0 = 22.$

1.3 Σ NOTATION

You will soon need to learn seven Greek letters. The first of these is the Greek (capital) S, called **sigma,** and written Σ. We shall use Σ as an abbreviation for the word "sum" or for the phrase "the sum of _____."

> Notation: Σ = Sum.

If X values are 1, 2, 3, 5, then $\Sigma X = 1 + 2 + 3 + 5 = 11$. Add all the X values.

If the weights W of five students are 110, 100, 160, 200, 190 pounds, then $\Sigma W = 110 + 100 + 160 + 200 + 190 = 760$ pounds.

Remember the rule for order of operations when Σ (standing for addition) is combined with exponents, multiplication, or parentheses.

Example 1

If Y values are 4, 3, 1, 6, 2, find (a) ΣY, (b) ΣY^2, (c) $(\Sigma Y)^2$.

(a) $\Sigma Y = 4 + 3 + 1 + 6 + 2 = 16$.

(b) $\Sigma Y^2 = 4^2 + 3^2 + 1^2 + 6^2 + 2^2 = 16 + 9 + 1 + 36 + 4 = 66$.

(c) $(\Sigma Y)^2 = 16^2 = 256$.

There are three rules for working with summations:

> **Rule 1.** $\Sigma(X + Y) = \Sigma X + \Sigma Y$. (This rule only makes sense if the number of X values is the same as the number of Y values.)
> **Rule 2.** If k is a constant, $\Sigma k X = k \Sigma X$.
> **Rule 3.** If A is a constant, $\Sigma A = nA$, where n is the number of values that are being added.

Example 2

X values: 4, 1, 6, 3; corresponding Y values: 2, 3, 7, -1. (a) $\Sigma(X + Y) = ?$ (b) $\Sigma 2X = ?$ (c) $\Sigma 5 = ?$ (d) $\Sigma(X - Y) = ?$

(a) $\Sigma(X + Y) = \Sigma X + \Sigma Y = (4 + 1 + 6 + 3) + (2 + 3 + 7 - 1) = 14 + 11 = 25$.

(Rule 1)

(b) $\Sigma 2X = 8 + 2 + 12 + 6 = 28$, but also
$\qquad = 2\Sigma X = 2(4 + 1 + 6 + 3) = 2(14) = 28$.

(Rule 2, which says a constant can be factored out)

(c) $\Sigma 5 = 4(5) = 20$. (Rule 3) It is assumed that the number of values being added (here, 4) is known from the context of the problem. A more sophisticated summation notation, using subscripts, can be adopted, but we shall not need it.

$$(d) \ \Sigma(X - Y) = \Sigma[X + (-1)Y]$$
$$= \Sigma X + \Sigma(-1)Y \quad \text{(Rule 1)}$$
$$= \Sigma X + (-1)\Sigma Y \quad \text{(Rule 2)}$$
$$= \Sigma X - \Sigma Y$$
$$= (4 + 1 + 6 + 3) - (2 + 3 + 7 - 1)$$
$$= 14 - 11$$
$$= 3.$$

1.4 EXERCISES

1. $4.38 + .12 + .02 =$

2. $6.21(.01) =$

3. $5(0) =$

4. $\dfrac{27(75)}{25(9)} =$

5. $\dfrac{25 + 75}{25 + 50} =$

6. Add: $-10, 5, -3, 2, 3, -5, 2, 10$.

7. Add: $-10, -2, 5, 3, 4, 12, -8, -6$.

8. $3(5^2) + 5 =$

9. $3(5^2 + 5) =$

10. $4 - \dfrac{10^2}{20} =$

11. $3(1 - 2)^2 - 4^2(2^2 - 3) =$

12. $2^2(2^2 + 1)^2 + 3(1 + 3)^2 =$

13. X values: 2, 5, 3. Find (a) ΣX, (b) ΣX^2, (c) $(\Sigma X)^2$, (d) $\Sigma 3X$, (e) $\Sigma 2X^2$, (f) $\Sigma 4$.

14. Y values: 4, 1, 1, 3. Find (a) ΣY, (b) $\Sigma 4Y$, (c) $(\Sigma Y)^2$, (d) ΣY^2, (e) $\Sigma 3$.

15. Z values: 2, -1, 0, 6, 3. Find (a) ΣZ, (b) ΣZ^2, (c) $\Sigma 5$, (d) $(\Sigma Z)^2$, (e) $\Sigma(-Z)^2$.

16.

f	X
3	2
4	5
2	3

Find (a) ΣfX, (b) ΣfX^2, (c) $(\Sigma fX)^2$.

17. X: 2, -1, 0, 3. Find $\Sigma X^2 - \dfrac{(\Sigma X)^2}{4}$.

18.

f	Y
2	0
9	1
3	4
1	8

Find (a) ΣfY, (b) $(\Sigma fY)^2$, (c) ΣfY^2.

19. X values are 3, 5, 8, 2, 0, 3, 0; corresponding Y values are 2, 3, 1, -1, 0, 2, 4. Find (a) ΣX, (b) ΣY, (c) $\Sigma(X + Y)$, (e) $(\Sigma X)^2$.

ANSWERS

1. 4.52   **2.** .0621   **3.** 0

4. $3(3) = 9$ ("Cancel" first; not $\dfrac{2025}{225} = 9$.)

5. $\dfrac{100}{75} = \dfrac{4}{3}$ (Don't try to cancel the initial 25's.)

6. Cancel -10 and 10, -5 and 5, -3 and 3. $2 + 2 = 4$.

7. Cancel -10, -2, and 12; 5, 3, and -8. $4 - 6 = -2$; OR add the positive numbers and negative numbers separately: $24 - 26 = -2$.

8. $3(25) + 5 = 75 + 5 = 80$.   **9.** $3(25 + 5) = 3(30) = 90$.

10. $4 - \dfrac{100}{20} = 4 - 5 = -1$.   **11.** $3(-1)^2 - 16(4 - 3) = 3(1) - 16(1) = -13$.

12. $4(4 + 1)^2 + 3(4)^2 = 4(5)^2 + 3(16) = 4(25) + 48 = 100 + 48 = 148$.

13. (a) 10, (b) $4 + 25 + 9 = 38$, (c) $10^2 = 100$, (d) $3(10) = 30$, (e) $2(38) = 76$, (f) $3(4) = 12$.

14. (a) 9, (b) $4(9) = 36$, (c) $9^2 = 81$, (d) $16 + 1 + 1 + 9 = 27$, (e) $4(3) = 12$.

15. (a) 10, (b) $4 + 1 + 0 + 36 + 9 = 50$, (c) $5(5) = 25$, (d) $10^2 = 100$, (e) 50.

16. (a) (Add a ΣfX column) $6 + 20 + 6 = 32$, (b) (Add a ΣfX^2 column) $12 + 100 + 18 = 130$, (c) $32^2 = 1024$.

17. $14 - \dfrac{4^2}{4} = 10$.

18. (a) $0 + 9 + 12 + 8 = 29$, (b) $29^2 = 841$, (c) $0 + 9 + 48 + 64 = 121$.

19. (a) 21, (b) 11, (c) $21 + 11 = 32$, (d) $21 - 11 = 10$, (e) $21^2 = 441$.

1.5 FORMULAS

A number of formulas will be developed in this course; they simply give shorthand directions for carrying out operations. Study the following examples, and learn how to translate from English into mathematical language and vice versa.

Example 1

Translate into English: $\Sigma(X - 4)$.
Subtract 4 from each X value, then add the new set.

Example 2

Write a formula that gives directions to subtract 4 from each X value, square each number in the new set, and then add the squares.

$\Sigma(X - 4)^2$

Example 3   Write a formula that gives directions to subtract 4 from each X value, add the new set of values, and square the sum.

$$[\Sigma(X - 4)]^2$$

Now test yourself on these two. Remember the order of operations!

Example 4   Translate into English: $\Sigma(X + 3)$.

Example 5   Write a formula that gives directions to subtract 5 from each X value, square each number in the new set, add the squares, and divide the sum by 3.

Answers   4. Add 3 to each X value, then add the new set of scores.

5. $\dfrac{\Sigma(X - 5)^2}{3}$

If there are three X values, 3, 5, 7, then the operations you have carried out are the following:

X	Example 1 $X - 4$	Example 2 $(X - 4)^2$	Example 4 $X + 3$	Example 5	
				$X - 5$	$(X - 5)^2$
3	-1	1	6	-2	4
5	1	1	8	0	0
7	$\underline{3}$	$\underline{9}$	$\underline{10}$	2	$\underline{4}$
	3	11	24		8

Example 3: $[\Sigma(X - 4)]^2 = 3^2 = 9.$

Example 5: $\dfrac{\Sigma(X - 5)^2}{3} = \dfrac{8}{3}.$

You may be given X values and an f corresponding to each X value, as shown in the first two columns below (f will be used in later chapters for the **frequency** with which X values appear); the other four columns will be needed in Examples 6–9.

(1) X	(2) f	(3) Xf	(4) $X - 5$	(5) $(X - 5)f$	(6) $(X - 5)^2 f$
3	1	3	-2	-2	4
5	4	20	0	0	0
7	2	$\underline{14}$	2	$\underline{4}$	$\underline{8}$
		37		2	12

Example 6   Translate into English: $\Sigma Xf = 37$.

Multiply each X value by the corresponding f and add the products; the result is 37 (Column 3).

Example 7   Write a formula that gives directions to subtract 5 from each X value, multiply the new set by the corresponding f, and add the products; the sum is 2.

$$\Sigma(X - 5)f = 2.$$

Again, try two examples yourself.

Example 8   Write a formula that gives directions to subtract 5 from each X value, multiply the new set by the corresponding f, add the products, and then square the sum; the result is 4.

Example 9   Translate into English: $\Sigma(X - 5)^2 f = 12.$

Answers   8. $[\Sigma(X - 5)f]^2 = 2^2 = 4$ (Column 5).
9. Subtract 5 from each X value, square each of the numbers in the new set, multiply each square by the corresponding f, and add the products; the result is 12 (Column 6).

1.6 EXERCISES

1. Explain (in words) the difference between ΣX^2 and $(\Sigma X)^2$.

2. You are given four X values. Write a formula that says
    (a) A is the sum of all of them.
    (b) B is obtained by subtracting 3 from each of them, then adding the set of differences.
    (c) C is the sum of the squares of the X values.
    (d) D is the square of the sum of the X values.
    (e) Find A, B, C, and D if the X values are 2, -1, 3, and 1.

3. Write a formula that says, "Subtract 5 from each of 10 Y values, add the differences, and divide the sum by the number of Y values; then add 5 to the quotient. The result is labeled M."

4. Write a formula that says, "Subtract 3 from each X value, square each new number, add the new set, divide the sum by 6, and then take the square root of the sum; call the result s."

5. Translate into English: $A = \dfrac{\Sigma(X - 3)}{4}$.

6. Translate into English: $B = \dfrac{\Sigma(Y - 1.5)^2}{10}$.

7. Translate into English: $C = \dfrac{\Sigma(Z - 14.5)}{6} + 14.5$.

8. A set of 10 Y values is given. Translate into English the directions given in $D = \dfrac{\Sigma(2Y + 5)^2}{10}$.

In Exercises 9–13, use the following X and f values:

X	f
2	3
5	4
3	9

9. Translate into English: $E = \dfrac{\Sigma Xf}{\Sigma f}$.

10. Translate into English: $F = \dfrac{\Sigma(X - 4)f}{\Sigma f} + 4$.

11. Write a formula that says, "Subtract 3 from each X score, multiply each difference by the corresponding f, add the products, and divide by the sum of the f column. Add 3 to the quotient and label the result G."

12. Write a formula that says, "(a) Multiply each X by the corresponding f, add the products, square the sum, and divide by the sum of the f column; (b) Multiply the square of each X value by the corresponding f; (c) Subtract the number obtained in (a) from that obtained in (b) and divide the difference by a number that is 1 less than the sum of the f column; label the result s^2."

13. Carry out the directions given in (a) Exericse 9, (b) Exercise 10, (c) Exercise 11, (d) Exercise 12.

14. X values: 104.5, 114.5, 124.5, 134.5. Find (a) $\dfrac{\Sigma(X - 114.5)}{4} + 114.5$; (b) $\dfrac{\Sigma(X - 124.5)}{4} + 124.5$.

15. Y values: 1, 2, 4, 9. Find (a) $\bar{Y} = \dfrac{\Sigma Y}{n}$, where n is the number of Y values. Read $\bar{Y}$ as "Y bar." (b) $s^2 = \dfrac{\Sigma(Y - \bar{Y})^2}{n - 1}$.

16. Z values: 2.1, 3.2, 1.4, 5.3. Find (a) $\bar{Z} = \dfrac{\Sigma Z}{n}$ where n is the number of Z values, (b) $s^2 = \dfrac{\Sigma(Z - \bar{Z})^2}{n - 1}$.

ANSWERS

1. For ΣX^2, square each X value and then add. For $(\Sigma X)^2$, add first and then square the sum.

2. (a) $A = \Sigma X$, (b) $B = \Sigma(X - 3)$, (c) $C = \Sigma X^2$, (d) $D = (\Sigma X)^2$, (e) $A = 5$,
$B = -1 + (-4) + 0 + (-2) = -7$ or $B = \Sigma X - \Sigma 3 = 5 - 4(3) = -7$,
$C = 4 + 1 + 9 + 1 = 15$, $D = 5^2 = 25$.

3. $M = \dfrac{\Sigma(Y - 5)}{10} + 5$ **4.** $s = \sqrt{\dfrac{\Sigma(X - 3)^2}{6}}$

5. Subtract 3 from each X value and add the differences. Divide the sum by 4 and call the result A.

6. Subtract 1.5 from each Y value, square each difference, and add the squares. Divide the sum by 10 and label the result B.

7. Subtract 14.5 from each Z value, add the differences, and divide the sum by 6. Add 14.5 to the quotient, and label the result C.

8. Multiply each Y value by 2, add 5 to each product, and square each of the 10 sums. Then add all 10 squares, divide the sum by 10, and label the result D.

9. Multiply each X by the corresponding f, add the products, and divide this sum by the sum of the f column to get E.

10. Subtract 4 from each X value, multiply the difference by the corresponding f value, and sum the products. Divide this sum by the sum of the f column. Add 4 to the quotient to find F.

11. $G = \dfrac{\Sigma(X - 3)f}{\Sigma f} + 3$

12. (a) $\dfrac{(\Sigma Xf)^2}{\Sigma f}$, (b) $\Sigma X^2 f$, (c) $s^2 = \dfrac{\Sigma X^2 f - \dfrac{(\Sigma Xf)^2}{\Sigma f}}{(\Sigma f) - 1}$

13. (a) $E = \dfrac{6 + 20 + 27}{3 + 4 + 9} = \dfrac{53}{16}$, (b) $F = \dfrac{-6 + 4 - 9}{16} + 4 = \dfrac{53}{16}$,

(c) $G = \dfrac{-3 + 8 + 0}{16} + 3 = \dfrac{53}{16}$,

(d) $s^2 = \dfrac{12 + 100 + 81 - \dfrac{(6 + 20 + 27)^2}{16}}{15} = \dfrac{193 - 175.6}{15} = 1.2$

14. (a) $\dfrac{-10 + 0 + 10 + 20}{4} + 114.5 = 119.5$,

(b) $\dfrac{-20 - 10 + 0 + 10}{4} + 124.5 = 119.5$

15. (a) $\bar{Y} = \dfrac{16}{4} = 4$, (b) $s^2 = \dfrac{9 + 4 + 0 + 25}{3} = \dfrac{38}{3} = 12.7$

16. (a) $\bar{Z} = \dfrac{12.0}{4} = 3.0$, (b) $s^2 = \dfrac{(-.9)^2 + .2^2 + (-1.6)^2 + (2.3)^2}{4} = \dfrac{8.70}{3} = 2.9$

1.7 GREEK LETTERS

You have already met Σ, the Greek capital S. You will need six other Greek letters; come meet them now all at once if you like. (If you prefer, wait until you bump into them later on.) Σ is a capital letter; all the others are lower case.

Greek letter	Name	Pronounced like	English equivalent
μ	mu	a kitty's mew	m
σ	sigma	stigma minus the t	s
α	alpha	as in Alfa Romeo	a
β	beta	"abate" in reverse	b
ρ	rho	fish egg; ignore the h	r
χ	chi	kie to rhyme with tie	(German ch)

1.8 VOCABULARY AND SYMBOLS

summation Σ
$\mu, \sigma, \alpha, \beta, \rho, \chi$

1.9 REVIEW EXERCISES

In Exercises 1–12, use the following X and Y values:

X	Y
3	2
4	1
1	0
2	1
5	-3

1. $\Sigma X =$
2. $\Sigma X^2 =$

3. $(\Sigma X)^2 =$
4. $\Sigma(X - 2) =$

5. $\Sigma 3 =$
6. $\Sigma Y =$

7. $(\Sigma Y)^2 =$
8. $\Sigma Y^2 =$

9. $\Sigma(X - Y) =$
10. $\sqrt{\dfrac{\Sigma(X - 3)^2}{4}} =$ (1 decimal place)

11. $\Sigma XY =$
12. $(\Sigma X)(\Sigma Y) =$

13. Explain (in words) the difference between (a) $\Sigma(X - 2)$ and (b) $(\Sigma X) - 2$.

In Exercises 14–17 use the following values of y and f:

y	f
2	4
4	2
5	4

14. $\Sigma yf =$ (What column should be added to the table above?)

15. $\Sigma y^2 f =$ (Hint: Add column fy^2d to the table above Exercise 14.)

16. $(\Sigma yf)^2 =$

17. If $\bar{y} = 3.6$, $\Sigma(y - \bar{y})f =$ (Hint: Add columns $y - \bar{y}$ and $(y - \bar{y})f$ to the table above Exercise 14.)

18. Six quiz grades (Q) are 90, 72, 84, 85, 73, 92. Write a formula that says, "Find the average grade (A) by adding the grades on all six quizzes and dividing the sum by 6."

19. The scores (X) of 10 students on the Miller Personality Test are as follows: 22, 21, 16, 26, 23, 27, 23, 18, 23, 31.

    (a) Write a formula that says: "Add the scores of all 10 students and divide the sum by the number of students; label the result $\bar{X}$ (read 'X bar')."

    (b) Translate into a formula the following directions: "Subtract 23 from each score, square each result, sum the squares, and then divide by one less than the number of scores. Take the square root of the result and label it s."

    (c) Carry out the directions given in (b) to one decimal place.

20. The heights (H) in inches of six sophomore men are 75, 71, 74, 67, 68, 71.

    (a) Write a formula that says: "Subtract 71 from each height, add the new set of numbers, divide the sum by the number of students measured, and then add 71 to the quotient; label the sum $\bar{H}$ (read 'H bar')."

    (b) Translate into a formula the following directions: "Subtract $\bar{H}$ from each height, square each of the new values, and then find their sum; divide the sum by one less than the number of men whose height has been measured, take the square root of the quotient, and label the result s."

    (c) Carry out the directions given in (a) and in (b) to one decimal place.

21. Eight defendants have been found guilty and are sentenced to terms (T) of 6, 10, 8, 1, 3, 1, 4, and 15 years in prison, respectively.

    (a) Translate into English: $\bar{T} = \dfrac{\Sigma(T - 7)}{8} + 7$.

    (b) Translate into English: $s^2 = \dfrac{\Sigma(T - \bar{T})^2}{7}$.

    (c) Carry out the instructions given in (a) and in (b) to one decimal place.

22. Evaluate $\Sigma \dfrac{A(B - C)^2}{B}$ for the following:

A	B	C
3	2	1
8	4	1

23. In June, 1975, observations (O) are made of 14 robins, 5 wrens, 17 starlings, and 12 sparrows on a certain plot of land. Based on previous experience in other years, the expected counts (E) on this plot are 10 robins, 8 wrens, 20 starlings, and 10 sparrows.

(a) Translate into English: $X^2 = \Sigma \left[\dfrac{(O - E)^2}{E} \right]$

(b) Compute X^2 to one decimal place.

ANSWERS

1. 15

3. $15^2 = 225$

5. $5(3) = 15$ (since there are 5 scores)

7. 1

9. $\Sigma X - \Sigma Y = 15 - 1 = 14$ or $1 + 3 + 1 + 1 + 8 = 14$

11. $6 + 4 + 0 + 2 - 15 = -3$

13. (a) Subtract 2 from each X score and add the differences;
(b) Add the X scores and subtract 2 from the sum.

15. $16 + 32 + 100 = 148$

17. $-6.4 + .8 + 5.6 = 0.0$

19. (a) $\bar{X} = \dfrac{\Sigma X}{10}$ (b) $s = \sqrt{\dfrac{\Sigma(X - 23)^2}{9}}$ (c) $\bar{X} = 23.0$, $s = \sqrt{\dfrac{168}{9}} = 4.3$

21. (a) 7 is subtracted from each term and the new set is summed; the sum is divided by 8 (the number of prisoners), and 7 is added to the quotient. The result is labeled. $\bar{T}$.
(b) The number $\bar{T}$ is subtracted from each term, these differences are squared, and then the squares are added. The sum is divided by 7, and the resulting number is labeled s^2.
(c) $\bar{T} = \dfrac{-8}{8} + 7 = 6$ years.

$s^2 = \dfrac{0^2 + 4^2 + 2^2 + (-5)^2 + (-3)^2 + (-5)^2 + (-2)^2 + 9^2}{7} = \dfrac{164}{7} = 23.4.$

23. (a) Subtract the Expected from the Observed count of each bird, square each of the differences, and divide each difference by the Expected count for that bird. Then sum the quotients, and label the result X^2.

(b) $\dfrac{4^2}{10} + \dfrac{(-3)^2}{8} + \dfrac{(-3)^2}{20} + \dfrac{2^2}{10} = 3.6$

2
INTRODUCTION TO STATISTICS

An introductory course in statistics consists of two parts: descriptive statistics and statistical inference. Here you will find a brief discussion of both, so you may have some idea of where you are going. To understand statistical inference, you will need to know the difference between a population and a sample. Then there is a very brief overview of how to set up an experiment using statistics. In most such experiments, one does not take a census (that is, get data on the whole population), but rather one takes a sample and collects data on just this portion of the population. In reading about how this sample is chosen, concentrate on simple random samples.

2.1   WHAT IS STATISTICS?

The word "statistics" is used with two different meanings:

1. **Statistics are classified facts (especially numerical facts) about a particular class of objects.** You have certainly heard of accident statistics, (49,400 deaths by motor vehicles in the United States in 1977), baseball statistics (Ted Williams had a batting average of .406 in 1941), and statistics on students entering college (the average SAT score of freshmen entering Drew University this year is 568). In this definition, statistics is a plural noun: "Statistics **are** . . . ." This is the meaning of the word when used by the general public.

   The term **descriptive statistics** is used for the part of the course that deals with the presentation of numerical facts (data) in either tables or graphs, and with finding numbers that summarize data, for example, by giving the center and the spread about the center. ("My average on quizzes in Economics is 86, and my grades range from 74 to 96.")

2. **Statistics is the area of science that deals with the collection of data on a relatively small number of cases so as to form conclusions about the general case;** the conclusion you draw is called a **statistical inference.** In this meaning, statistics is a singular noun; the question at the beginning of this section ("What is statistics?") should shout loudly and clearly that this is the meaning of particular interest in this course. You are familiar with Presidential polls: 2,000 voters are asked their preference for President. On the basis of these results, claims (and remarkably accurate claims) are made about how 75,000,000 voters will cast their ballots on election day. A conclusion is

made about how a large group will react from knowledge of reactions of a small part of the group.

In general can you draw conclusions about the whole if you can gather only part of the information needed to answer a question? The answer, of course, is "yes"—otherwise, few people would study statistics—but the "yes" is qualified by many limitations. In descriptive statistics—definition 1 above—errors may occur in recording data, or in arithmetic or rounding off computations, but otherwise the results are exact. In statistical inference, however, the situation is very different. As Wallis and Roberts write in *The Nature of Statistics,* "Statistics is a body of methods for making wise decisions in the face of uncertainty."* In order to make wise decisions, you must learn what kinds of information and how much of it is needed, what kinds of conclusions can be drawn, and how accurate they are likely to be.

2.2 POPULATIONS AND SAMPLES

A main part of statistics is learning how to make wise decisions about a large group (a population) after studying detailed information about a small part of it (a sample).

> **Definitions: A population** consists of all the individuals or objects in a well-defined group about which information is desired to answer a question.
>
> A **sample** is the part of a population about which information is gathered.

"Population" as used in statistics does not necessarily refer to all the individuals in a particular community. It may be the salary of each school teacher in Monmouth County in 1980, or it may be those residents of Arkansas who are registered voters in 1981.

We shall often be interested in measures or scores of different members of a population. We might be interested in the life span of each of four different strains of mice, or the amount each family of four pays for groceries each week in 1981. But, in either case, different members of the population may have the same measures or scores. Thus, if a population consists of the IQ of five students, A,B,C,D,E, the scores these five students receive on a standardized IQ test may be 105, 120, 120, 115, and 148, respectively; two scores are the same, even though the five students are different.

A sample is a relatively small group chosen so as to represent the population. Data for the sample are secured, but it is the whole population that the statistician is interested in. Soon (see Sections 2.5 and 2.6) we'll look into how a sample should be chosen.

If you know the IQ of every seventh grader in the junior high school in Madison, do you have a population or a sample? It is impossible to answer until

---

*Wallis, W. A., and Roberts, H. V., *The Nature of Statistics* (Glencoe, Illinois: The Free Press, 1965), p. 11.

you know the question that is being asked. You have a population if the question is "What is the average IQ of seventh graders in Madison Junior High School at the present time?" because you have all the relevant information about each member of the group about which the question is raised. You have a sample, however, if the question you are trying to answer is "What is the average IQ of present seventh graders in the United States?" or "What is the average IQ of seventh graders in Madison Junior High School over the past 10 years?" What if the question is "What is the average IQ of present seventh graders in Madison Junior High School who are taking typing?" Here you have a population—you know the IQ of every member of the group about whom the question is asked—plus some irrelevant information which you would ignore, namely the IQ of seventh graders not taking typing.

Example 1   What is the population and what is the sample in the following statistical problem?

A doctor tests a new drug on 100 patients with leukemia, chosen at random. After 6 years, 58 patients are alive. What proportion of all leukemia patients will be alive after 6 years of treatment with this drug?

The sample consists of the 100 patients on whom the new drug is tested. The population (the larger group of which the sample is a part) is all leukemia patients now living whose disease has been identified; the population about which the doctor wishes to draw conclusions consists of all leukemia patients treated with the drug now **or in the future.** Often, the population which you can sample and the population about which you hope to draw conclusions will differ, but the error made by extending conclusions about the sample population to a larger group is often hard to estimate.

Example 2   "Sullivan Company orders 10,000 light bulbs from a manufacturer who claims the mean life of his bulbs is at least 1,000 hours. Sullivan Company tests 15 bulbs from the shipment. Using the results of these tests, should the shipment be accepted?" Describe the population and sample.

Here the population consists of 10,000 bulbs; the sample, of the 15 bulbs which are tested.

Example 3   "In a random sample of 1,000 Democrats, 500 favor capital punishment, whereas 300 of a random sample of 500 Republicans favor capital punishment. Find the difference in the proportions of all Democrats and Republicans who favor capital punishment." Describe the populations and samples.

Two different populations (one consisting of all Democrats, the other of all Republicans) are to be compared after studying two samples (1,000 Democrats and 500 Republicans). For the comparison to be meaningful, the terms "Democrat" and "Republican" should be carefully defined.

2.3 EXERCISES

1. Which of the following use descriptive statistics, and which use statistical inference? Look at each sentence separately.

(a) Over the past 5 years, Jack Porter's monthly medical bills have varied from nothing to $860.

(b) The average IQ of 20 students, drawn at random from the freshman class, is 121. It is estimated that the average IQ of all freshmen is between 118 and 124.

(c) 40% of the plumbers in Austin made over $19,600 last year.

(d) Bill Abrams made more than 16 baskets in each of the last 14 basketball games in which he has played.

(e) Of 200 adults questioned in Atlanta, 120 believe that drunken drivers should lose their driving licenses for at least 4 months. It is estimated that between 53% and 67% of all adults in Atlanta agree.

2. The time each Brothers College student waits in line for dinner is measured on each of the nights from February 1 through February 15. For which of the following questions do you have information about the population, and for which only about a sample (that is, for which do you have all data necessary to answer the question, and for which do you have only part of the data needed)?
What is the average waiting time in line

(a) for dinner on February 3?

(b) for dinner on Saturdays in February?

(c) for dinner during the second semester?

(d) for all meals on February 3?

(e) for dinner on February 1–7?

3. For each of the following, describe the population. If time and cost were of no concern, could you collect information about the entire population?

(a) You are trying to decide whether a coin is fair (that is, equally likely to come up heads and tails when tossed). You toss it a large number of times, counting the number of heads and of tails that come up.

(b) A doctor has a new treatment for coreopsis. He tests it on all people who have this disease in 1980.

(c) An economist is interested in the number of unemployed men over 17 in the United States who are seeking employment in January.

4. Assume that you know the income from interest and dividends of each resident of Bangor, Maine. Give examples of situations in which you have information about (a) a population, (b) a sample.
In Exercises 5–9, describe the population(s) and sample(s).

5. An economist working for the Bureau of Labor Statistics knows the percentage of workers unemployed in Newark, New Jersey, last month. This month he surveys 5,000 Newark workers to discover whether the unemployment rate has changed.

6. A political scientist is investigating the difference in the proportion of registered Republicans among well-to-do and poor voters in Nassau County, New York. He finds that among 1,000 voters whose families have incomes over $26,000 the proportion of registered Republicans is 25.3%, while this proportion is 22.0% in 1,200 families whose income is less than $12,000.

7. Four identical packets of tomato seeds are treated with four different chemicals; then each of these packets, plus a fifth packet of untreated seeds, is planted. The average yield per plant of 35 different plants from each of the five packets is determined to see whether chemical treatments affect yield.

8. It is known that 1-year-old Eskimo dogs gain in weight, on the average, 2 pounds per month. A special diet supplement, HELTHPUP, is given to 35 1-year-old Eskimo dogs, and their gains in weight are measured to find out whether HELTHPUP affects their weight.

9. The registrar at Peterson University knows that the grade point average (GPA) of married students has been .3 higher than that of unmarried students in the past 4 years. This semester he compares the average GPA of 40 married and 50 unmarried students to see if there is still the same difference.

ANSWERS

1. Descriptive: (a), (c), (d), and the first sentences of (b) and (e). Statistical inference: the second sentences of (b) and (e).

2. Sample: (b), (c), (d). Population: (a), (e).

3. (a) The population consists of the results of all tosses of the coin, and cannot be determined. (b) "Does the new treatment help all who have coreopsis now or in the future?": Since the population includes those who will suffer from this dread disease in the future, the doctor cannot now collect all necessary information. "Does the new treatment help those who now have coreopsis?": Complete information is available— if you can recognize coreopsis. (c) Population: All unemployed men over 17 in the United States seeking employment in that January. Complete information could be determined by a census.

4. (a) What is the average income from interest and dividends of residents of Bangor, Maine? (b) What is the average income from interest and dividends of residents of Maine? (It is an exceedingly poor sample, of course.)

5. Population: those in the labor force in Newark, New Jersey, this month; sample: 5,000 chosen from the Newark labor force this month.

6. Two populations: registered voters in Nassau County whose families make over $26,000 and under $12,000, respectively. He does badly, however, in characterizing these as "well-to-do" and "poor" without giving family size: a family with 14 children and an income of $26,500 may have fewer luxuries than a single person with an income of $11,000; samples: 1,000 voters from the first population and 1,200 from the second.

7. Five populations: all seeds identical to those in the packets and planted in the same soil and grown under the same conditions that have been treated with one of the chemicals or left untreated; samples: 35 plants from each population.

8. Population: all 1-year-old Eskimo dogs with HELTHPUP in their diets; sample: 35 such dogs.

9. Two populations: married and unmarried students, respectively, at Peterson University this semester; samples: 40 students from the first population and 50 from the second.

2.4 FORMULATING A STATISTICAL EXPERIMENT

Five steps are necessary:

1. You must decide very carefully what question you want answered. In particular, you must identify the population about which the question is asked.

If you plan a study of the relationship between smoking and lung cancer, for example, one of the very many questions you will have to answer is "who is a smoker?" You might consider only cigarette smokers; are cigar and pipe smokers classified with non-smokers or not included in the study at all? You would probably need to classify cigarette smokers by number of cigarettes per day and the number of years of smoking. Is a smoker a person who smokes over two packs a day? over a pack a day? or anyone who has smoked this year? Do you want to compare the incidence of lung cancer in all smokers and all non-smokers, or only between those in certain occupations or in certain areas or in particular kinds of environment? The questions are endless; the answers depend on the time and money you have available, as well as on the aims of your study.

Usually you must restrict your aims because it takes so much effort to collect and analyze data, but then you must also restrict the generality of your conclusions. Counting hawks as they fly over Hook Mountain in New York will not tell you anything about the hawk population of Minnesota, for example. But a restricted experiment with results that are meaningful is worth much more than an experiment with scope but meaningless results. Plan carefully before you start collecting data.

2. You must decide whether the experiment is to be conducted on the whole population or on only part of it.

Can you take a census, or must you deal with a sample? If you can take a census, the statistical techniques you will need are those of descriptive statistics: you must present the data in tables and graphs, and you may summarize it by finding a number about which the data center, and a second number which indicates the spread of data around that center.

If the question you are asking is simple enough ("What is the average height of third graders in Meadowlands Elementary School?") and the population small enough, a census may be easily carried out. But for a large population, a census is ordinarily so costly and time-consuming that it is carried out only for the most important questions ("What is the population by age in each town in the United States?"). To answer some questions ("What percentage of patients who have this disease now **or in the future** will be cured with this drug?") a census of the whole population is impossible regardless of money and effort now available.

For most experiments you will use a sample. From the sample data you will make inferences about the population; most of this book will deal with the techniques by which this is done. How should the sample be chosen? This question will be answered in the rest of this chapter. How large should the sample be? Some tentative answers will be given in Chapter 9, and there will be further discussion in later chapters. For the moment, assume that you have advice on sample size from your friendly neighborhood statistician. One point should be emphasized here, however: the whole experiment, including the sample size, should be planned before data are collected.

3. You must determine what data are to be collected, how they are to be collected, and what criteria are to be used to interpret the results.

Suppose, for example, you are determining the IQ of 40 pairs of identical twins separated soon after birth and raised in different environments. You will need to decide not just which IQ test is to be given, but also whether the same person or two different people should give it to each pair (will the use of one person assure uniformity, or will this person's knowledge of the result of testing one twin unconsciously affect the results of the second test?); if you are interested in the effect of different environments on identical twins, you will have to consider very carefully what factors in their environment are to be measured, how to get the necessary information, and how comparisons are to be made.

It is essential to decide before an experiment is carried out what criteria will be used to interpret the results. If you are trying to decide whether a coin is fair, for example, you might decide to experiment by tossing it 100 times. If it comes up heads 50 times you will certainly decide the coin is fair; if it comes up heads only 3 times, you will be quite certain it is not a fair coin. But what if it comes up heads 20 times? 40 times? 48 times? You will have to learn not only what criteria to use to test your results, but also how to determine the probability of error in using those criteria. Much more will be said about this later, but again you should realize at the beginning of your study of statistical inference that criteria for interpreting the results should be set up **before** data are collected.

4. You must select the sample, if one is to be chosen, and collect the data.

This process will take much of your time and energy, and possibly much money. It will be easier and certainly your results will be more meaningful if you have laid careful plans first.

5. You must analyze the results, draw conclusions, and estimate the precision of the results.

The last requirement is related, of course, to the criteria you set up for interpreting the results before data were collected. This estimate is sometimes not included in reports of experiments that seem otherwise respectable; such reports are usually meaningless.

2.5 RANDOM SAMPLES

Suppose that you plan to measure the heights of 30 people, a sample of all the inhabitants of a small town. Clearly you would not stop the first 30 children entering an elementary school, nor would you stand outside the high school gym and measure 30 candidates for the basketball team. Our aim will be to take the results of a sample and generalize them to statements about the whole population from which the sample has been taken. Obviously, the two samples suggested above would not be representative of the heights of residents of the town.

A famous example of an unrepresentative or biased sample is a poll of voting opinion conducted by the *Literary Digest* magazine in 1936 before the presidential election. The *Digest* poll showed Alfred Landon winning by a wide margin, but on election day Franklin Roosevelt won by a landslide. The *Literary Digest* soon folded, partly because readers lost confidence in it after this fiasco. The sample was taken from those who had a telephone or owned a car. Why were the conclusions so erroneous? The poll was taken at the height (or should I say "bottom"?) of a great depression; only relatively well-to-do people had a telephone or owned a car. Also, ballots were returned by mail. Probably subscribers to the *Literary Digest* were more likely to return them than those unfamiliar with the *Digest*. But the sample was very unrepresentative of all voters, most of whom supported Roosevelt.

How should a sample be chosen? Intuitively, one wants the sample to be "representative." But how is this to be put into practice? If the population is completely uniform, you can choose any sample of it. This assumption is made, for example, when a sample of your blood is tested. But suppose you want to choose 30 residents of the small town and measure their heights. Can you decide to include a quota of 2 tall men, 3 infants under 2 years old, and allot the other choices in similar ways? No matter how the quotas for each group are allotted, it is done well only if you already know well the characteristics of the population—and, in that case, why bother getting data on a sample? The sample must not be chosen because of a subjective decision about what is typical of the population, since this causes the results of the experiment to be biased by the prejudices of the selector.

Similarly, a sample should not be chosen by convenience. The interviewer who stops people in front of a department store to ask their preference for governor may avoid hippies or blacks, or may get too many suburban housewives and too few farmers or miners. The *Literary Digest* had convenient telephone directories and lists of car owners, but failed miserably in forecasting the election. If you are experimenting on 12 white mice, you may find it necessary to use mice ordered at the same time from one supplier—but you must be aware that you may have litter-mates in your sample, and must ask yourself whether this may affect the outcome of your experiment.

A sample should be chosen objectively, and should not be determined by the convenience of the experimenter. What method of selection should be used? As you will see, valid conclusions about the population can be made if the sample taken is a **simple random sample.**

> **Definition:** A simple random sample is chosen so that each sample of that size in the population is equally likely to be chosen. Such a selection is said to be made **at random.**

The techniques of statistical inference that you will study in this book are based on the assumption that your sample is a simple random one. There are other kinds of random samples, but we shall work always with simple random samples, and therefore the word "simple" will not be used in the future.

It isn't necessary to list all possible samples and then choose one of them at random from the list. Instead, the sample is a random one if each additional member added to the sample is chosen so that all remaining members of the population are equally likely to be chosen. A sample can, theoretically, be chosen either with or without replacement in the population of the sample member that has just been chosen. In the first case, the same person can be chosen twice, but if sampling is without replacement all persons or objects in the sample are different. In practice, however, it is almost always taken without replacement. If you are interviewing students, for example, you would feel foolish indeed—and would probably not get a warm reception—if you presented yourself again to repeat the same interview. If you are a doctor testing a drug, you cannot try it on the same patient twice and hope to get the same results as when it is tried independently on two different patients.

✓ How can a random sample of 30 students be chosen from a freshman class of 400 students? One method is to write the name of each freshman on a slip of paper, put the identical slips of paper in a hat, mix well, and blindly draw 30 names. Another method is to assign each student a number and then choose 30 numbers from a table of random numbers. Such a table can be made up by writing the 10 digits 0 through 9 on separate cards and mixing them in a box. One card is drawn and the digit noted, then that card is replaced and the cards mixed again. A space is left after the process has been carried out five times for ease in reading the table. Table 3, Appendix C, is such a table.

Selection of a sample of 30 from 400 students is straightforward. Each student is assigned a number from 001 to 400. A starting point is chosen at random in Table 3, and any direction you wish: left, right, up, down, or move like a knight on a chess board if you prefer. You may take only the first 3 digits of each 5-digit group, or the middle 3, or ignore the spaces in the table and take each successive set of 3 numbers (3 because in this example 400 has 3 digits; 2 if you were choosing from a class of 99 or less, 4 if from a class of 1,000 to 9,999, etc.). If the 3 digits form a number over 400 or a number previously chosen, ignore it; otherwise, assign the student with that number to the sample.

p.419

Example 1

Select a sample of size 12 without replacement from a population of size 80, starting with row 18, column 5 of Table 3 and reading horizontally, continuing with row 19. Part of this table is reproduced here:

```
18                        . . . 62542  30536  14777  72360
19   70791  39030 . . .
```

Assume the population is assigned numbers 01, 02, 03 . . ., 80. Starting with 62 in the random number table, the numbers marked A are chosen for the sample:

```
 A   A   A   D   A        A   A   A   A   A   A   A   B   C   A
70  79  13  90  30       62  54  23  05  36  14  77  77  23  60
```

The second 77 (marked B) is omitted because it has appeared before; the second 23 (marked C) is also omitted. The number 90 (marked D) is omitted because the population is of size 80, and no member of it was assigned the number 90.

Is a random sample a representative sample? Unfortunately, it may not be. A fair coin, tossed 50 times, **might** come up heads 10 times instead of the 25 you expect. A random sample of 10 students from a class of 200 **might** include only those with A averages. And there are certainly fluctuations to be expected between samples. As a result, the conclusions you draw about a population after studying a sample cannot be of the form, "At least 53% of the voters will vote for Bray for governor tomorrow" or "The hypothesis that the average birth weight of full-term babies is 120 ounces or more is rejected," but rather "I am 95% certain that at least 53% of the voters will vote for Bray for governor tomorrow" or the phrase "but there is 1 chance in 20 that it should be accepted" is added to the sentence about the birth weight of babies. So you hope your random sample is representative, but—even if it is not—you will learn how to find out the probability that your conclusions are correct, so long as the sample you have used is a random one.

Choosing a random sample either by writing names on slips of paper or by assigning numbers and using a random number table may be too time-consuming, too expensive, or simply impossible. Would you look forward to choosing a sample of 8,000 voters from all the registered voters in the United States by either of these methods? So other methods have been devised.

2.6   OTHER METHODS OF SAMPLING

> **Definition:** In **stratified sampling** the population is divided into strata having common characteristics, and then a random sample is chosen from each stratum.

You might want to do this because data on separate strata are easier to assemble, or because you want information about separate strata as well as about the combined population.

The strata must not be determined by the characteristic you are measuring. If you are polling a sample on political preference, for example, you may divide the population by sex and have 48% males and 52% females if that is the proportion of men and women in the population, but you cannot say "In the last election 49% voted for the Republican candidate, so 49% of my sample will be Republicans."

A stratified sample is not a simple random sample. If, for example, a population of 6 is divided into 2 strata and 1 member of the sample is chosen from each stratum:

Stratum 1	Stratum 2
A B C	D E F

then the samples might consist of A and E. A and E are equally likely to be chosen, but the sample cannot include both A and B. In simple random sampling, A and B are as likely to be chosen as A and E.

Stratified sampling is indeed important in statistics, but sufficiently complicated so that it will not be considered in this book.

> **Definition:** In **cluster sampling** the population is divided into geographical clusters, and certain clusters are picked at random. Then sample members are chosen at random from each selected cluster (or the whole of each one may be included).

To choose a sample of 30 students from 400 freshmen living in four dormitories, you might pick at random one floor in each dormitory, and then choose randomly from the students on the selected floors. Cluster sampling is relatively inexpensive, since there is less travel time between interviews, but cluster sampling gives less accurate predictions about the population if individuals in one cluster are alike. If you wish to predict how New York City voters will vote for President, you would hardly be wise to pick all your samples from one block, since rich and poor generally live in separate neighborhoods, and family income may affect voting patterns. Cluster sampling is a particular form of stratified sampling, in which the strata are geographical areas.

> **Definition:** In **systematic sampling** the sample is chosen systematically from the population: every tenth name on the list, every fourth house on a street, etc.

To choose your sample of 30 students from the freshman class of 400, you might pick a starting place at random in the freshman directory and then list every thirteenth name after that until you have 30 names, continuing with the A's after you finish the Z's (thirteenth because $400/30 \approx 13$; read $\approx$ as "approximately equals"). This method of sampling is simple and frequently used.

2.7   EXERCISES

1.   Which of the following are random samples from the suggested population? Explain the reason if not a random sample.

   (a) A chemical society is interested in the number of technical papers written by those who have earned a Ph.D. in chemistry in the past 5 years. It sends a questionnaire to all those awarded such a degree in the past 5 years, and bases its estimate on the replies received.

   (b) A magazine takes a presidential preference poll by sending a questionnaire to one subscriber from each zip code district on its mailing list.

   (c) A manufacturer of automobile horns sends every hundredth horn it manufactures to quality control for testing.

(d) A psychologist chooses 30 students from a large class by assigning each member of the class a number, 001, 002, ..., 485. He uses the middle three digits in a group of 5 numbers in a random number table, but skips every other group of 5. He skips every number that repeats one previously chosen, and every number over 485.

(e) A sample of 50 is to be taken from telephone subscribers in Buffalo. You choose numbers at random from the Buffalo telephone directory and spend every evening calling selected numbers on the telephone until you have 50 answers.

(f) You are a doctor carrying out research at a famous cancer hospital in New York. You have developed a new chemical for treatment of lung cancer. You give the chemical to every third patient who enters with lung cancer if the patient is willing, leaving the others for controls for comparison of treatment.

2. In a table of random numbers, what proportion of numbers do you expect to be 4's?

3. Choose a sample of 10, without replacement, from a population numbered 0001 to 8432, starting with the third row, second column of Table 3, Appendix C, and reading horizontally from left to right with each succeeding line above the previous one.

4. Choose (without replacement) a sample of size 4 from a population of size 10, starting with the tenth row, fifth column of Table 7 and reading the first digit in each group of 5 vertically.

5. Criticize the following:
(a) A market researcher asks people entering a supermarket, "Do you use Kleenex?"
(b) A sociologist asks a random sample of 75 students in his Introduction to Sociology class, "How many times have you stolen something valued over $10?"

6. One interviewer asks each of 200 randomly selected students whether he or she smokes Marlboros. Another interviewer asks each of 200 randomly selected students what brand of cigarettes he or she smokes. Do you expect the proportion of students who smoke Marlboros to be the same in the two surveys? Explain.

In Exercises 7–10, use these scores of 36 students on the Miller Personality Test:

22	22	20	27	30	23	22	21	26
21	23	25	29	18	22	31	20	28
16	28	33	25	23	31	23	18	24
26	25	17	22	25	28	19	24	20

7. Choose a random sample of 8 scores on the Miller Personality Test, using the table of random numbers. Explain very carefully how you do it.

8. Treat each row as a stratum. Select a stratified sample of size 8 by randomly choosing 2 scores from each stratum. Explain carefully how you do it.

9. Select a systematic sample of size 8 of scores on the Miller Personality Test. Explain carefully how you do it.

10. It has been found that poor people are less likely to return a mail survey than people with middle or upper incomes. Give some reasons why you think this occurs.

ANSWERS

1. (a) Not random. Those who have written no papers are less likely to answer.
   (b) Not random. Rural areas are represented more than big-city areas. Nor does the description claim that the one subscriber is chosen at random from his district. Also, a voluntary response is needed; this may further bias the results (toward people with a better education?).
   (c) This is systematic, not random, sampling. It may not be objectionable, but what if a machine that makes part of the horn has a cyclic defect that causes trouble on every hundredth or thousandth horn?
   (d) This is a random sample. The psychologist will probably need a longer table of random numbers than that given in Table 7.
   (e) Not random. You will almost certainly get few business telephones, and will get answers from individuals who stay at home and not those who go out in the evening.
   (f) It is a very poor sample. Those who are willing to try the new drug may be hopeless of cure by any other means. It is not even a systematic sample, much less a random sample, of those entering that hospital, and certainly not a random sample of all patients with lung cancer—more "late" cases would probably be sent on to this hospital. And you are about to lose your medical license; a doctor can't just try out new drugs freely and ignore established methods of treatment.

2. 1/10, since 10 digits are equally likely to be chosen.

3. If the first 4 numbers in each group of 5 are used, then members of the population numbered 5617, 0983, 3422, 4389, 3851, 1161, 3033, 1792, 2402, 5093, and 0801 are chosen. If the last 4 numbers are used, then the sample consists of those numbered 6179, 4227, 3897, 8517, 1617, 0338, 7929, 4021, 0932, 8012, and 7925. If all 5 digits are used, then the sample consists of 5617, 3334, 2274, 3897, 3851, 7116, 1730, 3381, 7929, 2402, and 1509.

4. Number the population 0 to 9 rather than 1 to 10, so you need choose only 1-digit numbers. The numbers that appear vertically in the random number table starting at the designated spot are 8, 9, 8, 9, 0, 3, 1, ... The numbers chosen for the sample are 8, 9, 0, 3.

5. a) In spite of efforts by the manufacturer to preserve the trademark, many people refer to all paper tissues as "Kleenex."
   b) The sample is presumably random, but the answers would be biased by those willing to admit (or boast) that they had stolen something valued over $10.

6. It is probably higher in the first interviewer's survey. It is easier to say "yes" than to think of the brand one is smoking. And a smoker who switches brands but includes Marlboro as one possibility is more likely to appear in the first survey.

7. Assign numbers 01 to 36 to each score, from left to right. Starting, say, in column 4, row 19 of the random number table, choose scores numbered 20, 13, 07, 09, 31, 30, 08, and 21 for the sample. The corresponding scores are 28, 29, 22, 26, 22, 17, 21, and 33.

8. Assign number 1 to 9 to each stratum. Start with row 7, column 1, say, and choose single numbers. From the first stratum, choose numbers 8 and 6 (scores of 21 and 23), from the second, 9 and 2 (scores 28 and 23), from the third, 2 and 4 (scores 28 and 25), and from the last, numbers 8 and 9 (scores 24 and 20).

9. There are 36 scores in all, and 8 are to be in the sample. 36/8 = 4.5. Randomly choose a starting point (say, the fifth score in the third row). Choose that score and every fourth after it, starting over at the beginning when necessary. The scores chosen by this method are 23, 24, 22, 24, 20, 22, 23, and 22.

10. Poor people may be less likely to (a) read and write easily, (b) feel that the survey is needed, (c) have leisure time to fill out the questionnaire.

2.8 VOCABULARY

descriptive statistics	representative sample
statistical inference	random sample
population	stratified sampling
sample	cluster sampling
census	systematic sampling

2.9 REVIEW EXERCISES

In Exercises 1–6, which use descriptive statistics and which use statistical inference?

1. After sprouting 48 bean plants, I conclude that the chances are 9 to 1 that the average germination time for this type of bean seed is 112 to 117 hours.
2. The average annual rainfall is 7 inches in Arizona, 8 inches in New Mexico, and 67 inches in Alabama.
3. After tossing a die 1,200 times and noting the number of times 1, 2, . . . , 6 come up, I've decided this die is loaded—but there's a 5% chance that my conclusion is wrong.
4. The average score of 16 college students on the Colorado Personality Test is 26.
5. The earnings per share of American Express Company stock increased from $2.09 in 1974 to $4.31 in 1978.
6. A poll of 1,634 voters show that 68% plan to vote for Mrs. A for governor next week; Mrs. A is almost certain to win the election.

In Exercises 7–11, describe (a) the population(s), and (b) the sample(s).

7. A psychologist tests the reaction time of 36 rats to a 30-volt electric shock in order to estimate the average reaction time of all rats of the same strain to the same shock.
8. To estimate the average number of miles driven per week by 1½-ton trucks, an economist studies the weekly mileage records of 36 1½-ton trucks.
9. 324 male chemistry majors and 106 female chemistry majors who started working for chemical companies this year are questioned about their salaries

in order to estimate the difference in initial salaries of male and female chemistry majors.

10. The proportion of defective axles produced on two different machines from January 2–5 are compared in order to test the hypothesis that one machine is more reliable than the other.

11. The average number of years of school completed by adults over 30 in the United States is 11.1. The Board of Education in Madison surveys 200 adult residents over 30 in order to test the hypothesis that Madison adults differ from the national average in years of schooling completed.

12. What is a representative sample? a random sample? Is a random sample representative? Comment.

13. You have a list of names, addresses, and college class of all commuters to your college. Describe briefly how you would choose (a) a random sample, (b) a stratified sample, (c) a cluster sample, (d) a systematic sample.

14. Give one or more advantages and disadvantages of each kind of sample (random, stratified, cluster, systematic).

15. Describe in detail how to choose a random sample of 12 students from the freshmen at your college, using Table 3, Appendix C.

16. You suspect that emotional stress triggers juvenile rheumatoid arthritis. Outline briefly the steps that you would follow in conducting a study to test whether this is true.

17. "I interviewed 50 people over 18 who live on my street, and found that 60% think the state income tax should be cut in half. I conclude that a majority of all people over 18 in my state agree."
(a) What is (i) the population? (ii) the sample?
(b) Comment on the method of sampling.

18. Choose a random sample of 18 students from a list of 462. Describe carefully how you do it.

ANSWERS

1. Statistical inference.

3. Statistical inference.

5. Descriptive statistics.

7. (a) All rats of the same strain; (b) the 36 tested rats.

9. (a) 2 populations: all male and all female chemistry majors who started working for chemical companies this year; (b) 324 male and 106 female chemistry majors who started jobs this year.

11. (a) Adult residents over 30 in Madison. (The question is whether or not this population is like the population of all adults over 30 in the United States.) (b) The 200 Madison adults over 30 who are surveyed.

13. (a) Assign each commuter a number, then use the random number table to choose which of these numbers are to be in the sample.
(b) Classify by sex and college class, for example. Then if 14% of commuters are male freshmen, choose 14% of your sample at random from male freshmen commuters.

(c) Mark commuters' homes on a map, divide the map into areas, each including approximately the same number of commuters, randomly choose some of these areas, and then randomly choose commuters from the selected areas.

(d) Divide the number of commuters by the size of the sample; suppose the quotient $= p$. Then choose at random a starting place in a list of commuters, and every p^{th} name after that, starting over at the beginning when necessary.

15. Suppose there are 5,467 freshmen. Assign each a number 0001, 0002, . . . , 5,467 to each. Starting in column 4, row 7, say, and reading down, use the first 4 numbers in each group of 5. Omit any numbers over 5,467 or any repetitions. Students whose assigned numbers are 1496, 5005, 2235, 3320, 3069, 2612, 4112, 0458, 4642, 1249, 1945, and 1313 would be selected.

17. (a) (i) All over 18 in my state; (ii) the 50 on my street.

    (b) Those living on the same street tend to have similar incomes. What if the range on my street is $20,000 to $40,000? The sample is neither random nor representative.

3

PRESENTING DATA IN A TABLE OR GRAPH

In carrying out a statistical investigation, what you can do depends on the kind of data you have. This chapter starts by describing three types of data: categorical, ranked, and metric. You may collect masses of data; most of this chapter deals with how to present it in a table or a graph.

3.1 TYPES OF DATA

Data may be classified as categorical, ranked, or metric.

✓ **Categorical Data.** Individuals are simply placed in the proper category or group, and the number in each category is counted. Each item must fit into exactly one category.

Example 1

Sex of Essex College Students
1980

Sex	Number
Male	706
Female	678

In Example 1, categories (Male, Female) are listed in a column. Sometimes information is classified in two ways at once. Then a row listing categories in the second set is added. Be especially careful with your tally if you are making a two-way classification.

Example 2

Eye Color of Essex College Students by Class
1980

Color	Class			
	FR	SOPH	JR	SR
Blue	124	86	82	98
Brown	150	170	135	136
Green	15	4	7	2
Hazel	103	85	95	92

CELLS MAKE UP TABLE

29

In the first example there were two categories; in the table above there are 16 (brown-eyed sophomores, for example). If a pink-eyed albino freshman turns up at Essex College, another category would be added to the column under Color, and four more to the table; the line "Pink 1 0 0 0" would be added.

What's wrong with the following categories for types of publications? "Newspapers, books, magazines, *The New York Times*."*

✓ **Ranked Data** have order among the categories. If 6 horses race, the statement "bets were paid off on A, B, and C but not on D, E, and F" presents two categories (winners and losers). The statement "A won, followed by B, C, D, E, and F in that order" presents ranked data. Note that there are not necessarily equal intervals or differences between ranks. A may win by a nose over B, but E may come in 10 lengths ahead of F.

✓ **Metric Data** involve measurement. To continue the horse race analogy, the times the 6 horses take to run the race are metric data. Units are assigned (seconds or minutes for the horse race), with equal intervals between the units. This makes it possible to carry out arithmetical operations. With ranked data, we can only remark "A came in ahead of B." With metric data, we can say "A's time was 62.3 sec., B's 62.4 sec.; the difference is .1 sec."

Other examples of metric data are incomes of residents of Milwaukee, times for rats to solve a given maze, and SAT scores of students who applied for admission to Cornhill College in the class of 1983.

In general, separate techniques are used for the different types of data. Much of the data dealt with in this book will be metric. Techniques developed for metric data can seldom be used with categorical or ranked data. At other times, techniques especially suited to ranked or categorical data will be presented. Most of these could be applied to metric data, but they usually are not, since more powerful methods specific to metric data are preferred.

3.2 PRESENTATION OF DATA IN A TABLE

Eighty-two students obtain the following scores on the Miller Personality Test:

22	22	20	27	30	23	29	21	26
21	23	25	29	18	22	31	30	28
16	28	33	25	23	31	23	18	24
26	25	17	22	25	28	19	24	20
23	26	21	31	25	24	33	29	20
27	21	25	28	24	23	25	30	27
23	26	22	24	17	33	26	24	19
18	33	25	28	31	29	27	28	24
26	24	22	26	24	18	21	29	22
31								

---

*The New York Times is a newspaper and therefore fits into two categories.

These are raw data (that is, data that have not been numerically organized). It is difficult to digest such a mass of figures, and indigestion becomes more acute as the number of figures is increased. A **tally** is made by listing the different scores in order from the smallest to the largest and adding a vertical tally mark as each score appears until a multiple of 5 is reached; then a tally mark is made through the previous four.

Score	Tally	Frequency
16	|	1
17	||	2
18	||||	4
19	||	2
20	|||	3
21	卌	5
22	卌 ||	7
23	卌 ||	7
24	卌 ||||	9
25	卌 |||	8
26	卌 ||	7
27	||||	4
28	卌 |	6
29	卌	5
30	|||	3
31	卌	5
32		0
33	||||	4
		82

√√ A **score** is any relevant numerical measurement (any piece of metric data), regardless of how obtained or the kind of unit used. After the scores are tallied, the number in each class is written in a column labeled **frequency**. We shall commonly use X or Y for scores, f for frequency. A **frequency distribution** is a record of the number of scores that fall in each score class. The frequency distribution made from the 82 scores on the Miller Personality Test would start, then, as follows:

X	f
16	1
17	2
18	4
.	.
.	.
.	.

Each frequency distribution must include classes of observations (X's) and the frequency in each class (f's). Each score must fit into exactly one class.

Ignore the middle column of tally marks and look at the frequency distribution for scores on the Miller Personality Test; compare it with the raw data given at the beginning of this section. The frequency distribution is easier to assimilate; it is easy to determine the lowest score (16) and the highest score (33); it is easy to determine how many people scored 27 (4). There are still 18 different score classes. It is often advantageous to **group** data by combining score classes. Here are two ways in which this may be done:

(a)		(b)	
X	*f*	*X*	*f*
16-17	3	16-18	7
18-19	6	19-21	10
20-21	8	22-24	23
22-23	14	25-27	19
24-25	17	28-30	14
26-27	11	31-33	9
28-29	11		82
30-31	8		
32-33	4		
	82		

ANSWER ON P.??

As the number of score classes decreases, the information given becomes easier to assimilate. Grouped frequency distribution (b) above can be absorbed in about 5 seconds, while (a) takes longer, and the original distribution longer still. But simplicity has a price: detailed information is lost by grouping. (b) tells us there are 7 scores between 16 and 18, but we don't know how these 7 are distributed; the original data show 1 score of 16, 2 of 17, and 4 of 18. So grouping is not an unmixed blessing; a compromise must always be found between clarity and loss of information.

The **class width** is the difference between the lowest score that fits into one class and the lowest score that fits into the next higher class. (It is not the difference between the largest and smallest scores in one class.) In (a) above, each class is of width 2, and in (b) each has width equal to 3.

The **midpoint** of a class is the average (later you will learn to be fancier and call this the "mean") of the smallest and largest scores that could go into a class. The midpoint of the 16–17 class in (a) is $\dfrac{16 + 17}{2} = 16.5$; the midpoint of the 16–18 class in (b) is $\dfrac{16 + 18}{2} = 17$.

3.3   HOW TO GROUP SCORES

Here are some guidelines for grouping.

1.   Every score must fit into exactly one class.

This guideline is never ignored.

2.   Use 10 to 20 classes.

This guideline is ignored often—but only for a good reason. Use as few as 5 classes if you are willing to sacrifice a lot of information in order to gain extreme simplicity; use many more than 20 classes if you are presenting masses of information to a sophisticated audience.

MORE DATA → MORE CLASSES
<100 OBSERVATIONS
10-20 CLASSES

3.   Classes should be of the same width.

This is the conventional method of grouping. Sometimes it is impossible for all classes to be of the same size. A psychologist who times 40 rats in a maze, but removes any rat who has not found its way through the maze in 3 minutes, might present his data as follows:

Time (minutes)	f
0 and under 1	10
At least 1 and under 2	18
At least 2 and under 3	4
At least 3 and over	8

The width of the last class cannot be determined. Such a class is called **open-ended.** It has no upper limit and no midpoint.

4.   Consider customary preferences in numbers (we count by 5, 10, 20, 25, . . . , most easily), choose your class width to reflect this, and start (or if necessary end) your classes with "nice" numbers to emphasize this.

Good choices:	100–124, 125–149, . . .
	10–19, 20–29, . . .
	30–34, 35–39, . . .
Poor choices:	100–122, 123–145, 146–168, . . . (class width is 23)
	106–115, 116–125, . . . (class width is 10 but this is not
	immediately apparent)

Start the classes with "nice" numbers when you can, but instead end them this way if necessary. If for example one student answers all 140 questions correctly on a T-F quiz, then classes of width 10 would be 61–70, 71–80, . . . , 131–140; scores that are impossible should not be included in any class.

The last three guidelines are not firm rules; they are ignored if the purpose for which the data are used so demands.

A mass of raw scores is organized by using the following procedures:

1.   Tally the scores.

2.   Determine the range (= the difference between the highest and lowest scores.)

3.   Select the class width and the number of classes so that the four guidelines are followed and so that (class width) × (number of classes) covers the range. Once you estimate the number of classes, you can divide the range by this number to approximate the class width.

4. Determine the classes; that is, state which scores are to be included in each class.
5. Make a table showing the classes and the frequency in each class. In any table, always include a title, give a date if meaningful, and give the source of data if you did not collect it yourself.

Example 1

The lowest of 100 scores is 11 and the highest is 54. How should they be grouped in about 10 classes?

The range $= 54 - 11 = 43$. Dividing the range by the approximate number of classes, we obtain $43/10 = 4.3$. The nearest "nice" number is 5. The lowest class must include 11, the highest 54. Classes should be 10–14, 15–19, . . . , 50–54. Nine classes, each of width 5, cover the range.

Now try two examples yourself.

Example 2

About 15 classes are to be used to group 300 scores which range from 207 to 592. How should classes be chosen?

Example 3

900 scores, of which the largest is 418 and the smallest 112, are to be grouped in about 18 classes. How should this be done?

Answers

2. The range is $592 - 207 = 385$. $\frac{385}{15} \approx 26$; 25 is a convenient choice for class width. Classes: 207–231, 232–256, . . . , 582–606? The "nice" class width of 25 is hardly apparent; a better choice is 200–224, 225–249, . . . , 575–599.

3. The range is $418-112 = 306$. $\frac{306}{18} = 17$. The nearest "nice" number is 20. Classes: 100–119, 120–139, . . . , 400–419. $\frac{420 - 100}{20} = 16$; 16 classes are needed. Other answers: class width of 25, classes 100–124, . . . , 400–424 is not as good because only 12 classes are used. Class width of 15 is possible (100–114, 115–129, . . . , 415–429) but is not as good because we count more easily by 20 than by 15, and 22 classes are needed.

Definition: In a **relative frequency** distribution, the frequency in a score or class is replaced by the ratio of that frequency to the total number of scores.

In a **cumulative frequency** distribution, add the frequencies starting with lowest or highest scores or classes. It can be a "less than," "or less," "or more," or "more than" distribution.

Example 4

For the scores in the box below, find (a) a relative frequency distribution, and (b) an "or less" cumulative frequency distribution.

Grades on Statistics Quiz

Grade	Number of students	(a) Relative frequency	(b) Grade	Number of students
50–59	3	.041	50 or less	3
60–69	6	.081	69 or less	9
70–79	28	.378	79 or less	37
80–89	26	.351	89 or less	63
90–99	11	.149	99 or less	74
	74	1.000		

3.4   EXERCISES

1. Classify as metric, ranked, or categorical data:
    (a) Number of cases of each reportable disease reported by doctors in New York City in 1979.
    (b) The average number of years of school completed by adults over 25 in the United States is 11.1.
    (c) James Pollock came in first in the National Ice Skating Championship, Lee Rohrs was second, and Harold Nagel came in third.
    (d) Eight percent of the fruit flies used in Experiment 4 of a genetics class had banded wings.
    (e) The average weight gain of 6 1-year-old collies at the Crispin Kennel was 1.8 pounds last month.

2. (a) Tally the following scores: 17, 17, 15, 10, 12, 15, 17, 11, 10, 15, 12, 11, 15, 15, 11, 10, 11, 14, 12, 15.
    (b) What is the highest score?
    (c) Which score occurs most frequently? How frequently does it occur?
    (d) Group the scores in classes of width 2.
    (e) Make (i) a relative frequency, and (ii) a "less than" cumulative frequency distribution for the (ungrouped) scores in (a).

3. Grades on a statistics quiz in a small recitation section are 77, 69, 81, 90, 52, 98, 72, 83, 85, 77, 80, 83, 59, 83, 63, 87, 60, 94, 96, 75, 67, 70.
    (a) Tally the grades.
    (b) What is the lowest grade? the highest?
    (c) Group the grades in 5 classes.
    (d) Another student gets a grade of 100, and this grade is added to those above. How should the 5 classes now be chosen?
    (e) What is the relative frequency of the grades in your second class in (d)?

4. How should 400 SAT scores be grouped in approximately 15 classes if the lowest score is 458 and the highest score is 782?

5. How should 400 SAT scores be grouped in approximately 12 classes if the lowest score is 326 and the highest is 800? (Note: 800 is the highest possible SAT score.)

6. At Perkins College, grade-point averages (GPA) can vary from 0.00 to 4.00. Among 1,200 seniors the lowest GPA is 1.98 and the highest is 3.69. How should their GPA's be grouped in approximately 17 classes? How many classes are actually needed?

7. Which (if any) of the guidelines for grouping are not followed?

(a) X	f	(b) Y	f	(c) Z	f
10–18	2	100–109	26	27–41	5
19–27	3	110–119	41	42–61	7
28–36	7	120–129	60	62–71	9
36–45	4	130–139	34	72–101	14
46–54	0	140–149	20	102–121	11
55–63	9	150–159	25	123–141	8
		160–169	28	142–161	3
		170–179	12	162–181	4
		180–189	21	182–201	5
		190–199	8	202–221	2

8. The lowest of 728 scores is 16 and the highest is 204. (a) How should the scores be grouped in approximately 10 classes? (b) Exactly how many classes are needed?

9. The price of 42,000 commuter tickets sold on the Dover line of ConRail on September 28 ranged from $.54 to $3.82. (a) How should the prices be grouped in approximately 13 classes? (b) Exactly how many classes are needed?

10. (a) Make a tally for the following scores:

126	132	121	149	130	139	127	136	128	129
121	134	139	135	138	123	133	136	124	130
127	136	132	126	145	139	131	133	142	131
134	130	141	144	136	124	136	136	133	128
123	125	139	145	148	141	126	145	138	139
133	147	136	134	132	142	149	122	131	139
130	139	136	148	132	147	121	124	148	133
139	127	147	124	148	135	142	142	133	142
121	146	145	148	127	136	130	144	143	124
148	140	136	136						

(b) What is the smallest score? the largest? How many times does 139 appear?

(c) Group the scores in 6 to 8 classes.

11. Find the (a) "or less," (b) "less than," (c) "or more" cumulative frequency distribution for the following:

Y	f
20–29	10
30–39	18
40–49	26
50–59	12

ANSWERS

1. Metric: (b), (e); ranked: (c); categorical: (a), (d).

2. (a) Scores from 10 to 17, f = 3, 4, 3, 0, 1, 6, 0, 3; (b) 17; (c) 15 appears 6 times; (d) 10–11, 12–13, 14–15, 16–17; f = 7, 3, 7, 3; (e) Scores still 10–17, (i) Relative frequencies: .35, .15, .35, .15; (ii) Cumulative frequencies: 3, 7, 10, 10, 11, 17, 17, 20 less than 11, 12, . . . , 18, respectively.

3. (a) Frequency = 1 except where shown in parentheses: 52, 59, 60, 63, 67, 69, 70, 72, 75, 77(2), 80, 81, 83(3), 85, 87, 90, 94, 96, 98; (b) 52, 98; (c) 50–59, 60–69, . . . , 90–99 with f = 2, 4, 5, 7, 4; (d) 51–60, 61–70, . . . , 91–100, with f = 3, 4, 5, 7, 4; (e) $\dfrac{4}{22}$ = .18 (or .182).

4. 782 − 458 = 324 and $\dfrac{324}{15}$ = 22. 450–469, 470–489, . . . , 770–789 (17 classes), or 450–474, 475–499, . . . , 775–799 (14 classes).

5. 800 − 326 = 474 and $\dfrac{474}{12}$ = 39.5; 321–360, 361–400, . . . , 761–800.

6. 3.69 − 1.98 = 1.71, $\dfrac{1.71}{17}$ = .10; 1.90–1.99, 2.00–2.09, . . . , 3.60–3.69 (18 classes).

7. (a) Classes are not the same width, 36 appears in 2 classes, and class width of 9—most common width—is not recommended; (b) all guidelines are followed; (c) not of same width, not "nice" numbers, score of 122 doesn't fit in any class.

8. (a) $\dfrac{204 - 16}{10}$ = 19; 10–29, 30–49, . . . , 190–209; (b) 10 classes.

9. (a) $\dfrac{3.82 - .54}{13}$ = 25; \$.50–\$.74; \$.75–\$.99, . . . , \$3.75–\$3.99; (b) 14.

10. (a) X column: 121 to 149; f column: 4, 1, 2, 5, 1, 3, 4, 2, 1, 5, 3, 4, 6, 3, 2, 11, 0, 2, 8, 1, 2, 5, 1, 2, 4, 1, 3, 6, 2.
    (b) 121, 149, 8.
    (c) 120–124, 125–129, . . . , 145–149, with f = 12, 10, 21, 24, 11, 16.

11.

(a)		(b)		(c)	
Y	f	Y	f	Y	f
29 or less	10	Less than 30	10	20 or more	66
39 or less	28	Less than 40	28	30 or more	56
49 or less	54	Less than 50	54	40 or more	38
59 or less	66	Less than 60	66	50 or more	12

3.5 GRAPHICAL PRESENTATION OF DATA

To many people, data are more meaningful if presented graphically rather than in a table. Every graph, just like a table, should include a **title,** a **date** if this is meaningful, and the **source of data** if you did not collect it yourself.

Bar Charts. These are used for categorical data. Categories are usually shown on the horizontal axis, and frequency, proportion, or per cent is shown on the vertical axis, unless the names of the categories are long. The bars are usually separated from each other to emphasize the distinctness of the categories. The bars must all be of the same width, and the length of each bar must be proportional to the number in that category. Labels are needed on both axes.

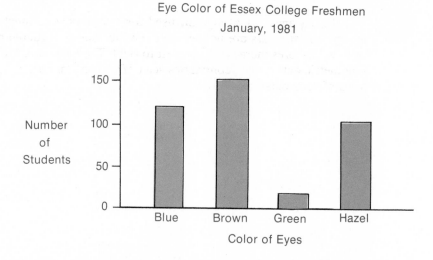

Pie Charts. These are also used for categorical data. The pie is cut into wedges, with the central angle of each wedge proportional to the size of the category or scores which it represents. Since there are 360° in a circle, multiply the relative frequency by 360 to find the angle at the center of the pie. Use a protractor if you have one, or judge by eye if you don't.

Sources of Pike College income (in millions of dollars)		Relative frequency	Angle
Tuition	9.8	.856	308
Endowment	1.2	.104	37
Other	.5	.043	15
	11.5		

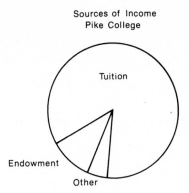

Sources of Income
Pike College

Histograms. A **histogram,** used for metric data, is similar to a bar chart except that the bars are not separate, and classes are ordered on the horizontal axis, with scores increasing from left to right. The bars are adjacent to emphasize that metric data is on a continuous scale. Either the midpoint or the class boundaries of each class are shown.

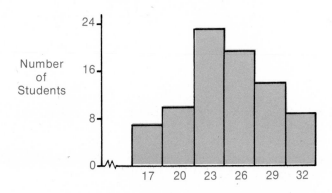

Miller Personality Test
82 Students in Psychology 21
January, 1981

A histogram cannot be used if one of the classes is open-ended.

Frequency Polygons. A **frequency polygon** is a line graph made by connecting the midpoints of the bars in a histogram by straight lines. If it makes sense, a class of zero frequency is assumed at each end to bring the eye down to the axis again.

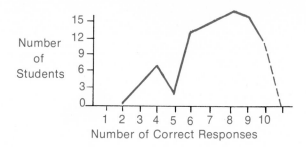

Math 3
Correct Responses on Ten-Question T-F Test

Note: Should the dotted line on the above figure be filled in as part of the graph, or should it have been omitted? It should have been omitted, since including it would imply that 11 is a possible score. The eye is not brought down to the axis on the right end of this frequency polygon.

A histogram is very confusing if two distributions are shown on the same graph; a frequency polygon is usually preferable.

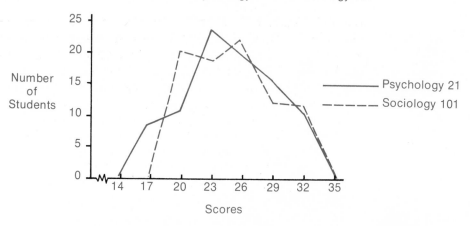

Miller Personality Test
Students in Psychology 21 and Sociology 101

Symmetrical and Skewed Distributions in Graphs. A histogram (or a frequency polygon) may be **symmetrical** (if you cut it vertically down the middle, one half can be flipped over exactly onto the other half), **skewed to the left** (has a longer tail on the left), or **skewed to the right.** Study these examples:

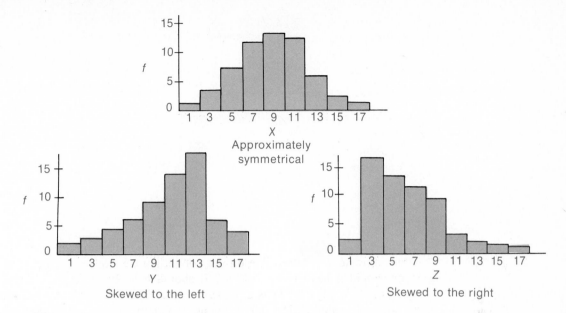

Approximately symmetrical

Skewed to the left

Skewed to the right

A **variable** is a quantity which can take on numerical values. Usually X or Y will be used to denote a variable.

✓ **Discrete and Continuous Variables.** In the distributions you have been considering, the variable (scores on the Miller Personality Test, for example) has taken on a finite number of values (16, 17, 18, . . . , 33). Such a variable is called **discrete.** A variable is called **continuous** if it can take on all the values of a continuous scale. For example, if X is any number between 1 and 4 (such as 1.42 or $\sqrt{3}$ or 3.9117), then X is a continuous variable. It is hard to give practical examples involving measurements: the height of a child is a continuous variable, in that a child passes through every intermediate height in growing from 3 feet to 4 feet; but ordinarily in measuring a child's height we measure to the nearest quarter inch, and therefore our measurements are discrete variables. Time flows on continuously (continuous variable), but is measured to the nearest second (discrete variable). And yet, as you will see in later chapters, continuous distributions are of tremendous theoretical importance in statistics. The graph illustrating a continuous distribution will be a curve rather than a frequency polygon.

3.6 EXERCISES

1. Marc Bernstein in Miss Van Vleet's biology class at North Plainfield High School measured the diameter of 20 fresh garden peas (in mm.): 3, 3, 4, 4, 5, 5, 6, 6, 7, 8, 8, 9, 9, 9, 10, 10, 10. Draw a histogram of this distribution.

2. The rest of Miss Van Vleet's class also measured peas, and their results were as follows:

Length (mm.):   2   3   4   5   6   7   8   9   10   11
Number of peas: 2   1   8   8   18   19   18   28   47   18

Draw a histogram.

3. Illustrate with two pie charts:

Education of American Workers over 25 (1977)*

| Education | Number (in millions) | |
	Females	Males
Elementary	9.7	11.4
High School	25.9	29.9
College	12.5	18.9

4. What is misleading in each of the following?

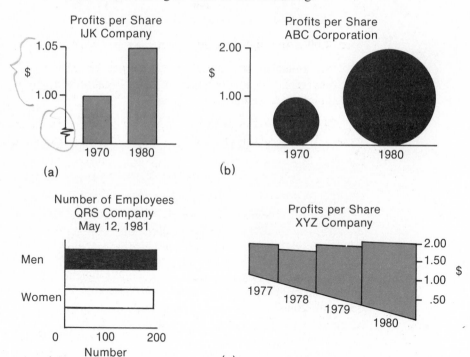

(a)   (b)   (c)   (d)

5. Illustrate the following data for the number of tables with 1, 2, . . . , 6 people that Waitress Nancy and Waitress Maria waited on at Marchant's Restaurant on St. Valentine's Day:

Number at table: 1   2   3   4   5   6
Number } Nancy: 2   35   9   3   0   1
of tables } Maria: 4   25   7   3   3   0

*Ginzburg, E., "The Professionalization of the U.S. Labor Force," *Scientific American,* March 1979, pp. 48–53.

6. In 1974, earnings per share of American Express Company were $2.09; in 1975, $2.29; in 1976, $2.70; in 1977, $3.65; and in 1978, $4.31. Illustrate.

7. The following are the concentrations (mg./100 ml.) of serum phosphate in patients with breast cancer, from a thesis by Iris Weisman: 4.0, 2.1, 3.6, 3.2, 3.8, 2.8, 2.7, 3.8, 3.5, 3.4, 5.0, 3.2, 4.3, 3.2, 3.6, 4.0, 3.5, 4.3, 2.5, 3.9, 3.3, 4.1, 3.7, 4.0, 4.1, 3.1, 3.2, 2.8, 3.7, 3.0, 3.2, 3.2, 3.5, 2.9, 3.7, 3.2, 3.4, 3.4, 3.7. Group in 7 classes and draw a histogram.

8. Illustrate with a single frequency polygon:

Grade Point Averages, Second Semester

GPA	Men	Women
2.00–2.39	140	120
2.40–2.79	840	732
2.80–3.19	1224	1410
3.20–3.59	686	720
3.60–3.99	110	140

Note: 4.00 is the highest grade-point average that can be obtained.

9. Illustrate the grade-point averages for men given in Exercise 8 with (a) a relative frequency polygon, (b) a cumulative frequency polygon, (c) a cumulative relative frequency polygon.

10. The number of blacks, other minorities, women, and white men working as professionals in firms with at least 100 employees are given for the years 1966 and 1975:

	1966	1975
Blacks	22,000	78,000
Other minorities	37,000	105,000
Women	237,000	731,000
White men	1,396,000	1,526,000
Total	1,692,000	2,440,000

Illustrate with pie charts.

11. Here are the current yield and volume of some of the bonds sold on the New York Stock Exchange on February 15: (Multiply sales by $1,000.)

Yield	Sales	Yield	Sales	Yield	Sales
11	12.1	9.25	5	6.75	7.1
9	9.2	7.4	7.6	7	7.3
8.0	9.0	9.5	11.4	10.0	11.1
4.8	5.8	7.5	8.4	8.5	9.0
8.6	9.1	8.5	10.0	7.0	9.4
9.4	10.0	11.3	10.4	7.5	8.9
11.7	12.1	9.5	11.4	9.5	12.5

(a) Tally and then draw a histogram for the yields of these bonds.
(b) Do the same for sales.

ANSWERS

1.

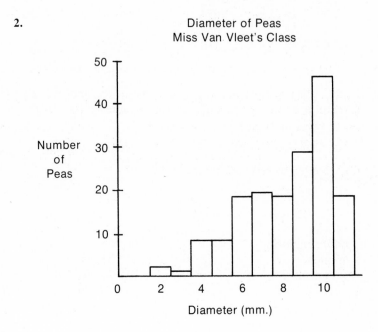

Diameter of Peas
Marc Bernstein

Number of Peas

Diameter (mm.)

2.

Diameter of Peas
Miss Van Vleet's Class

Number of Peas

Diameter (mm.)

3.

Education of American Workers over 25
1977

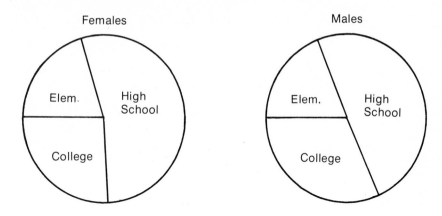

Females

Males

4. (a) The break in the vertical scale is misleading. It looks as though the second bar is twice the height of the first, but it should only be 5% more.
(b) The area of the second circle is four times that of the first, and this impresses the eye more than the height, which is twice as great.
(c) The darker color is more impressive, although it is almost the same length.
(d) More recent years are emphasized unfairly.

5.

Marchant's Restaurant
Tables Waited on by Nancy and Maria
February 14

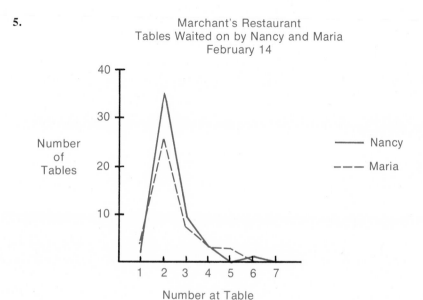

6.

American Express Company
Earnings per Share

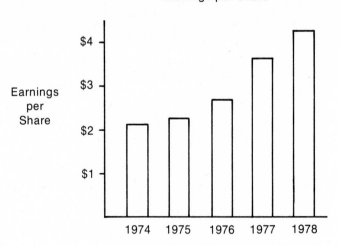

Earnings
per
Share

7.

Serum Phosphate Concentration
in Patients with Breast Cancer

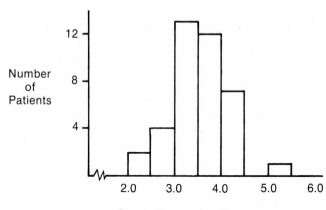

Number
of
Patients

Serum Phosphate Concentration
(mg./100 ml.)

8.

Grade Point Averages
Second Semester

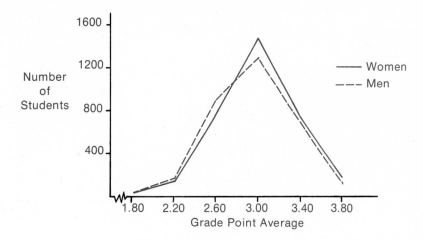

9. Grade Point Averages for Men
Second Semester

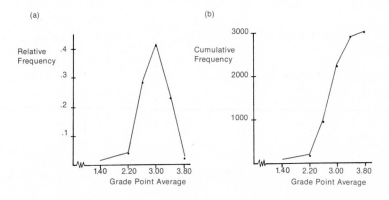

(a)

(b)

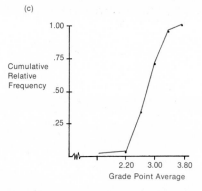

(c)

10.   Professionals at Work

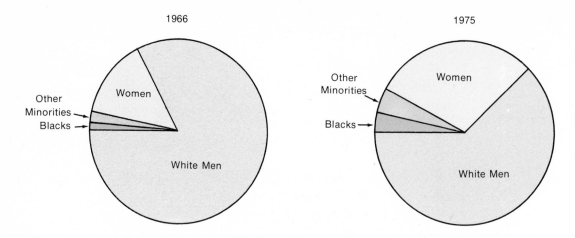

1966

1975

11.   Some Bonds Sold on the New York Stock Exchange
February 15

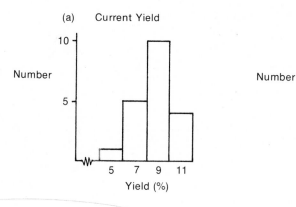

(a)   Current Yield

(b)   Sales (in thousands)

3.7 VOCABULARY AND SYMBOLS

categorical data	range
ranked data	≈
metric data	bar chart
raw data	histogram
tally	frequency polygon
score	symmetrical distribution
frequency	skewed to the left
frequency distribution	skewed to the right
class width	discrete variable
midpoint of a class	continuous variable
open-ended class	

3.8 REVIEW EXERCISES

1. Classify as categorical, ranked, or metric data:
(a) At an art show, a judge lists his preferences for paintings.
(b) A chemist makes 10 different solutions by adding 100 cc. of sulfuric acid to 10, 20, . . . , 100 cc. of water, and then measures the boiling point of each solution.
(c) Dresses are inventoried by style number and size.
(d) Four hundred firemen are queried about their ethnic backgrounds.
(e) Ten bars of chocolate are rated by an expert taster.
(f) Thirty students are given IQ tests.
(g) Consumption of alcohol per capita in 1976 is determined for each of 46 countries.

2. Make a two-way classification (by sex and political affiliation) of the members of the Rhodes Political Union (R = Republican, D = Democrat, I = Independent, M = Man, W = Woman):

	1	2	3	4	5	6	7	8	9	10	11	12	13	14	15	16	17	18	19	20	21	22	23	24	25	26	27	28	29	30	31	32
R	x				x		x										x				x		x								x	
D								x		x				x			x	x	x						x		x		x			
I		x	x	x	x		x			x		x	x	x		x		x						x		x		x		x		x
M	x	x			x	x	x		x	x			x	x		x	x		x					x		x	x	x				
W		x	x				x			x	x	x		x		x		x	x	x	x		x							x	x	x

3. Follow the guidelines to find the class width and the number of classes for each of the following distributions. What scores should the class with the lowest scores include?

	Number of scores	Highest score	Lowest score	Approximate number of classes
(a)	10,000	208	22	20
(b)	600	737	112	14
(c)	120	114	52	10

4. (a) The diameters of a sample of 120 half-inch bolts are measured, and found to range from .4954 to .5072 inches. How should the data be grouped?
(b) What are the midpoints of the two classes with the smallest scores?

5. The registrar at Dixon College computes grade-point averages with A = 4.0, B = 3.0, C = 2.0, D = 1.0, F = 0.0. The grade-point averages of 490 students range from 0.42 to 4.00. How should these averages be grouped in about 15 classes? (Note: It is impossible for a student to have a grade-point average greater than 4.00.)

6. The highest grade on an hour exam in statistics is 97 and the lowest is 62; 100 students took the exam. How should the grades be grouped in classes whose width is 5? How many classes are needed?

7. Find the class width, and the midpoint of the **first** and **second** classes in the following distributions.

(a) Trials for rats to learn discrimination		(b) Heights of 188 American males	
Number of trials	Number of rats	Height (in inches)	Number
1–3	7	60–64	10
4–6	5	65–69	150
7–9	3	70–74	120
10–12	3	75–79	8

8. Use a histogram and a frequency polygon (if suitable) to present graphically each of the distributions shown in Exercise 7.

9. Little did friends arriving for Peter's Groundhog Eve party on February 1 realize that the first thing they would be served was—a questionnaire! Here are the heights reported by the guests as one of their answers: 66, 69, 66, 69, 63, 66, 53, 72, 66, 71, 74, 74, 75, 62, 66, 66, 62, 64, 62, 58, 63, 70, 73, 71, 67, 67, 68, 62, 65, 68, 64 inches.

    (a) Tally and group the heights in classes of width 3.
    (b) Draw a histogram.
    (c) Do you notice anything very striking?

10. Miss Van Vleet's biology class (see Exercise 2, p. 41) also measured the lengths of pine needles in mm.:

Length:	1	2	3	4	5	6	7	8	9	10	11	12	13	14
Number:	8	11	9	7	14	22	27	26	41	56	43	30	96	2

    (a) Draw a frequency polygon to illustrate.
    (b) Is the distribution of pine needle lengths symmetric or skewed to the left or to the right?

11. College Women's basketball scores on February 17 were:

Game:	1	2	3	4	5	6	7	8	9	10	11	12	13	14	15	16	17
Winning Score:	73	94	63	50	71	89	108	62	76	71	79	57	98	89	66	87	85
Losing Score:	62	82	61	48	63	64	42	57	69	53	73	51	46	78	35	57	50
Win Margin:	11	12	2	2	8	25	66	5	7	18	6	6	52	11	31	30	35

    (a) Tally and group each row in classes of width 10.
    (b) Draw a histogram for each.
    (c) Which are (approximately) symmetric, and which are skewed?

12.   Here are the scores for men's winning and losing basketball scores on the same day:

Score	Number of Wins	Number of Losses
20–39	1	2
40–59	6	33
60–79	72	75
80–99	40	23
100–119	12	1
120–139	3	0

Draw a histogram of the winning scores, and another under it for the losing scores. Use the same scales for both so you can compare the two.

13.   Based on the distributions of winning and losing scores given in Exercise 12, find the distribution of win margins (winning minus losing score) and draw a histogram.

14.   A student (who is about to flunk the course) wishes to illustrate the information given in the table at the left below with a frequency polygon. He comes up with the object shown at the right. Find two major and a number of (relatively) minor errors in his work. ("Major" means "so fundamental it cannot be easily patched up.") List his errors, and indicate which are the major ones.

Ages of Residents of
Concord, New Hampshire
September 1, 1980

Age	Number of Residents (in thousands)
>0, but less than 10	4
>10, but less than 20	5
>20, but less than 30	7
>30, but less than 40	5
40 or over	9

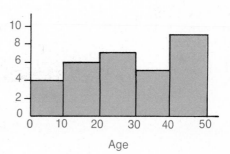

ANSWERS

1.   Categorical: (c), (d); ranked; (a), (e); metric: (b), (f), (g).

3.   (a) 10, 19, 20–29.
(b) 50, 13, 100–149.
(c) 5, 13, 50–54 (better than 51–55).

5. Fifteen classes of width .25: .26–.50, .51–.75, . . . , 3.76–4.00. Since no GPA is greater than 4.00, you must either have unequal classes (.25–.49, . . . , 3.50–3.74, but the last one, 3.75–4.00), or else end the classes on ''nice'' numbers instead of beginning them so.

7. (a) 3, 2; 3, 5; (b) 5, 62; 5, 67.

9. 51–53, 54–56, . . . , 75–77 with f = 1, 0, 1, 4, 5, 9, 5, 4, 1 (see histogram); or 52–54, 55–57, . . . , 73–75 with f = 1, 0, 1, 6, 8, 6, 4, 4; or 53–55, . . . , 74–76 with f = 1, 1, 0, 8, 9, 5, 4, 3.

(b)

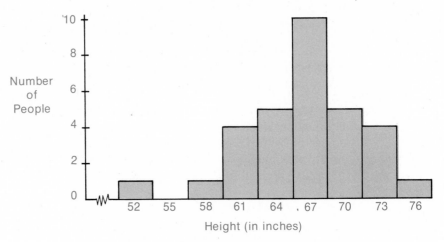

Heights of Guests
at Groundhog Eve's Party
February 1

(c) Two children were at the party: Sonja, 53 inches and Ismael, 58 inches.

11. (a) Winning scores: 50–59, 60–69, . . . , 100–109. f = 2, 3, 5, 4, 2, 1. Losing scores: 30–39, 40–49, . . . , 80–89, f = 1, 3, 5, 5, 2, 1. Winning margins: 0–9, 10–19, . . . , 60–69, f = 7, 4, 1, 3, 0, 1, 1.

(b)

College Women's Basketball Scores
February 17

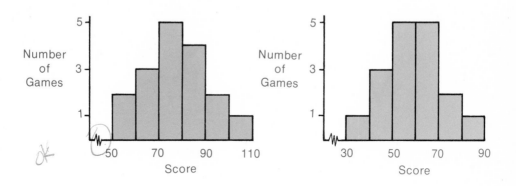

Winning Scores

Losing Scores

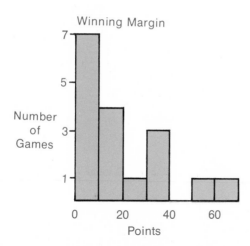

Winning Margin

(c) Winning and losing scores are almost symmetric, winning margins are skewed to the right.

13. Of course you can't; you can only find win margins if you know the win-lose scores for individual games (as in Exercise 11). You can't subtract histograms to get histograms of the differences of scores.

4
MEASURES OF CENTER, LOCATION, AND VARIABILITY

To summarize a set of scores, it is customary to give two numbers: one to measure the center or "average" score, and the other to give some idea of the amount of variability or the spread of the scores. The mode, median, and mean are three different measures of center. The mean especially is essential for further work in statistics. Become very familiar with it. Percentiles, quartiles, and percentile rank are discussed in this chapter also before taking up the second measure; these are easy once you understand the median.

Then we shall look at three measures of variability: the range, the variance, and the standard deviation. The last two are more difficult to find than the range, but concentrate on them; they are used so much in the rest of the book that you must have a clear understanding of them.

The chapter concludes with a discussion of standard scores, in which the unit of measurement is a standard deviation.

4.1   INTRODUCTION

We are looking first for a single number that will best represent a whole set of scores. There are three ways of doing this in common use: The **mode**, the **median**, and the **mean**. These three terms are lumped together as "measures of center" because each, in its own way, gives an idea of the score about which a whole set centers.

If you look at the coins in your wallet and say, "there are **more** pennies than anything else," you are using the mode. "My verbal aptitude score was well above the **middle** score of 500" uses the median. "The **average** of my quiz grades is 80" almost certainly states that the mean grade is 80, since "average" is most commonly used for "mean" by the general public—although some colleges which claim an average SAT score for entering freshmen of, say, 625 give the median rather than the mean, because their median SAT score is higher and therefore more impressive.

4.2   THE MODE

> ✓ **Definition:** The **mode** is the score which appears most frequently.
> **Notation:** *Mo*

The mode is the only measure of center that can be used for categorical data.

Example 1   *X:* 1, 1, 1, 2, 2, 2, 2, 2, 3, 3, 3, 3, 3, 3, 4, 4, 5, 6, 8.

A score of 3 appears 6 times, more frequently than any other score. *Mo* = 3.

Example 2   *Y:* 1, 1, 1, 2, 2, 2, 2, 2, 3, 3, 4, 4, 4, 4, 4, 5, 6, 6

Scores of 2 and 4 each appear 5 times. *Mo* = 2, 4. This set of scores is called **bimodal.** In contrast, the *X* scores above are **unimodal.** Occasionally the words **trimodal** (3) and **multimodal** (many) are used. As an adjective, mode becomes "modal": "For the *X*'s, the modal score is 3."

Example 3   *Z:* 1, 7, 8, 10, 12, 14, 17, 23, 24.

Each score appears the same number of times (once); the mode doesn't make sense. Often this will be the case when the number of scores is small.

Example 4

Magazine	Circulation (in thousands)
Mademoiselle	850
Psychology Today	835
New Yorker	485
Scientific American	540

The modal category is *Mademoiselle.*

The mode is the most common score. It is the easiest of the three measures of center to compute, necessitating only a tally of the scores, but no arithmetic beyond adding up the tally marks. It is not affected by extreme scores: The sets 1, 3, 5, 5, 5, 6 and 1, 3, 5, 5, 5, 213 have the same mode. One aberration—for example, a rat, being timed in a maze, who hasn't listened to the explanations and takes a nap on the way—doesn't affect the results. The mode is seldom used, however, except with categorical data. The modal frequency may vary slightly or very much from the frequency of other scores or classes.

4.3 THE MEDIAN

> **Definition:** The **median** is the middle score after the scores have been arranged in either increasing or decreasing order.
>
> **Notation:** *Md*

The median can be used for ranked or metric data, but not for categorical data.

Example 1
X: 8, 1, 10, 7, 17, 23, 12, 15, 24.
There are 9 scores here. After ranking them (1, 7, 8, 10, 12, 15, 17, 23, 24), the middle score is 12. *Md* = 12.

Example 2
Y: 1, 7, 8, 10, 12, 15, 17, 23, 24, 29.
Here there are 10 scores; the middle score is halfway between 12 and 15, so *Md* = 13.5. Note that the median of a set of scores need not be one of the scores.

With just a few scores, it is very much better to know what the median means and to use common sense in finding it. Some people like rules, however, and rules are extremely useful with a large number of scores, so here they are: Let n stand for the number of scores. If $n/2$ is an integer, the median is halfway between the $(n/2)$th score and the next higher score after they have been ranked in increasing order. If $n/2$ is not an integer, then round up to find the number of the median score; count that many scores from the smallest to find the median.

Example 3
n scores have been ranked in increasing order. What is the median if (a) $n = 200$, (b) $n = 201$?
(a) $200/2 = 100$. The median is halfway between the 100th and the 101st scores.
(b) $201/2 = 100.5$. The median is the 101st score. (But don't conclude that *Md* must equal 101!)

The median is about as easy to understand as the mode, but more work to find: you have to arrange the scores in order, and count how many there are. The arithmetic involved is of the simplest sort, however. Like the mode, the median is not affected by extreme scores: the sets 1, 3, 4, 6, 9, 12, and 1, 3, 4, 6, 9, 213 have the same median (5). The median cannot be used for categorical data. It **can** be used for ranked data, while neither the mode nor the mean can. If samples of coffee are tasted and ranked for bitterness, the middle sample can be said to be of "average" bitterness.

4.4   THE MEAN

> **Definition:** The **mean** is the sum of the scores divided by the number of scores.

The mean is used only for metric data.

> **Notation:** μ = mean of a population
> N = number of scores in a (finite) population.
> $\bar{X}$ (read this as "X bar") = mean of a sample of X scores.
> $\bar{Y}$ ($\bar{Z}$, etc.) = mean of a sample of Y (Z, etc.) scores.
> n = number of scores in a sample.
>
> **Formulas:** $\mu = \dfrac{\Sigma X}{N}$, $\bar{X} = \dfrac{\Sigma X}{n}$

Example 1

Quiz grades: 70, 75, 80, 80, 85, 90.

$$\text{Mean} = \frac{70 + 75 + 80 + 80 + 85 + 90}{6} = \frac{480}{6} = 80.0$$

> **Round-off:** Let us agree that computations will, in general, be carried to one more decimal place than in the given scores, and that = rather than $\approx$ will be used for the answer. Thus $\dfrac{-20}{60} + 185 \approx 184.7$. But don't round off until your final answer; keep at least one more decimal place until then.

CARRY AS MANY AS CAN TILL END

Example 2

Number of inches of rain each month in Albany, New York: 2.2, 2.0, 2.8, 2.7, 3.3, 3.0, 3.1, 2.9, 3.1, 2.6, 2.8, 2.9. What is the mean monthly rainfall in Albany?

$$\bar{X} = \frac{2.2 + 2.0 + 2.8 + 2.7 + 3.3 + 3.0 + 3.1 + 2.9 + 3.1 + 2.6 + 2.8 + 2.9}{12}$$

$$= \frac{33.4}{12} = 2.78 \text{ inches.}$$

To understand the mean better, think of it as the balance point of a balance beam with a weightless board; each weight on the board represents one score. For example, the scores 70, 75, 80, 80, 85, 90 might look like this:

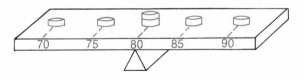

The balance point is at 80, which is the mean of the scores. The balance point will always be between the smallest and greatest weights. Check that the mean you compute is always between the smallest and largest scores.

What happens when each score in a set is decreased by the same amount? Look at the balance beam and think what happens when each weight is moved a distance of 5 units to the left:

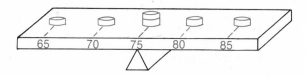

Is it clear that if the beam is to balance, the fulcrum must be moved 5 units to the left? But if you know where the balance point is **after** the weights have been moved 5 units to the left, you can tell where it must have been originally. Similarly, a constant A can be subtracted from each score to give easily handled numbers. After the mean of the new set is found, the constant A is added to it to find the mean of the original set.

Example 3   Find the mean of the X sample scores in the box below.

	X	$A = 5$ $X - 5$	$A = 20$ $X - 20$	$A = 70$ $X - 70$	$A = 80$ $A - 80$
	70	65	50	0	−10
	75	70	55	5	− 5
	80	75	60	10	0
	80	75	60	10	0
	85	80	65	15	5
	90	85	70	20	10
Sum:	480	450	360	60	0
Mean:	80	75	60	10	0
		$\overline{X} = 75 + 5$	$\overline{X} = 60 + 20$	$\overline{X} = 10 + 70$	$\overline{X} = 80 + 0$

✓   Note that any constant A can be used; choose an A which gives easily handled numbers. Ordinarily, the arithmetic will be simpler if you estimate the mean and let A be some convenient number near this estimate.

Example 4   Find the mean of the Y sample scores in the following box, letting (a) $A = 40$, (b) $A = 200$, (c) $A = 204.5$.

	(a)	(b)	(c)
Y	$Y - 40$	$Y - 200$	$Y - 204.5$
197	157	−3	−7.5
198	158	−2	−6.5
199	159	−1	−5.5
201	161	1	−3.5
203	163	3	−1.5
207	167	7	2.5
212	172	12	7.5
219	179	19	14.5
	1316	36	0.0

$$\overline{Y} = \frac{1316}{8} + 40 \qquad \overline{Y} = \frac{36}{8} + 200 \qquad \overline{Y} = \frac{0}{8} + 204.5$$

$$= 164.5 + 40 = 204.5 \qquad = 204.5 \qquad = 204.5$$

Clearly, 40 is a miserable choice for A (although the correct mean is found). $A = 200$ is probably the obvious choice. (c) was included to show you that if the constant subtracted from each score equals the mean, then $\Sigma(Y - \overline{Y}) = 0$. This will always be the case. **For any set of scores, the sum of the differences from the mean equals zero. If the sum of the differences from some constant is zero, then the constant is the mean.** See Exercise 12, page 63.

Sometimes a few scores are repeated many times. Then it's quicker to tally them and use one of the following formulas:

$$\overline{X} = \frac{\Sigma Xf}{n} \quad \text{or} \quad \mu = \frac{\Sigma Xf}{N}$$

Example 5   (a) Find the mean of X scores using $\overline{X} = \frac{\Sigma X}{n}$; (b) tally the scores and then find

the mean using $\overline{X} = \frac{\Sigma Xf}{n}$.

X: 1, 1, 1, 1, 2, 2, 2, 2, 2, 2, 2, 2, 2, 3, 3, 4, 4, 4, 4, 4 grams.

(a) $\overline{X} = \dfrac{48}{20} = 2.4$ grams.

(b)

X	f	Xf
1	4	4
2	9	18
3	2	6
4	5	20
	20	48

$\overline{X} = \dfrac{48}{20} = 2.4$ grams.

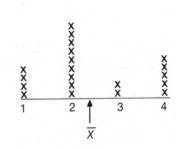

The formula $\overline{X} = \dfrac{\Sigma Xf}{n}$ "works" because multiplication is repeated addition:

$(1 + 1 + 1 + 1) + (2 + 2 + 2 + 2 + 2 + 2 + 2 + 2 + 2) + (3 + 3) + (4 + 4 + 4 + 4 + 4) = 1(4) + 2(9) + 3(2) + 4(5) = \Sigma Xf$.

It isn't worth doing this, however, if only one or two scores are repeated: if Y scores are 1, 2, 2, 3, 4, 6, 9, 9, then $\overline{Y} = \dfrac{36}{8} = 4.5$.

If you have only a few scores, the mean is the hardest measure of center to compute. If you have 1,000 scores, it's touch-and-go whether it's more difficult to add them for the mean or to arrange them in increasing order for the median. Be grateful that a computer can do either of these tasks for you, and arrange to use one (or at least a calculator) if you must find the mean of a great many scores. The mean can be found only for metric data, not for categorical or ranked data. But even for metric data the mean may not exist. Assume, for example, that a psychologist times five rats in a maze, but removes any rat which does not solve the maze in 5 minutes and simply records the time as "over 5." What is the mean time for the rats if their recorded times are 2, 3, 3, 4, over 5 minutes? The first four numbers can be added easily enough, but is "over 5" best represented by 6 or by 17 minutes?

The mean, in contrast to the mode and median, does depend on each of the scores. Sometimes this is not an advantage. For the scores 7, 9, 9, 31, is the median (9) or the mean (14) more representative? If you are an economist reporting on wages (in thousands of dollars) in a small shop, you would probably prefer the median, since the mean is raised so much by one high wage (the owner's).

So far, the mean does not seem to have a clear-cut advantage over the mode and the median, and yet it is used enormously more by statisticians. The first reason for this is that dependence of the mean on all the scores is usually a tremendous advantage. If your quiz grades in a course are 50, 50, 50, 90, 100, you would certainly prefer the mean (68) to the median (50)—and not just because it is higher, but because it fits your sense of justice. The second reason is that the mean is determined algebraically, and is amenable to algebraic operations. We shall find that when we look at a sample and make inferences about the population from which it is taken, it is the mean that will be of greatest use to us. You will appreciate the mean more fully as you learn to use it—and you will have lots of practice.

A minor algebraic advantage of the mean can be seen immediately. Consider the following yields of bushels of beans in experimental plots:

Field *X:* 1, 3, 3, 3, 4, 4, 5, 5, 6, 6 bushels. $Mo = 3$, $Md = 4$, $\overline{X} = 4$ bushels, $n = 10$.

Field *Y:* 1, 1, 2, 2, 4, 4, 4, 6, 6, 6, 6, 7, 7, 9, 10 bushels. $Mo = 6$, $Md = 6$, $\overline{Y} = 5$ bushels, $n = 15$.

Now suppose the data from plots in both fields are joined together to give a set of 25 scores. To find the new mode and median it is necessary to rearrange the scores as follows, and start over:

Fields X and Y: 1, 1, 1, 2, 2, 3, 3, 3, 4, 4, 4, 4, 4, 5, 5, 6, 6, 6, 6, 6, 6, 6, 7, 7, 9, 10 bushels. $Mo = 6$, $Md = 4$ bushels.

To find the mean of the joint set, however, it is not necessary to interweave the scores in this fashion; we can reason as follows: $\overline{X} = \dfrac{\Sigma X}{10} = 4$, so $\Sigma X = 40$, and $\overline{Y} = \dfrac{\Sigma Y}{15} = 5$, so $\Sigma Y = 75$. Therefore the sum of all 25 scores is $\Sigma X + \Sigma Y = 40 + 75 = 115$, and the mean of the joint set is $\dfrac{115}{25} = 4.6$ bushels. This example may not be persuasive, since the number of scores is so small; you can probably count or re-order them mentally to find the mode and median, without writing down the joint set—but imagine doing the same with a hundred or more scores in each set!

4.5 EXERCISES

1. Find the mode, median, and mean of each of the following:
    (a) X: 0, 0, 1, 1, 1, 2, 4, 5, 5, 10 microvolts.
    (b) Y: 1, 3, 2, 4, 0, 1, 0, 6, 0 chickens.
    (c) Z: 498, 502, 503, 501 pints.

2. Translate from English into mathematical language: "To find the mean of X scores in a sample of size n, subtract a constant A from each score, find the mean of the new set, and then add A; label the result $\overline{X}$."

3. Twelve students in a statistics class receive the following quiz grades:

Quiz 1: 72, 75, 75, 97, 54, 72, 86, 72, 63, 79, 82, 91
Quiz 2: 78, 42, 72, 88, 86, 97, 91, 79, 82, 86, 91, 74

    (a) Find the mode, median, and mean for each quiz.
    (b) Find the mode, median, and mean for both quizzes together (24 grades).
    (c) Which measure of center best represents the grades on the second quiz?

4. Find the mode, median, and mean for each of the following sets of scores, if possible; if not, explain why.
    (a) Weekly pay (in dollars) of cafeteria workers: 70, 60, 70, 70, 6, 4, 90, 30, 16, 80, 12, 70.
    (b) Grades of students on a statistics quiz: 50, 100, 90, 100, 50, 60, 80, 70, 40, 100, 90, 95, 75, 50, 74, 81, 76, 69, 75.

(c) A scale used for weighing letters measures up to 10 ounces. Twenty-four pieces of first class mail have the following weights (in ounces): .7, .9, .8, 1.1, .5, .8, .5, 1.1, .7, .6, .4, .9, 1.3, .4, .6, .5, 1.2, .4, .3, .5, over 10, 3.0, .2, .4.

(d) Number of cars parked in the Student Center parking lot at noon each day, April 17–23: 212, 650, 435, 672, 510, 621, 287.

(e) Number of each make of car parked in the Student Center parking lot at noon on April 22: Ford 131, Chevrolet 142, VW 183, other 165.

5. The annual salaries of 20 men are given below.

$8,000	$ 9,800	$11,200	$11,700	$12,400
8,500	10,000	11,400	11,800	12,800
8,700	10,400	11,600	11,900	13,500
9,200	10,800	11,700	11,900	43,000

(a) Find the median and the mean salary.

(b) Is the median or the mean more representative? Why?

6. Which measure of central tendency would each of the following be likely to use to summarize his data?

(a) The manager of a shoe store contemplating the shoes in his shop listed by size.

(b) An economist studying the income of heads of families in low-cost housing in Pawtucket, Rhode Island.

(c) An engineer studying the maximum load supported by each of 100 cables produced by his company.

7. Find the mean of the following scores. Subtract a constant A from each score to simplify the arithmetic; choose A wisely.

(a) X: 15, 17, 18, 19, 20, 22, 18, 16, 21, 17.

(b) Y: 514, 514, 520, 536, 518, 504, 510, 520, 521.

(c) Z: .0021, .0027, .0019, .0018, .0023, .0026, .0029, .0018.

8. The rainfall in inches each month in Miami, Florida, is shown in the following table. Find (a) the mean monthly rainfall, (b) the median.

Month	Jan	Feb	March	April	May	June	July	Aug	Sept	Oct	Nov	Dec
Inches of rain	2.0	1.8	2.5	3.7	6.8	7.5	6.5	7.1	9.7	8.0	2.6	1.8

9. Thirty families live in an apartment house on Delancey Street in Brooklyn. The number of children in each family is as follows:

```
3 1 2 0 4 6 1 0 1 5 2 3 3 0 7
2 2 0 3 1 2 2 4 2 4 0 4 3 1 4
```

Tally and then find (a) the mean, (b) the median, and (c) the modal number of children.

10. In each of the following, decide whether it is possible for the given measure of center to have the given value if the other requirements are satisfied. If impossible, explain why. If possible, give a set of scores which satisfy all the requirements (For example: $n = 8$, lowest score $= 7$, $Mo = 5$. Answer: impossible, since every score is greater than 5.)

(a) $n = 5$, lowest score $= 4$, range $= 10$, $\overline{X} = 14$.
(b) $n = 5$, lowest score $= 4$, range $= 10$, $Md = 14$.
(c) $n = 5$, lowest score $= 50$, highest score $= 100$, $\overline{X} = 55$.
(d) $n = 5$, $\overline{X} = 6$, $Md = 5$.

11. In each of the following, give two examples of scores which meet the given requirements. (For example, $n = 6$, range $= 10$, $Mo = 4$; range $=$ difference between highest and lowest scores. Answers: 3, 4, 4, 4, 7, 13, or 0, 2, 4, 4, 4, 10, for example.)

(a) $n = 5$, range $= 7$, $Md = 54$.
(b) $n = 3$, range $= 10$, $\overline{X} = 90$.
(c) $n = 8$, range $= 20$, $Mo = 42$, $Md = 45$.
(d) $n = 4$, $\overline{X} = 10$, $Md = 8$.

12. Study this proof that $\Sigma(X - \overline{X}) = 0$:

$$\Sigma(X - \overline{X}) = \Sigma X - \Sigma\overline{X} \qquad \text{(Rule 1 for } \Sigma \text{, page 3)}$$
$$= \Sigma X - n\overline{X} \qquad \text{(Rule 3 for } \Sigma \text{, page 3)}$$
$$= \Sigma X - n\left(\frac{\Sigma X}{n}\right) \qquad \left(\overline{X} = \frac{\Sigma X}{n}\right)$$
$$= 0$$

Prove that if $\Sigma(X - A) = 0$, where A is a constant, then $A = \overline{X}$.

13. The mean grade point average (GPA) of 1,000 students in the School of Liberal Arts at Triple University during one semester was 2.85, while it was 2.90 for 200 students in the School of Engineering, and 3.10 for 400 students in the School of Education. What was the mean GPA of all the undergraduates in the three schools of Triple University that semester?

ANSWERS

1. (a) $Mo = 1$, $Md = 1.5$, $\overline{X}$ (or μ) $= 2.9$ microvolts; (b) $Mo = 0$, $Md = 1$, $\overline{Y}$ (or μ) $=$ 1.9 chickens; (c) no Mo, $Md = 501.5$ $\overline{Z}$ (or μ) $= 501.0$ pints.

2. $\overline{X} = \dfrac{\Sigma(X - A)}{n} + A$.

3. (a) Quiz 1: $Mo = 72$, $Md = 75$, $\overline{X}$ (or μ) $= \dfrac{918}{12} = 76.5$.

Quiz 2: $Mo = 86$ or 91 (bimodal), $Md = 84$, $\bar{Y} = \dfrac{966}{12} = 80.5$.

(b) $Mo = 72$, $Md = 79$, $\bar{Z} = \dfrac{918 + 966}{24} = 78.5$ (or $\bar{Z} = \dfrac{76.5 + 80.5}{2} = 78.5$, since the same number of students took each quiz).

4. (a) $Mo = \$70$; $Md = \$65$; $\bar{X} = \$48.20$. (b) $Mo = 50, 100$ (bimodal); $Md = 75$; $\bar{X} = 75.0$. (c) $Mo = .4$ and $.5$ (bimodal); $Md = .65$; no mean (open-ended). (d) Mo is meaningless; $Md = 510$; mean $= 483.9$. (e) $Mo = 183$; no median or mean (categorical data).

5. (a) $Md = \$11,500$; $\bar{X}$ (or μ) $= \dfrac{250,300}{20} = \$12,520$.

   (b) The median, since one high salary pulls up the mean so much, but does not affect the median.

6. (a) Mo; (b) $\bar{X}$ (no extraordinarily large income is allowed); (c) $\bar{X}$.

7. (a) $A = 20$: $\bar{X} = \dfrac{-17}{10} + 20 = 18.3$; or $A = 18$: $\bar{X} = \dfrac{3}{10} + 18 = 18.3$.

   (b) $A = 500$; $\bar{Y} = \dfrac{162}{9} + 500 = 518.0$, or $A = 520$; $\bar{Y} = \dfrac{-18}{9} + 520 = 518.0$.

   (c) $A = .0020$: $\bar{Z} = \dfrac{.00021}{8} + .0020 = .00226$.

8. (a) $\dfrac{60.0}{12} = 5.00$ inches; (b) 5.1 inches.

9. Number of children 0 to 7 with $f = 5, 5, 7, 5, 5, 1, 1, 1$.
   (a) $\bar{X} = \dfrac{72}{30} = 2.4$, $Md = 2$, $Mo = 2$.

10. (a) Impossible: highest score is 14, $\bar{X} < 14$.
    (b) 4, 7, 14, 14, 14.
    (c) Impossible: scores with the lowest mean are 50, 50, 50, 50, 100, so $\bar{X} = 60$ at least.
    (d) 3, 4, 5, 6, 12 or 0, 0, 5, 10, 15.

11. (There are many possible answers to each part.)
    (a) 48, 50, 54, 54, 55, or 54, 54, 54, 57, 61.
    (b) 85, 90, 95, or 86, 88, 96.
    (c) 42, 42, 42, 44, 46, 48, 50, 62, or 40, 42, 42, 42, 48, 50, 55, 60.
    (d) 7, 8, 8, 17, or 4, 6, 10, 20.

12. $\Sigma(X - A) = 0$ (Given)
    $\Sigma X - \Sigma A = 0$ (Rule 1 for Σ, page 3)
    $\Sigma X - nA = 0$ (Rule 3 for Σ, page 3)

    $$A = \dfrac{\Sigma X}{n}$$

    $$A = \bar{X}$$

13. $\dfrac{1,000(2.85) + 200(2.90) + 400(3.10)}{1600} = 2.919$.

4.6 QUARTILES, PERCENTILES, AND PERCENTILE RANK

The same reasoning that is used for the median may be used to find quartiles and percentiles.

> **Definitions:** The **first quartile** is the number that is larger than one quarter of the scores (and smaller than three quarters).
> The **second quartile** is the median.
> The **third quartile** is larger than three quarters of the scores.
> The **sixtieth percentile** is larger than 60% of the scores (and smaller than 40%).
>
> **Notations:** Q_1 = the first quartile.
> Q_3 = the second quartile.
> P_{60} = the sixtieth percentile.

Note that $P_{25} = Q_1, P_{50} = Md, P_{75} = Q_3$. Quartiles and percentiles are called **measures of location.**

If eight scores are ranked, Q_1 will be larger than two scores and smaller than six; the first quartile will be halfway between the second and third scores. If $\dfrac{n}{4}$ is an integer, then Q_1 is halfway between this score and the next higher. If $\dfrac{n}{4}$ is not an integer, it is customary to round up to find Q_1. Thus, if $n = 50$, Q_1 will be the thirteenth score after they are put in increasing order, since $\dfrac{50}{4} = 12.5$, which rounds up to 13. 12 scores are smaller and 37 are larger. But if you think of the thirteenth score as being half in each group, there are 12.5 scores below and 37.5 above Q_1.

Q_3 and P_a are determined in similar fashion: multiply n by $\dfrac{3}{4}$ and by $\dfrac{a}{100}$, respectively. If the result is an integer, choose the number halfway between this score and the next; otherwise, round up.

Example 1   Ninety scores are ranked in increasing order. Find (a) Q_1, (b) Q_3, (c) P_{60}, (d) P_{16}.

(a) $\dfrac{1}{4}(90) = 22.5$: Q_1 is the 23rd score; count up to find it. (*Not* $Q_1 = 23$).

(b) $\frac{3}{4}(90) = 67.5$; Q_3 is the 68th score; count up to find it.

(c) $\frac{60}{100}(90) = 54$; P_{60} is halfway between the 54th and 55th scores.

(d) $\frac{16}{100}(90) = 14.4$; P_{16} is the 15th score.

✓ | **Definition:** The **percentile rank** of a score is the percentage of scores which are smaller.

If 75 is the 349th score in a list of 400 scores arranged in increasing order, then 348 are smaller. $\frac{348}{400}(100) = 87\%$ are smaller than 75; the percentile rank of a score of 75 is 87.

Example 2   A high school student has an average grade of 92.5; he is sixteenth highest in a class of 300. What is his percentile rank?

He does better than $300 - 16 = 284$ in his class. His percentile rank is $\frac{284}{300}(100) = 94.7$. (His average grade of 92.5 is not relevant.)

4.7 EXERCISES

1.   For the following grades on a statistics quiz, find (a) the first quartile, (b) the third quartile, (c) the percentile rank of a grade of 88: 78, 42, 72, 88, 97, 56, 91, 79, 82, 86, 91, 74, 80.

2.   For the scores on the Miller Personality Test which are tallied on page 31, find (a) the first quartile, (b) the fortieth percentile, (c) the seventy-fifth percentile, (d) the percentile rank of a score of 21, (e) the percentile rank of a score of 30.

3.   The students in a statistics course at Hyde College obtain the following grades on a quiz:

24	30	32	40	48	50	58	60	60	62	63	64	64	67	67	67
70	70	71	72	74	74	75	75	76	77	78	78	78	78	79	80
81	83	83	83	84	84	84	84	85	86	87	88	88	88	89	90
92	92	93	95	95	97	98	99	100	100						

(a) What is the modal grade?
(b) What is the median grade?
(c) Find Q_1 and P_{10}.
(d) What is the percentile rank of a score of 93?

4. Comment on each of the following:
   (a) A high school principal said to one of his teachers, "All the classes in this school took the same test, but your class scored below the average. I don't want any of our classes to be below the average."
   (b) "I am in the first quartile of my class," Ed boasted.
   (c) "My percentile rank in a class of 90 students is 100," Ann boasted.

5. TV ratings (percent of viewers tuned in to that show) of the "Golden Boy" series are as follows over an 8-week period: 12.4, 14.5, 10.6, 18.6, 16.1, 15.2, 13.3, 10.0. What is (a) the median rating? (b) the mean rating? (c) the percentile rank of a 12.4 rating?

6. Find (a) Md, (b) Q_1, (c) P_{10}, (d) the percentile rank of a score of 106.0 for the following weights (in grams) of 22 mice: 70.4, 71.5, 72.3, 72.4, 76.8, 80.5, 84.3, 86.2, 87.1, 89.8, 90.6, 92.0, 94.5, 96.8, 100.2, 104.8, 105.7, 106.0, 109.0, 113.1, 113.4, 113.7.

7. Give 10 scores whose median is 54 and $P_{80} = 70$.

8. Give a set of scores for which the percentile rank of a score of 15 is 10 and $Q_3 = 32$.

ANSWERS

1. (a) $\frac{1}{4}(12) = 3$, so Q_1 is the mean of the third and fourth grades when in increasing order; $Q_1 = \frac{74 + 78}{2} = 76.0$.

   (b) $\frac{3}{4}(12) = 9$, so Q_3 is the mean of the ninth and tenth grades; $Q_3 = \frac{88 + 91}{2} = 89.5$.

   (c) 8 grades are less; the percentile rank of a grade of 88 is $\frac{8}{12}(100) = 66.7$.

2. (a) $\frac{1}{4}(82) = 20.5$. The 21st score is $Q_1 = 22.0$.

   (b) $\frac{40}{100}(82) = 32.8$. The 33rd score is $P_{40} = 24.0$.

   (c) $\frac{75}{100}(82) = 61.5$. The 62nd score is $P_{75} = Q_3 = 28$.

   (d) $\frac{12}{82}(100) = 15$.

   (e) $\frac{70}{82}(100) = 85$.

3. (a) 78, 84 (bimodal)
   (b) 78
   (c) $Q_1 = 67, P_{10} = 50$
   (d) $\left(\frac{50}{58}\right)100 = 86.2$

4. (a) "Average" probably signifies "mean," but it is not clear. If no class is below the mean, all have exactly the same result; this is highly unlikely.

   (b) He could be **at** the first quartile but not in it, and he shouldn't boast if he is better than just one quarter of the class. "I am in the top quarter of my class" is probably what he meant.

   (c) Even if Ann is first in her class, her percentile rank is $\frac{89}{90}(100) = 99$, not 100.

5. (a) 13.9 (b) $\frac{110.7}{8} = 13.8$ (c) $\frac{2}{8}(100) = 25$.

6. (a) 91.3 grams; (b) 80.5 grams; (c) 72.3 grams; (d) $\frac{17}{22}(100) = 77.3$.

7. There are many possible answers; one is 10, 30, 40, 45, 50, 58, 60, 65, 75, 90.

8. Again there are many possible answers, for example 10, 15, 20, 25, 30, 30, 31, 32, 33, 39.

4.8 INTRODUCTION TO VARIABILITY

"Oklahoma and the Canary Islands have the same mean annual temperature." Would you therefore be equally comfortable throughout the year in both places? Oklahoma may vary from 0° to 100° during the year, the Canary Islands from 60° to 80°, so during most of the year their climates are very different.

The following sets of scores have the same mode, median, mean, and number of scores, but differ markedly:

> X: 40, 50, 50, 50, 50, 60
> Y: 0, 30, 50, 50, 70, 100

These sets of scores differ in their spread about the center. We shall look now at three different measures of spread or variability.

4.9 THE RANGE

The range of a distribution is the difference between the highest and lowest scores.

Example 1

> X: 5, 12, 13, 13, 14, 15, 15, 15, 18, 20.
> The range is 20 − 5 = 15.

Example 2

> Y: 5, 6, 10, 11, 11, 19, 19, 19, 20, 20.
> The range is again 20 − 5 = 15.

The range is easy to understand and easy to compute. But the scores in Examples 1 and 2 have the same mean (14) and range (15), but "look" quite different.

4.10 VARIANCE AND STANDARD DEVIATION

The range depends only on the extreme scores. Let us search now for a measure which depends on **all** the scores. It should have other characteristics as well: it should be small when all the scores are close together and large when the scores are widely scattered; it should not get large just because the number of scores is large.

How about taking the sum of the deviations from the mean? The sum $\Sigma(X - \overline{X})$ is always equal to 0, as we proved in Exercise 12, page 63, and as the following example illustrates again:

$$X: 1, 2, 6; \quad \overline{X} = 3; \quad \Sigma(X - \overline{X}) = (1 - 3) + (2 - 3) + (6 - 3)$$
$$= -2 - 1 + 3 = 0$$

Next you might decide to ignore the minus signs: add the absolute values* of the deviations from the mean. In the example above, you would get $|1 - 3| + |2 - 3| + |6 - 3| = 2 + 1 + 3 = 6$. If there were 1,000 scores, this sum might be large even if all the scores were between 1 and 6. If the sum of the absolute values of the deviations from the mean is divided by the number of scores, we arrive at a measure of variability called the **mean absolute deviation**. It satisfies the criteria set up in the first paragraph, but is seldom used now. One reason for this is that absolute values are difficult to work with algebraically; the mean absolute deviation does not enter nicely into formulas indicating its relation to other statistics.

Any number, positive or negative, becomes positive when squared: $(+4)^2 = +16$, $(-2)^2 = +4$, $(-1.2)^2 = +1.44$. Instead of taking the mean of the absolute values of the deviations of scores from the mean, we take the mean of the squares of these deviations, arriving at the mean square deviation $\dfrac{\Sigma(X - \mu)^2}{N}$ for the variance of a finite population.

σ (sigma) is the Greek lower case s.

Notation: σ^2 = variance of a population

Formula: $\sigma^2 = \dfrac{\Sigma(X - \mu)^2}{N}$

There is a comparable formula often used for the variance of a sample, but it will cause only a little confusion now and save a lot more confusion later on if we define the variance of a sample differently:

Notation: s^2 = variance of a sample

Formula: $s^2 = \dfrac{\Sigma(X - \overline{X})^2}{n - 1}$

---

*The absolute value of a number is its magnitude without regard for sign; the notation is two vertical lines, one on either side of the number. Thus $|7| = 7$ and $|-7| = 7$.

Unfortunately, the term "variance" is used by statisticians whether the sample variance has $n - 1$ or n in the denominator, so check the usage if you are reading any other book or an article. Here, $n - 1$ will always be used.

Why is the denominator in the sample variance changed from n to $n - 1$? Later we shall want to estimate the population variance by using the variance of a sample. If n is the denominator for the sample variance, the estimate will be consistently too low. Note that a fraction is increased if the denominator is decreased (1/2 is greater than 1/3). Changing n to $n - 1$ in the denominator increases the fraction just enough so that the sample variance will, on the average, give a good estimate of the population variance. "On the average" means that if the variance of all possible different samples is computed, the average (mean) of all these variances will approximate pretty well the population variance.

The variance satisfies all the criteria set up at the beginning of this section for a measure of variability. It has one drawback, however: it measures in square units. If X scores are in inches, for example, then so are $\bar{X}$ and the difference $X - \bar{X}$; then $(X - \bar{X})^2$ and the variance are in square inches. To get back to the original unit, the (positive) square root is taken. The square root of the variance is called the **standard deviation.** It is another measure of variability, and it is measured in the same units as the scores.

> **Notations:** σ = standard deviation of a population.
> s = standard deviation of a sample.
>
> **Formulas:** $\sigma = \sqrt{\dfrac{\Sigma(X - \mu)^2}{N}}$, $s = \sqrt{\dfrac{\Sigma(X - \bar{X})^2}{n - 1}}$

Note that to find the standard deviation using this formula you must first find the mean. Be careful to square first and then add; $\Sigma(X - \bar{X})^2$ is not the same as $[\Sigma(X - \bar{X})]^2$. Use Table 1, Appendix C, or a calculator to find the square root.

Example 1

Find the variance and standard deviation of the X scores in pounds: 1, 2, 6, if these are scores in (a) population, (b) a sample.

(a) $\mu = \dfrac{9}{3} = 3.0$ pounds; (b) $\bar{X} = \dfrac{9}{3} = 3.0$ pounds.

X	$X - \mu$ or $X - \bar{X}$	$(X - \mu)^2$ or $(X - \bar{X})^2$
1	-2	4
2	-1	1
6	3	9
9		14

(a) $\sigma^2 = \dfrac{14}{3} = 4.67$ pounds squared, $\sigma = \sqrt{4.67} = 2.2$ pounds.

(b) $s^2 = \dfrac{14}{2} = 7.00$ pounds squared, $s = \sqrt{7.00} = 2.6$ pounds.

Example 2   Find the variance and standard deviation of Y: 10, 12, 14, 18 if these are eggs laid by Farmer Brown's chickens in a random sample of 4 days. $\overline{Y} = \dfrac{54}{4} = 13.5$ eggs.

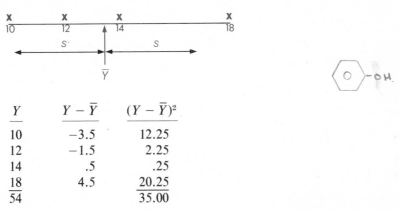

Y	$Y - \overline{Y}$	$(Y - \overline{Y})^2$
10	−3.5	12.25
12	−1.5	2.25
14	.5	.25
18	4.5	20.25
54		35.00

$s^2 = \dfrac{35}{3} = 11.67$ square eggs. $s = \sqrt{11.67} = 3.4$ eggs. (Do you see why the units in the standard deviation make more sense?)

Be careful not to round off too early. If you are finding σ or s to one decimal place, keep at least two until after you have taken the square root.

This last example involved more arithmetic than Example 2; it is not hard to find examples where the necessary computations are still more complicated. Fortunately, another form of the formula for variance or standard deviation can be found that frequently simplifies computations. These computational formulas are

$$\sigma^2 = \dfrac{\Sigma X^2 - \dfrac{(\Sigma X)^2}{N}}{N}, \quad \sigma = \sqrt{\sigma^2}$$

$$s^2 = \dfrac{\Sigma X^2 - \dfrac{(\Sigma X)^2}{n}}{n - 1}, \quad s = \sqrt{s^2}$$

Squares of numbers are given in Table 1, Appendix C; these formulas are also handy for a calculator. Be very careful to distinguish between ΣX^2 (square first and then add) and $(\Sigma X)^2$ (add first and then square the sum).

Where do these computational formulas come from? See if you can follow these steps, at least closely enough to realize that the computational formulas are legitimate (remember that $\overline{X}$ is a constant, so $\Sigma \overline{X}^2 = n\overline{X}^2$, by Rule 3 for Σ, page 3):

$$\Sigma(X - \bar{X})^2 = \Sigma(X^2 - 2X\bar{X} + \bar{X}^2)$$

$$= \Sigma X^2 - \Sigma 2X\bar{X} + \Sigma \bar{X}^2 \qquad \text{(Rule 1 for } \Sigma, \text{ page 3)}$$

$$= \Sigma X^2 - 2\bar{X}\Sigma X + n\bar{X}^2 \qquad \text{(Rules 2 and 3 for } \Sigma, \text{ page 3)}$$

$$= \Sigma X^2 - 2\frac{\Sigma X}{n}\Sigma X + n\left(\frac{\Sigma X}{n}\right)^2 \qquad \left(\bar{X} = \frac{\Sigma X}{n}\right)$$

$$= \Sigma X^2 - 2\frac{(\Sigma X)^2}{n} + \frac{(\Sigma X)^2}{n}$$

$$= \Sigma X^2 - \frac{(\Sigma X)^2}{n}$$

Now divide $n - 1$. This shows that the definition of the variance of a sample and the computational formula for it are equivalent. The proof for the variance of a population is similar.

Example 3   Find the standard deviation of (a) the population and (b) the sample resistances Z: 1.2, 2.1, 3.0 ohms.

Z	Z^2
1.2	1.44
2.1	4.41
3.0	9.00
6.3	14.85

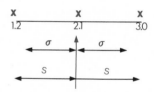

$$\text{(a) } \sigma = \sqrt{\frac{14.85 - 6.3^2/3}{3}} = .73 \text{ ohms.}$$

$$\text{(b) } s = \sqrt{\frac{14.85 - 6.3^2/3}{2}} = .90 \text{ ohms.}$$

Example 4   Compare the two methods (definition and computational formula) for finding the variance of the sample X: 1, 2, 3, 5, 10.

	Definition			Shortcut	
X	$X - \bar{X}$	$(X - \bar{X})^2$		X	X^2
1	−3.2	10.24		1	1
2	−2.2	4.84		2	4
3	−1.2	1.44		3	9
5	.8	.64		5	25
10	5.8	33.64		10	100
21		50.80		21	139

$\bar{X} = 21/5 = 4.2$

$s^2 = 50.80/4 = 12.7$

$$s^2 = \frac{139 - 21^2/5}{4} = \frac{139 - 88.2}{4}$$

$$= 12.7$$

In computing the mean, we found that arithmetic can often be simplified by subtracting a constant from each of the scores in a sample. The same device can be used here, and is even more effective because the constant doesn't have to be added again at the end. The "spread" of scores remains the same if a constant A is subtracted from each of them; since the mean is reduced by A also, the square deviations of the new scores about the new mean remain the same. Many times, A can be chosen wisely so that arithmetic is much easier.

Example 5   Find the standard deviation of X scores in a sample: 15.5, 17.5, 21.5, 23.5 pounds (a) directly, (b) after subtracting 17.5 from each score.

(a)		(b)	
X	X^2	$X - 17.5 = Y$	Y^2
15.5	240.25	−2.0	4.00
17.5	306.25	0.0	0.00
21.5	462.25	4.0	16.00
23.5	552.25	6.0	36.00
78.0	1,561.00	8.0	56.00

$$\text{Variance of } X \text{ scores} = s_X{}^2 = \frac{1{,}561 - (78)^2/4}{3} = \frac{40.00}{3} = 13.33, \quad s_X = 3.65$$

pounds.

$$\text{Variance of } Y \text{ scores} = s_Y{}^2 = \frac{56 - 8^2/4}{3} = \frac{40.00}{3} = 13.33, \ s_Y = 3.65 \text{ pounds.}$$

Note the use of subscripts here to distinguish between the standard deviation of the X scores, s_X, and of the Y scores, s_Y.

Example 6   Find the variance of Z scores in a population: 42, 47, 54, 53, 48, 51, 50, 49, 52 miles.

Z	$Z - 50 = X$	X^2
42	−8	64
47	−3	9
54	+4	16
53	+3	9
48	−2	4
51	1	1
50	0	0
49	−1	1
52	2	4
	−4	108

$$\sigma^2{}_Z = \sigma^2{}_X = \frac{108 - (-4)^2/9}{9} = \frac{108 - 1.78}{9} = 11.8 \text{ square miles.}$$

If many scores are repeated, it's easier to use the following formulas:

$$\sigma^2 = \frac{\Sigma X^2 f - \dfrac{(\Sigma X f)^2}{N}}{N}, \quad s^2 = \frac{\Sigma X^2 f - \dfrac{(\Sigma X f)^2}{n}}{n-1}$$

Since you need $\Sigma X f$ anyway, find $\Sigma X^2 f$ by multiplying each Xf by X, then add.

Example 7   Find σ for the scores in the box below.

X	f	Xf	$X^2 f$
1	10	10	10
2	12	24	48
3	3	9	27
	25	43	85

$$\sigma = \sqrt{\frac{85 - 43^2/25}{25}} = \sqrt{.44} = .7.$$

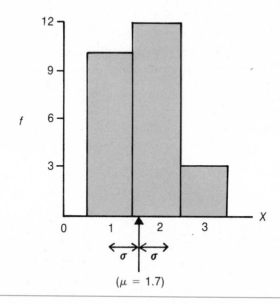

$(\mu = 1.7)$

As before, a constant may be subtracted from each score without changing the variance or the standard deviation.

Example 8   Find s for the scores in the box below.

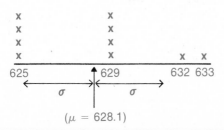

X	f	$Y = X - 629$	Yf	Y^2f
625	4	-4	-16	64
629	4	0	0	0
632	1	3	3	9
633	1	4	4	16
	10		-9	89

$$s = \sqrt{\frac{89 - (-9)^2/10}{9}} = \sqrt{8.99} = 3.0.$$

4.11   INTERPRETATION OF THE STANDARD DEVIATION

For any set of scores:

1. At least 3/4 of the scores are within 2 standard deviations of the mean.
2. The standard deviation is never more than one-half the range.

CHECKS

(a)

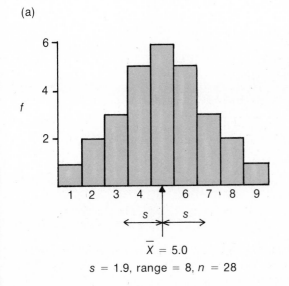

$\overline{X} = 5.0$

$s = 1.9$, range $= 8$, $n = 28$

(b)

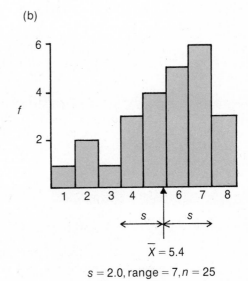

$\overline{X} = 5.4$

$s = 2.0$, range $= 7$, $n = 25$

(c)

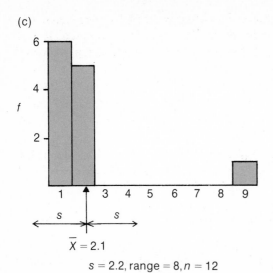

$\overline{X} = 2.1$

$s = 2.2$, range $= 8, n = 12$

(d)

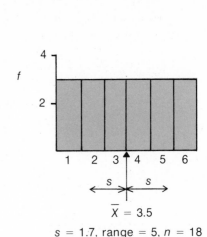

$\overline{X} = 3.5$

$s = 1.7$, range $= 5, n = 18$

Check this rule on the following examples.
If the histogram is mound shaped ⌒ or bell shaped ◠ :

CHECKS

1. Approximately 2/3 of the scores are within 1 standard deviation of the mean.
2. Approximately 95% of the scores are within 2 standard deviations of the mean.

Check this rule on (a) above, which is approximately bell shaped.

4.12 EXERCISES

1. Find (a) the range, (b) the mean, (c) the standard deviation using the definition, and (d) the standard deviation using the computational formula and whatever shortcut you can find, for each of these sets of scores in samples:

(i) *X:* 1, 1, 2, 4 bullets.
(ii) *Y:* 51, 54, 59, 62, 63, 63, 67, 69 pounds.
(iii) *Z:* 30, 24, 21, 25, 20, 30, 22, 28.
(iv) *W:* 14, 16, 17, over 18 seconds.

2. Find (a) the range, (b) the standard deviation for the melting points and then for the boiling points of these alcohols:

Alcohol	Melting point (°C)	Boiling point (°C)
Methyl	− 97	64.7
Ethyl	−114	78.3
Isopropyl	− 89	82.3
Allyl	−129	97.0
Isobutyl	−108	107.9
Isoamyl	−117	131.5

3. Find (a) the range, and (b) the standard deviation for these grades on a statistics quiz: 78, 42, 72, 88, 86, 97, 91, 79, 82, 86, 91, 74.

4. The systolic and diastolic blood pressures of 10 adults were taken and the difference d between the readings for each individual was computed. It was found that $\Sigma d = 300$, and that $\Sigma d^2 = 9{,}200$. Find the mean and the variance of these differences.

5. The Upton Calculator Company received 5 shipments of 1,000 batteries. The numbers of defective batteries in each shipment were 4, 12, 6, 4, and 8, respectively. What is the mean number of defective batteries per shipment, and what is the standard deviation?

6. An instructor in a calculus course gave the GLUG algebra test to his students at the beginning of the semester, with the following results:

78	72	84	81	78	76
83	89	81	71	72	78
79	83	81	72	84	68
86	82				

Find the standard deviation of these scores, assuming these are scores in a population.

7. In a psychological experiment, 10 subjects were given the same test after 5 of them had been told the test would be very difficult and 5 were told it would be very easy. The scores on the test were as follows:

"It is very difficult"	"It is very easy"
38	30
52	38
46	40
42	24
38	32

Find the standard deviation of each group of 5, and of all 10 subjects together.

8. Find three scores in a sample whose mean is 10 and whose variance is 1.

9. Find two scores in a sample whose mean is 40 and whose variance is 8.

10. The velocity of sound in meters per second for various metals is as follows:

Metal	Velocity (meters per second)
Aluminum	5,000
Cadmium	2,300
Copper	3,600
Iron	5,100
Lead	1,200
Nickel	5,000
Platinum	2,700
Silver	2,600
Tin	2,500

Find s. Hint: If a set of scores is divided by 100, and s for the new set is computed, it must be multiplied by 100 to find s for the original set.

11. Six coins are tossed 100 times; each time the number of coins that land heads up is noted. The results are as follows:

Number of heads	0	1	2	3	4	5	6
Frequency	2	7	24	34	22	10	1

Find the mean number of heads, and the standard deviation.

ANSWERS

1. (i) (a) 3 bullets; (b) 2.0 bullets; (c) $\sqrt{\dfrac{6}{3}} = 1.4$ bullets; (d) $\sqrt{\dfrac{22 - (8^2/4)}{3}} = 1.4$ bullets.

(ii) (a) 18 pounds; (b) $\dfrac{8.0}{8} + 60 = 61.0$ pounds; (c) $\sqrt{\dfrac{262}{7}} = 6.1$ pounds; (d) (Subtract 60.) $\sqrt{\dfrac{270 - (8^2/8)}{7}} = 6.1$ pounds.

(iii) (a) 10 inches; (b) 25.0 inches; (c) $\sqrt{\dfrac{110}{7}} = 4.0$ inches; (d) $\sqrt{\dfrac{110 - (0^2/8)}{7}} = 4.0$ inches.

(iv) Cannot be determined (open-ended).

2. Melting points: (a) 40°C; (b) (Add 100 to each temperature.) $\sqrt{\dfrac{1,520 - (-54)^2/6}{5}} = 14.4$°C. Boiling points: (a) 66.8°C; (b) (Subtract 90 from each temperature.) $\sqrt{\dfrac{2,927.93 - (21.7)^2/6}{5}} = 23.9$°C.

3. (a) $97 - 42 = 55$; (b) 14.2.

4. Mean $= \dfrac{300}{10} = 30$, $s^2 = \dfrac{9,200 - \left(\dfrac{300^2}{10}\right)}{9} = 22.2$.

5. Mean $= 6.8$, $s = 3.4$.

6. (Subtract 80 from each score.) $s = \sqrt{\dfrac{620 - (-22)^2/20}{19}} = 5.6$.

7. "Very difficult": $s = 5.9$; "Very easy": $s = 6.4$; all subjects: $s = 8.0$.

8. 9, 10, 11.

9. 38, 42.

10. $s = 1,414$ meters per second.

11. $\bar{X} = \dfrac{301}{100} = 3.0$, $s = \sqrt{\dfrac{1,047 - 301^2/100}{99}} = 1.2$.

4.13 STANDARD SCORES, OR SCORES IN STANDARD DEVIATION UNITS

The important question in this section is this: If you have a set of X scores and know their mean μ and their standard deviation σ, how can you find a related set whose mean is 0 and whose standard deviation is 1? The letter z will be used for scores in the related set; they will be called **z scores** or **standard scores** or **scores in standard deviation units.**

To find z scores: Subtract μ from each of the X scores, and then divide each of the new set by σ, thus arriving at the set of z or standard scores.

$$z = \frac{X - \mu}{\sigma} \qquad z = \frac{X - \bar{X}}{s}$$

$\sigma \; \vee \; S \;=\;$ STAN. DEV.

Example 1   Given the X scores in the box below, find the corresponding z scores.

DONE AS A POP.

X	$X - \mu$	$(X - \mu)^2$	$z = \dfrac{X - 8}{3}$
5	-3	9	-1
7	-1	1	$-1/3$
7	-1	1	$-1/3$
13	5	25	5/3
32		36	

$$\mu = \frac{32}{4} = 8.$$

$$\sigma = \sqrt{\frac{36}{4}} = 3.$$

The corresponding z scores are -1, $-1/3$, $-1/3$, $5/3$. Their mean is 0 and their standard deviation is 1. (Check these!)

Why does this work? When you subtract a constant from a set of scores, the mean is reduced by that constant but the spread of the scores is not changed. Here you subtract μ, so the mean of the new set is 0. Then you divide by the standard deviation; this doesn't change the mean again ($0/\sigma = 0$), but the new standard deviation becomes 1: if $\sigma^2 = \dfrac{\Sigma(X - 0)^2}{N}$, then $\dfrac{\Sigma(X/\sigma - 0)^2}{N} =$

$$\frac{\Sigma X^2}{\sigma^2 N} = \left(\frac{1}{\sigma^2}\right)\frac{\Sigma(X - 0)^2}{N} = \frac{1}{\sigma^2}(\sigma^2) = 1.$$

Example 2   For these X scores in a population, $\mu = 36.0$ and $\sigma = 19.8$ (check these): 0, 29, 31, 36, 41, 43, 72.

(a) Find the corresponding z scores.

(b) Check that the mean of the z scores is 0 and that the standard deviation is 1.

(c) An X score of 43 is how many standard deviations above the mean of the X scores? What is the corresponding z score?

(d) If an X score of 72 corresponds to a z score of 1.8, then 72 is how many standard deviations above μ?

(a)

X:	0	29	31	36	41	43	72
$X - \mu = X - 36$:	-36	-7	-5	0	5	7	36
$z = \dfrac{X - \mu}{\sigma} = \dfrac{X - 36}{19.8}$:	-1.82	$-.35$	$-.25$	0	.25	.35	1.82

(b) Mean of z scores $= \dfrac{-1.82 - .36 - .25 + 0 + .25 + .36 + 1.82}{7} = 0.00$

Standard deviation of z scores $=$

$$\sqrt{\frac{(-1.82)^2 + (-.35)^2 + (-.25)^2 + 0^2 + .25^2 + .35^2 + 1.82^2}{7}} = 1.00.$$

(c) $\dfrac{43 - 36.0}{19.8} = .35$ standard deviation above μ; the corresponding z score is .35.

(d) 1.8 standard deviation units above μ.

Example 3   In a population with mean 100 pounds and standard deviation 10 pounds, what weight is 1.3 standard deviation units below the mean?

1 z unit $=$ 1 standard deviation unit $=$ 10 pounds, so -1.3 z units $= -1.3$ standard deviation units $= -1.3(10) = 13$ pounds. The weight 13 pounds below the mean of 100 is 87 pounds.

Alternatively, solve $z = \dfrac{X - \mu}{\sigma}$ for X if $z = -1.3$, $\mu = 100$, $\sigma = 10$:

$$-1.3 = \frac{X - 100}{10}$$

$$-13 = X - 100$$

$$100 - 13 = X$$

$$X = 87 \text{ pounds}$$

Example 4   In a distribution with $\mu = 40$, $\sigma = 3.1$, what score corresponds to a z score of 2.1?

$$z = \frac{X - \mu}{\sigma}$$

$$2.1 = \frac{X - 40}{3.1}$$

$$2.1(3.1) = 6.5 = X - 40$$

$$X = 46.5$$

We shall use z scores many times in succeeding chapters. The formula will always give directions for subtracting the mean and then dividing by the standard deviation, but will have many varied forms. If Y scores in a sample have a mean

$\overline{Y}$, the formula will be $z = (Y - \overline{Y})/s$. There will also be much more astonishing forms, however. But always there is a fundamental relation:

 $$z = \frac{(\text{Score in distribution}) - (\text{Mean of the distribution})}{\text{Standard deviation of the distribution}}$$

Here are two examples of the ways in which z scores or scores in standard deviation units can be useful:

Example 5   A student gets 82 in a final examination in mathematics; the mean grade is 75, with a standard deviation of 10. In economics he gets 86 in a final examination on which the mean grade is 80 with a standard deviation of 14. Is his relative standing better in mathematics or in economics?

In mathematics, $z = \dfrac{82 - 75}{10} = .70$; in economics $z = \dfrac{86 - 80}{14} = .43$. His grade in mathematics is .70 standard deviation units above the mean, while in economics his grade is .43 standard deviation units above the mean. His relative standing is higher in mathematics than in economics.

> Note: No assumption is made here that the grades in either mathematics or economics are normally distributed; z scores have many other uses besides comparing any normal distribution to a standard normal distribution.

Example 6   On a mathematical aptitude test, Philip Nolan gets a raw score of 88.3. The mean of all scores on that test is 75, with a standard deviation of 8. Each year scores are standardized, with a mean of 500 and a standard deviation of 100, before reports are sent to those taking the test. What score is reported to Philip Nolan?

Philip's score is $\dfrac{88.3 - 75}{8} = 1.66$. In a distribution with $\mu = 500$, $\sigma = 100$, a score of 666 corresponds to a z score of 1.66: solve $1.66 = \dfrac{X - 500}{100}$. The score reported to Philip Nolan is 666.

4.14   EXERCISES

1.   For each of the following populations, find (a) μ and σ; (b) the corresponding z scores; (c) the mean and standard deviation of the z scores.
  (i) X: 0, 6, 6, 8 pounds; (ii) Y: 0, 0, 0, 0, 5 grams.

2.   In a population with $\mu = 16$ inches and $\sigma = 3$ inches,
  (a) What z score corresponds to 22 inches?
  (b) How many inches correspond to 1 standard deviation unit? to 2.4 standard deviation units?
  (c) What score (in inches) is 1.5 standard deviation units below the mean?

3.   The mean temperature in Hawaii in January is 72°, with a standard deviation of 2°. On January 12 the temperature is 2.5 standard deviations above the mean. What is the temperature on January 12?

4.   The club batting averages for the National League in 1978 were Chicago .264, Los Angeles .264, Houston .258, Philadelphia .258, Pittsburgh .257, Cincinnati .256, Montreal .254, San Diego .252, St. Louis .249, San Francisco .248, New York .245, Atlanta .244.

  (a) Chicago was how many standard deviations above the mean?
  (b) San Francisco was how many standard deviations below the mean?

5.   A psychologist gives a personality test to a sample of 6 students, who obtain the following scores: 28, 34, 32, 31, 30, 31. In writing a paper he wishes to report these as "standardized" (not **standard**) scores with a mean of 50 and a standard deviation of 10. What scores does he report?

6.   John Bradford's grades on hour exams in physics during one semester are as follows:

Exam	John's Grade	Class Mean	Class Standard Deviation
1	86	85	5
2	90	85	10
3	60	55	5
4	74	72	5

  (a) What is John's average grade?
  (b) What is John's grade in the course if the teacher averages z scores for each exam, and assigns final grades with a mean of 78 and a standard deviation of 6?
  (c) Does (a) or (b) give a more just way of determining John's grade in the course?

7.   The times of Brooklyn runners in the 60-yard dash in the Borough Championships were 6.5, 6.8, 6.8, 6.8, and 6.8 seconds. Calculate (a) $\overline{X}$; (b) s; (c) the z score of the 6.5-second runner; (d) the z score of the 6.8-second runners. Is it the same except for sign as the 6.5-second runner?

8.   1,000 yard running times for Brooklyn in the Borough Championships were 140.7, 142.5, 146.5, 147.6, and 150.6 seconds. Find the z score of the fastest runner on the Brooklyn team in this race.

9.   Combine the times for the 10 runners given in Exercises 7 and 8. What is the z score of 6.5 seconds relative to the combined scores? What do you learn about the 6.5 time from this?

10.   Scores in a distribution are proportions p. Their mean, $P = .20$, and their standard deviation $= \sqrt{\dfrac{P(1-P)}{n}}$. If $n = 9$, find the z score when (a) $p = .17$; (b) $p = .30$; (c) What proportion p is 2 standard deviation units above the mean?

ANSWERS

1. (i) (a) = 5.0, = 3.0; (b) −1.67, .33, .33, 1.00; (c) mean = .00, standard deviation = 1.00.

   (ii) = 1.0, = 2.0; (b) −.50, −.50, −.50, −.50, 2.00; (c) mean = .00, standard deviation = 1.00.

2. (a) $\dfrac{22 - 16}{3} = 2.00$; (b) 3 inches; 3(2.4) = 7 inches; (c) $-1.5 = \dfrac{X - 16}{3}$, $X = 11.5$ inches.

3. $2.5 = \dfrac{X - 72°}{2}$, $X = 77°$.

4. $\mu = .254$, $\sigma = .0064$; (a) $\dfrac{.264 - .254}{.0064} = 1.56$; (b) $\dfrac{.248 - .254}{.0064} = -.94$.

5. $\overline{X} = 31.0$, $s = 2.0$. (ROUNDED)
   z scores: −1.5, 1.5, .5, 0, −.5, 0.
   Standardized scores: 35, 65, 55, 50, 45, 50.

6. (a) 77.5. (b) Average of z scores $= \dfrac{+ .2 + .5 + 1 + .4}{4} = .53$; $.53 = \dfrac{X - 78}{6}$, so $X = 81$. (c) Using standard scores gives a more accurate reflection of his relative ability— but whether he thinks (b) is more just may depend on the teacher's choice of mean and standard deviation for the class in assigning final grades.

7. (a) 6.74 seconds; (b) .13 seconds; (c) −1.846; (d) 0.46. No, because a number of 6.8-second times are bunched together.

8. $\overline{X} = 145.58$ seconds, $s = 3.98$ seconds, $z = -1.23$.

9. $\overline{X} = 76.16$, $s = 73.22$, $z = -0.95$. Garbage. You don't average 60-yard and 1,000-yard running times; you don't judge someone's 60-yard running time relative to how long it takes someone else to run 1,000 yards.

10. $z = \dfrac{p - .20}{.133}$; (a) −.23; (b) .75; (c) $p = .47$.

4.15 VOCABULARY AND SYMBOLS

mode	Mo	range	Q
unimodal		squared deviations from the mean	
bimodal		mean absolute deviation	
multimodal		variance	σ^2, s^2
median	Md	standard deviation	σ, s
mean	$\overline{X}$, $\overline{Y}$	standard score	
quartile	Q^1, Q^3	z score	
percentile	P_a	score in standard deviation units	
percentile rank			

LESS
IMP

4.16 REVIEW EXERCISES

The following data are to be used in Exercises 1–8. Tony Rotatori tried a reinforcement schedule to help obese retarded adults lose weight. The group on this regimen was compared with a group of 8 controls. To see whether treated and comparison groups were comparable to begin with, initial height, weight, percentage overweight, and IQ were recorded:

		Study Group						Control Group		
	Height	Weight	% Overweight	IQ			Height	Weight	% Overweight	IQ
A	68	201	32	65		O	64	159	20	52
B	68	169	11	78		P	66	176	23	58
C	65	151	10	56		Q	65	165	20	65
D	66	162	13	68		R	67	162	22	59
E	69	211	34	72		S	60	174	59	66
F	64	195	58	52		T	65	154	12	78
G	63	161	35	58		U	61	162	45	57
H	63	172	45	55		V	62	156	34	52
I	63	162	36	64						
J	62	138	10	55						

1. Find the mean height and the standard deviation for (a) the study group and (b) the control group.

2. Find the mean height of all 18 of these people.

3. Find (a) the median, (b) the mode of the heights for (i) the study group, (ii) the control group, (iii) all 18 people.

4. Compare the mean percent overweight of the study group and of the control group.

5. What is the first quartile for the weights (a) in the study group, (b) in the control group, (c) in both groups together?

6. What is the mean weight of (a) the study group, (b) the control group? What are the standard deviations of weights in these two groups?

7. Find the percentile rank of a weight of 172 pounds in (a) the study group, (b) both groups together.

8. What is the z score of an IQ of (a) 55, (b) 78 relative to the study (i) the study group, (ii) the entire group of 18?

9. A test of anxiety state was given by C. Smoszyk to patients who knew they had cancer both before and after a talking-it-out treatment to relieve anxiety, with the following results (a lower score shows less anxiety):

Patient	A	B	C	D	E	F	G	H	I	J
Before	23	48	69	58	48	44	57	20	26	34
After	20	26	67	57	33	24	44	21	26	27
Change	−3	−22	−2	−1	−15	−20	−13	+1	0	−7

    (a) Find the mean for each row. Do you see any relation between the three means?

    (b) Find the standard deviation for the changes.

10. Winning margins (winner's score − loser's score) in college women's basketball on February 17 were 11, 12, 2, 2, 8, 25, 66, 5, 7, 18, 6, 6, 52, 11, 21, 30, 35. The 66 and 52 look conspicuous. Are they really out of line with the rest?

    (a) Calculate μ and σ for the whole group.

    (b) Find the z score of (i) 66, (ii) 52.

    (c) Then find the z score for the lowest difference, 2.

11. How many blankets do you use in winter? And how long do you sleep at night, typically? Here are the statistics for some people I know; the first figure gives the number of blankets and the second the usual number of hours of sleep: Eugene, 2 blankets, 7 hours; Everett 2, 9; Win 4, 8; Eleanor 3, 8; John 3, 7½; Vicki 2, 10; Roger 7, 8½; Jean 3, 10; Peter 5, 6½.

    (a) Calculate the mean number of blankets and the mean hours of sleep.

    (b) Find the standard deviations for both variables.

    (c) Calculate Roger's and Win's standard scores on each variable.

12. In Exercise 5, page 42, you made histograms of the number of tables of different sizes that Waitress Nancy and Waitress Maria waited on. Here's the data again:

		Number at Table					
		1	2	3	4	5	6
Number	Nancy	2	35	9	3	0	1
of							
Tables	Maria	4	25	7	3	3	0

    (a) For Nancy's tables, what is the percentile rank of a table with 3 people?

    (b) What is the mean number of people at a table waited on by (a) Nancy, (b) Maria?

    (c) What is the z score for that table of 6 relative to Nancy's tables?

13. The heights of adults at the groundhog's eve party (see Exercise 9, page 50) were 66, 69, 66, 69, 63, 66, 72, 66, 71, 74, 74, 75, 62, 66, 66, 62, 64, 62, 63, 70, 73, 71, 67, 67, 68, 62, 65, 68, 64, 64, 73, 69, 72 inches.

    (a) Calculate the mean and variance of the heights. (If you subtract 67 from each height, you'll have smaller numbers to work with.)

(b) If you had included the heights of 2 children, Sonja 53″ and Ismael 58″, what effect would this have on the values obtained for $\overline{X}$ and s^2?

Use the following data in Exercises 14–16. Rutile—titanium oxide—is one of 14 known compounds with a certain tetragonal crystal structure with apical height from metal to non-metal atoms equal to the side length of the equatorial square except for a little distortion. The distances are measured in Angstrom units $(1Å = 10^{-7} \text{ mm.})$.

Distortion of AB_6 octahedra in AB_2 compounds with rutile structure*

AB_2	Apical A-B	Equatorial A-B	Ratio apical/equatorial
TiO_2	1.98Å	1.949Å	1.016
GeO_2	1.902	1.872	1.016
SiO_2	1.810	1.757	1.030
SnO_2	2.057	2.052	1.002
PbO_2	2.16	2.17	0.99
RuO_2	1.942	1.984	0.979
OsO_2	1.962	2.006	0.978
CrO_2	1.88	1.92	0.98
MgF_2	1.979	1.998	0.990
MnF_2	2.104	2.131	0.987
FeF_2	1.998	2.118	0.943
CoF_2	2.027	2.049	0.989
NiF_2	1.981	2.022	0.980
ZnF_2	2.012	2.046	0.983
ΣX	27.794	28.074	13.863
ΣX^2	55.2827	56.4484	13.7332

14. Find (a) the range, and (b) the mean μ of apical heights.

15. Find (a) the mean μ and (b) the standard deviation σ of equatorial heights.

16. The difference between the ratio of apical to equatorial height and 1 is a measure of distortion. Find (a) the mean, (b) the variance, and (c) the standard deviation of the ratios.

17. Sleep efficiency scores of 11 men who had suffered combat trauma as Israeli soldiers in the Yom Kippur war were: 94.8, 95.4, 93.0, 89.3, 89.2, 77.5, 85.6, 87.2, 77.1, 67.8, 75.9. Find (a) the median, (b) Q_1, (c) the mean; (d) How many standard deviation units above the mean is a score of 93.0?

---

*Abrahams, S. C., and Bernstein, J. L., "Rutile; Normal Probability Plot Analysis and Accurate Measurement of Crystal Structure," *The Journal of Chemical Physics*, 55, 1971, pp. 3206–3211.

ANSWERS

1. (Subtract 65 from each height.) (a) $\bar{X} = \dfrac{1}{10} + 65 = 65.1$ inches, $s = \sqrt{\dfrac{57 - 1^2/10}{9}} = 2.4$
   inches; (b) $\bar{Y} = \dfrac{-10}{8} + 65 = 63.8$, $s = \sqrt{\dfrac{56 - (-10^2)/8}{7}} = 2.5$ inches.

3. (i) $Md = 64.5$ inches, $Mo = 63$ inches; (ii) $Md = 64.5$ inches, $Mo = 65$ inches; (iii)
   $Md = 64.5$, $Mo = 63$ or 65 inches.

5. (a) 161 pounds; (b) 157.5 pounds, (c) 159 pounds.

7. (a) $\dfrac{6}{100}(100) = 60$; (b) $\dfrac{12}{18}(100) = 67$ (or 66.7).

9. (a) Before, 42.7; after, 34.5; change -8.2; $42.7 - 8.2 = 34.5$. The difference of means
   equals the mean of the differences. (b) 8.6.

11. (a) 3.4 blankets, 8.3 hours of sleep; (b) 1.7 blankets, 1.2 hours of sleep; (c) Blankets:
    Roger, $z = \dfrac{7 - 3.4}{1.7} = 2.12$, Win, $z = .35$; hours of sleep: Roger, $z = \dfrac{8.5 - 8.3}{1.23} = .16$,
    Win, $z = -.25$.

13. (a) $\bar{X} = 67 + \dfrac{18}{33} = 67.5$, $s^2 = \dfrac{508 - 18^2/33}{32} = 15.6$.
    (b) $\bar{X}$ would be decreased and s^2 would be increased.

15. (a) $\mu = \dfrac{28.074}{14} = 2.0053$; (b) $\sigma = \sqrt{\dfrac{56.4484 - 28.074^2/14}{14}} = .1042$.

17. (a) $Md = 87.2$; (b) $\dfrac{11}{4} = 2.75$, so the third score after they are arranged in increasing
    order is Q_1; $Q_1 = 77.1$; (c) $\bar{X} = 80 + \dfrac{52.8}{11} = 84.8$; (d) $z = \dfrac{93.0 - 84.8}{8.98} = .91$.

$$\frac{1722}{4} \; *$$

5
PROBABILITY

Up to this point you have been studying descriptive statistics, but now you are ready to begin the study of statistical inference. On the basis of data from a sample, you will learn how to draw conclusions about the population from which the sample is drawn. Probability is basic to statistical inference; you will meet the word "probability" dozens of times in every chapter after this.

You could study probability for years and years before getting to the boundaries of what is known about it. Fortunately, for this book you need master only a very few basic facts. Concentrate on (1) the probability that event A or event B will happen, when they cannot occur at the same time, and (2) the probability that event A and event B will both take place when the occurrence of one has no effect on the occurence of the other.

5.1   WHY STUDY PROBABILITY?

Probability and statistics are very closely related. In statistical inference you may take the list of registered voters in your county, randomly choose 100 names, and ask these people how they will vote in the election next week; then you draw conclusions about who will be elected mayor next week when everybody votes. In probability, you know how many voted for each of candidates A, B, and C last week; then you randomly choose 100 voters and draw conclusions about how many of these 100 probably voted for candidate A. In statistics, you put your hand into a black box of colored and white marbles, examine your handful, and try to answer the question "What is in the box?" In probability, you look into a transparent box, count the colored and white marbles in it, mix them up well, and then blindly take out one handful; without opening your eyes, you

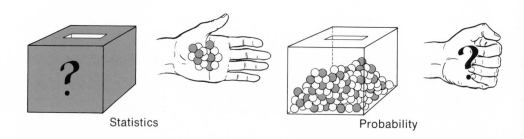

Statistics   Probability

predict how many marbles of each kind are in your hand. You will learn how to forecast the election or answer the question about the black box only after you have studied probability, since you will use this study in choosing the sample, in making your estimate about the population (the contents of the black box), and in deciding what degree of confidence you have in your estimate.

5.2   INTRODUCTION TO PROBABILITY

A penny has been tossed 800 times. Here is a record of the number of times it came up heads, and the proportion of heads in the throws already made:

Number of tosses	Number of heads in last 80 tosses	Cumulative number of heads	Proportion of heads in total number of tosses
80	38	38	.475
160	40	78	.488
240	47	125	.521
320	39	164	.513
400	40	204	.510
480	45	249	.519
560	32	281	.502
640	40	321	.502
720	38	359	.499
800	42	401	.501

As the result of this experiment, we say that the probability of heads [notation: $\Pr(H)$] on one toss of this coin is about .50: $\Pr(H) = .50$.

> **Definition:** The **probability** that something occurs is the proportion of times it occurs when exactly the same experiment is repeated a very large (preferably infinite!) number of times in independent trials; ''independent'' means the outcome of 1 trial of the experiment doesn't affect any other outcome.
>
> **Notation:** $\Pr(A)$, if A represents the ''something'' that occurs.

Heads come up from 0 to 800 times in 800 tosses, so the proportion of heads is between $\frac{0}{800} = 0$ and $\frac{800}{800} = 1$. A probability is defined as a proportion, and a proportion is never less than 0 or more than 1. Furthermore, since heads came up 401 times, tails came up $800 - 401 = 399$ times, so $\Pr(T) = \frac{800 - 401}{800} = 1 - \frac{401}{800} = 1 - \Pr(H)$. This is true in general: the proportion of times an outcome A of an experiment occurs = 1 − (the proportion of times A does not occur), so $\Pr(A \text{ occurs}) = 1 - \Pr(A \text{ doesn't occur})$.

> If A is an outcome of 1 trial of an experiment that is carried out many times in independent trials, then
>
> $0 \leqslant \Pr(A) \leqslant 1$
> $\Pr(A) = 0$ if A never occurs.
> $\Pr(A) = 1$ if A always occurs.
> $\Pr(A \text{ occurs}) = 1 - \Pr(A \text{ doesn't occur}).$

Example 1 Of 100,000 men aged 40, past experience indicates that 99,382 will be alive a year later and 618 will have died. What is the probability a 40-year-old is alive a year later?

$$\Pr(\text{alive a year later}) = \frac{99,382}{100,000} = .99382$$

We have tossed a penny 800 times, and estimated that $\Pr(H) = .50$. This doesn't mean that if we toss the penny again, half a head will come up; it will be either heads or tails. It may even come up heads seven times in the next seven tosses. But when we say $\Pr(H) = .50$, we are trying to describe the "randomness" of the real world; if the penny is tossed again and again, we expect tht half the time it will come up heads.

Carrying out an experiment a great many times may be very difficult and may be very dull. So let's try a different approach to probability, based on what we expect would happen if an experiment were carried out many times.

A bag contains three balls, all perfectly round and all of the same size and weight. Two of the balls are red and one is green. The bag is well shaken, one ball is chosen blindly and its color is noted. Then that ball is replaced in the bag, and the process is repeated until 300 balls have been drawn. (A quicker way of describing this is to say "300 balls are drawn **at random** and **with replacement**"; "at random" means on any one choice each ball is equally likely to be chosen, "with replacement" means the ball is replaced in the bag before the next choice is made.) In 300 choices, we expect each of the balls to be chosen about the same number of times (100 each). Since there are two red balls, we expect a red ball will be chosen 200 times out of 300. The probability of getting a red ball in a single draw is the proportion of red balls in the 300 choices; $\Pr(\text{Red}) = \dfrac{200}{300} = .67$.

We would arrive at the same result by looking at the proportion of red balls in the bag:

$$\Pr(\text{Red}) = \frac{\text{number of red balls}}{\text{total number of balls}} = \frac{2}{3} = .67$$

This leads us to what is called the "classical" probability model because it was developed in the sixteenth century when probability first began to be studied; it is only useful if all outcomes are equally likely.

REPLACEMENT NOT AB. NECESSARY

> If there are n equally likely outcomes of an experiment and if one outcome A can occur in s ways, then $\Pr(A) = \dfrac{s}{n}$.

Example 2   The population of Jasper County, Iowa, is 35,010, and of Lucas County, Iowa, is 10,031 (1970 census). If one inhabitant of these two counties is chosen at random, what is the probability he lives in Jasper County?

$$\Pr(\text{from Jasper}) = \frac{\text{number in Jasper}}{\text{total number}} = \frac{35,010}{35,010 + 10,031} = .78$$

> **Round-off:** Let's agree that, in general, the answer to a probability problem is rounded to 2 decimal places.   TRY 3

Example 3   In an election district in New Jersey there are 2,000 Republicans, 2,500 Democrats, and 5,500 Independents. If one voter is chosen at random, what is the probability (a) a Republican, (b) a Republican or a Democrat is chosen?
(a) Note that each voter is equally likely to be chosen.

$$\Pr(\text{Republican}) = \frac{2,000}{2,000 + 2,500 + 5,500} = \frac{2,000}{10,000} = .20$$

(b) $\Pr(\text{Republican or Democrat}) = \dfrac{2,000 + 2,500}{10,000} = .45.$

TRIAL   In statistics anything that results in a count or a measurement is called an experiment. It may be the number of acres planted with corn in Iowa or the blood count of each patient entering Overlook Hospital or measurements of social awareness among mentally disturbed children.

> **Definitions:** An **outcome** is one particular result of an experiment.
> An **event** is a set of outcomes.
> A **sample space** is the set of all possible outcomes of an experiment. These outcomes are also called the **elements** of the sample space.
>
> **Notation:**   Sample space $= S$

A sample space may be given by a list, a tree diagram (mathematicians cut down the tree so its branches fall to the right), or pictorially by a rectangle or circle. In the last case, an event is indicated by a circle or a rectangle inside S. This drawing is called a **Venn diagram.**

CLOSED
CONTOUR

Example 4 Indicate the outcomes, some of the events, and the sample space if a single die is tossed.

(a) The outcomes are 1, 2, 3, 4, 5, 6.

(b) Examples of events: 1; 4; even numbers from 2 through 6; 1, 3, 6; 1, 2, 3, 4, 5, 6.

(c) List: $S = \{1, 2, 3, 4, 5, 6\}$. Note that curly brackets are put around the items in the list.

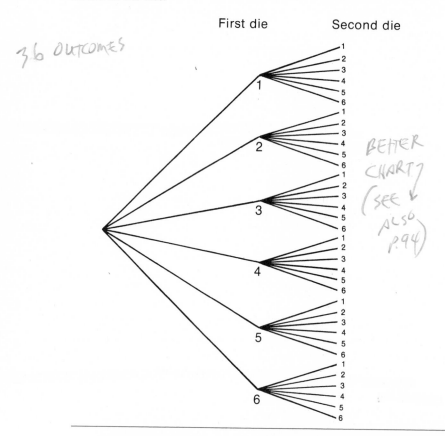

If S contains only a finite number of elements, then to each element we assign a non-negative number, called a probability, so that the sum of all the numbers assigned to all the elements of S is 1. Since the numbers are non-negative and their sum is 1, each of the numbers must be between 0 and 1. If we think an outcome of an experiment is quite likely, we assign to that outcome a number close to 1. If an outcome is quite unlikely to occur, we assign to it a number close to 0. To outcomes which are totally impossible (including those outside the sample space) we assign the number 0. In many experiments, all the outcomes are equally likely to occur; in this case the same number is given to each.

Example 5 In Montgomery County Court, 150 defendants are convicted of misdemeanors in October; 90 are fined and put on probation, and 60 receive prison terms of 1 year or less. If one of those convicted is chosen at random, what is the probability that he is (a) fined and put on probation? (b) sentenced to 1 year or less? (c) sentenced to more than 1 year?

Let A, B, and C be, respectively, the events {fined and given probation}, {sentenced to 1 year or less}, and {sentenced to more than 1 year}.

(a) $\Pr(A) = 90/150 = .60$.
(b) $\Pr(B) = 60/150 = .40$.
(c) $\Pr(C) = 0/150 = 0$.

A	B	C	S
$\frac{90}{150}$	$\frac{60}{150}$	$\frac{0}{150}$	

> To find the probability of any event A [$= \Pr(A)$], add all the numbers assigned to the elements of A. If an experiment takes place and S is the set of all the outcomes of the experiment, then S is certain to take place, so $\Pr(S) = 1$.

Example 6

If a die is tossed, what is the probability that a 1, 2, or 3 will come up?

$S = \{1, 2, 3, 4, 5, 6\}$, $A = \{1, 2, 3\}$. 1/6 is assigned to each element of S. The probability that a 1, 2, or 3 will come up $= \Pr(A) = 1/6 + 1/6 + 1/6 = 1/2$.

1	2	3	4	5	6	S
1/6	1/6	1/6	1/6	1/6	1/6	

A

Example 7

In Beach Bluff there are 4,112 Protestants, 1,332 Catholics, and 556 Jews. If one person is picked at random from these, what is the probability that he is (a) Protestant? (b) Catholic or Jewish?

$B = \{$Protestant$\}$, $C = \{$Catholic or Jewish$\}$.

(a) $\Pr(B) = \dfrac{4{,}112}{4{,}112 + 1{,}332 + 556} = \dfrac{4{,}112}{6{,}000} = .69$.

(b) $\Pr(C) = \dfrac{1{,}332 + 556}{6{,}000} = \dfrac{1{,}888}{6{,}000} = .31$.

Protestants	Catholics	Jews	S
.69	.22	.09	

B · · · · C

Example 8

If two dice are tossed, one red and one green, what is the probability that (a) 11 (total on tops of both dice) will come up? (b) 7 will come up?

(a) Here is it more difficult to list the elements in the sample space. Each element of S is a pair of numbers such as (4, 3), where the first number is the result of the toss on the red die, the second is the result on the green die. $S =$ $\{(1, 1), (1, 2), \ldots, (1, 6), (2, 1), \ldots, (2, 6), \ldots, (6, 6)\}$. It is easier to see from a tree or Venn diagram that there are 36 elements in S; the probability assigned to each is 1/36 if both dice are "fair." The event "11 comes up" includes the elements indicated by a cross. There are two of them, so Pr(11) = Pr(6 on the first die and 5 on the second, or 5 on the first and 6 on the second) = Pr[(6,5) or (5, 6)] = 1/36 + 1/36 = 1/18 = .06.

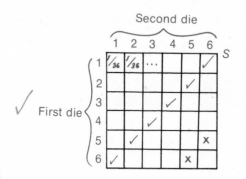

(b) The elements of the event "the total is 7" are indicated by checks. Pr(7) = Pr[(6, 1) or (5, 2) or (4, 3) or (3, 4) or (2, 5) or (1, 6)] = 6(1/36) = 1/6.

Example 9

One card is drawn at random from an ordinary (bridge) deck of 52 cards. What is the probability that it is (a) a king? (b) a heart? (c) a spade or a heart? What probability is assigned to each element of S?

(a) Pr(K) = ?

(b) Pr(♥) = ?

(c) Pr(♠ or ♥) = ?

1/52 is assigned to each element of S; (a) Pr(K) = 4/52 = 1/13; (b) Pr(♥) = 13/52 = 1/4; (c) Pr(♠ or ♥) = 26/52 = 1/2.

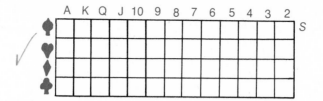

Example 10

A rather extraordinary die is manufactured. It is like the ordinary die only in that it is six-sided, and the sum of the numbers on each pair of "opposite" sides (that is, sides without a common edge) is 7. In the illustration, side 1 is on top and side 6 on the bottom, side 2 is facing you and side 5 is away from you; side 4 is to the left and side 3 is slanting upwards on the right. Because of its unusual construction, this die will not balance on side 3 (it immediately falls on side 6) so side 4 cannot come up, and it will not balance on side 1 (it falls on side 4), so side 6 cannot come up on top.

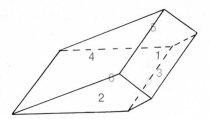

The die is tossed 1,000 times, with the following results.

Side	Number of times on top
1	470
2	110
3	310
4	0
5	110
6	0
	1,000

What is the probability that, when it is tossed once, (a) an even number will turn up? (b) a 1 or a 3 will turn up? (c) a 2 or a 5 will come up?

$S = \{1, 2, 3, 4, 5, 6\}$. Probabilities can be assigned here to the elements of the sample space only on the basis of the experimental results. Since 4 and 6 are impossible outcomes in practice, they are each assigned the number 0; 2 occurs 110 times in 1,000 trials, and so is assigned the probability 110/1,000 = .11; and so on. (Always check that the sum of the probabilities assigned to all the elements of the sample space is 1.) Now the problem should be simple for you.

(a) Pr(2, 4, or 6) = .11 + 0 + 0 = .11.

(b) Pr(1 or 3) = .47 + .31 = .78.

(c) Pr(2 or 5) = .11 + .11 = .22.

1	2	3	4	5	6	*S*
.47	.11	.31	0	.11	0	

5.3 EXERCISES

1. A fair die is tossed 1,200 times.
   (a) How many times do you expect (i) a 2 to come up? (ii) a 2 or a 4 to come up?
   (b) What proportion of times do you expect (i) a 2 to come up? (ii) a 2 or a 4 to come up?
   (c) If the die is tossed once, what is the probability that (i) a 2 comes up? (ii) a 2 or a 4 comes up?

2. From a group of three Republicans and five Democrats, one person is chosen at random and with replacement 400 times.
  (a) If Mrs. A is one of the Republicans, how many times do you expect that she is chosen?
  (b) How many times do you expect a Republican to be chosen?
  (c) What proportion of those chosen do you expect to be Republicans?
  (d) If one person is chosen at random, what is the probability she is
    (i) a Republican? (ii) a Democrat?

3. Assign one of the statements (i)–(vi) to each of the probabilities (a)–(g):

	Pr(A)		Statement
(a)	−.40	(i)	Event A will never occur.
(b)	0	(ii)	Event A is highly likely to occur, but not
(c)	.10		always.
(d)	.50	(iii)	A will always occur.
(e)	.80	(iv)	Something's wrong.
(f)	1.00	(v)	A is possible but unlikely
(g)	1.02	(vi)	A is as likely as not to occur.

4. A market analyst claims that the probabilities that stock in United States Rubber will go up more than 5 points, stay the same (within 5 points), or go down more than 5 points this year are .40, .20, and .30, respectively. A second analyst claims these probabilities are .50, .30, and .30, and a third that they are .40, .30, and .30, respectively. Comment on these claims.

5. If two (fair) dice are tossed, what is the probability that the sum of the two dice is (a) 10? (b) 9? (c) 9 or more? (d) less than 9?

6. A coin is tossed 1,000 times, and comes up heads 440 times. If the coin is tossed again, what is the probability that it will come up (a) heads? (b) tails?

7. A quiz consists of three true-false questions. A very poor student answers each question at random (by sheer guess).
  (a) Use a tree diagram to show the sample space.
  (b) Use a Venn diagram to show the sample space. (Let $T_1 F_2 T_3$, for example, stand for answers of True on the first and third questions, False on the second.)
  (c) What is the probability that the student answers the first question correctly?
  (d) What is the probability that the student answers all three questions correctly?
  (e) What is the probability that the student answers two or less questions correctly?

8. If one card is drawn at random from a deck of 52 cards, what is the probability it is (a) a 9? (b) a diamond? (c) a diamond or a club? (d) a diamond or a 9?

9. "The probability of rain on any day in February in Portland, Oregon, is .75. It did not rain in Portland on February 15. Therefore, the claim made in the first sentence is wrong." Comment.

10. Of 640 cars driving past Exist 29 on Route 287 between 8 and 9 A.M., 360 are going more than 60 miles per hour. If one car is chosen at random, what is the probability that it is going (a) more than 60 miles per hour? (b) 60 miles per hour or less? (c) more than 65 miles per hour?

ANSWERS

1. (a) (i) 200, (ii) 400; (b) (i) $\frac{200}{1200} = \frac{1}{6}$, (ii) $\frac{400}{1200} = \frac{1}{3}$; (c) $\frac{1}{6} = .17$, (ii) $\frac{1}{3} = .33$.

2. (a) $\frac{1}{8}(400) = 50$; (b) 150; (c) $\frac{150}{400} = .38$; (d) (i) .38, (ii) $1 - .38 = .62$.

3. (a) - (iv); (b) - (i); (c) - (v); (d) - (vi); (e) - (ii); (f) - (iii); (g) - (iv).

4. The claims of the first and second analysts are impossible, since their probabilities do not add up to 1.

5. (See the Venn diagram or tree diagram on page 94.) (a) $3/36 = .08$; (b) $4/36 = .11$; (c) $10/36 = .28$; (d) $1 - 10/36 = 1 - .28 = .72$.

6. (a) $\frac{440}{1000} = .44$; (b) $1 - .44 = .56$.

7. (a)   (b)

Question

$$1 \qquad 2 \qquad 3$$

$$T_1\ T_2\ T_3 \mid T_1\ T_2\ F_3 \mid T_1\ F_2\ T_3 \mid F_1\ T_2\ T_3$$

$$T_1\ F_1\ F_2 \mid F_1\ T_2\ F_3 \mid F_1\ F_2\ T_3 \mid F_1\ F_2\ F_3$$

S

$P(A) = \frac{1}{8}$

$P(\text{ONE } F) = \frac{3}{8} = P(\text{TWO } T)$

$P(\text{AT LEAST } 2 \text{ T}s) = \frac{1}{2}$

(c) $\frac{1}{2} = .50$; (d) $\left(\frac{1}{2}\right)^3 = .13$; (e) $1 - \frac{1}{8} = .87$.

8. (a) $\frac{4}{52} = .08$; (b) $\frac{13}{52} = .25$; (c) $\frac{26}{52} = .50$; (d) $\frac{16}{52} = .31$.

9. Nonsense; if the first sentence is correct, then the probability it does not rain on any particular day is .25; on the average there is no rain 1 day in 4.

10. (a) $\frac{360}{640}$ = .56; (b) 1 − .56 = .44; (c) Cannot tell, but not more than .56.

5.4 MUTUALLY EXCLUSIVE EVENTS

If a single die is tossed, we already discovered (Example 6, page 93) that Pr(1, 2, or 3) = Pr(1) + Pr(2) + Pr(3) = 1/6 + 1/6 + 1/6 = 1/2. Soon we shall want to tackle problems such as this: What is the probability that heads will come up exactly twice if a fair coin is tossed three times? Here we would need a three-dimensional Venn diagram. We could use a tree diagram to show possible outcomes, as follows:

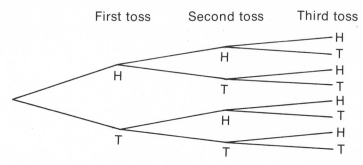

Or we could abbreviate "heads on all three tosses" by HHH, "heads on the first two and tails on the third" by HHT, and so on, and then list the elements of S: HHH, HHT, HTH, HTT, THH, THT, TTH, TTT. But if the coin is tossed seven times, both these methods become cumbersome too. Some shortcuts should be welcomed.

> **Definition:** Two events A and B are **mutually exclusive** if they have no elements in common. If A and B are thought of as outcomes of an experiment, this definition means that A and B cannot both happen.

Note that different paths from the root of a tree to a farthest branch represent mutually exclusive events. See the tree above for a single die tossed 3 times. HHH and HTH, for example, are mutually exclusive.

Example 1

One die is tossed. Let $S = \{1, 2, 3, 4, 5, 6\}$.
Let A = the event an odd number turns up; $A = \{1, 3, 5\}$.
Let B = the event a 1, 2, or 3 turns up; $B = \{1, 2, 3\}$.
Let C = the event a 2 turns up; $C = \{2\}$.
A and B have the elements 1, 3 in common, so A and B are not mutually exclusive.

It is impossible to throw an odd number **and** a 2 if one die is tossed, so *A* and *C* have no element in common; they are mutually exclusive. *B* and *C* have the element 2 in common, so *B* and *C* are **not** mutually exclusive.

Example 2   In the Sixth Congressional District, the normal election turnout is 180,000 if the weather is good; 100,000 are urban and 80,000 are rural. If the weather is bad, however, 10% of these urban voters and 30% of the rural voters stay home.

A is the event {urban voters who stay home in bad weather}.

B is the event {rural voters who stay home in bad weather}.

A and *B* are mutually exclusive.

If *C* is the event {urban voters}, then *A* and *C* are not mutually exclusive but *B* and *C* are.

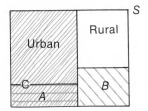

BETTER
RULE
P. 102

If *A* and *B* are mutually exclusive, then the probability that at least one of them will happen is the sum of their separate probabilities:

$$\Pr(A \text{ or } B) = \Pr(A) + \Pr(B).$$

But you can't add the probabilities of *A* and of *B* if they are not mutually exclusive.

Note: In general, *A or B* means *A* will occur or *B* will occur or both will occur. If they are mutually exclusive, *A or B* means *A* will occur or *B* will occur (but not both).

Example 3   If a fair coin is tossed twice, what is the probability that heads and tails will each come up once?

$S = \{HT, TH, HH, TT\}$. (Remember that HT means heads on the first toss, tails on the second; TH means tails on the first toss, heads on the second, and so on.) If the coin is fair, each outcome is equally likely and the probability 1/4 is assigned to each. Let $A = \{HT\}$, $B = \{TH\}$. Then *A* and *B* are mutually exclusive.

HH	HT	TH	TT	
1/4	1/4	1/4	1/4	*S*

Pr(heads and tails each come up once) = Pr(HT or TH) = Pr(HT) + Pr(TH)
$$= \frac{1}{4} + \frac{1}{4} = \frac{1}{2}.$$

Example 4

If a child is chosen at random from a fourth-grade class in which 6 children read below grade level, 12 read at grade level, and 7 read above grade level, what is the probability that he reads at or below grade level?

$S = \{\text{below grade level, at grade level, above grade level}\}.$

Pr(the chosen child reads below or at grade level) = Pr(he reads below grade level) + Pr(he reads at grade level) = 6/25 + 12/25 = 18/25 = .72.

Below		Above	
6/25	12/25	7/25	S

If A, B, and C are all mutually exclusive, then no element in A is in B or C, and none in B is in C, so Pr(A or B or C) = Pr(A or [B or C]) = Pr(A) + Pr(B or C) = Pr(A) + Pr(B) + Pr(C). In the same way, probabilities of more than three events which are mutually exclusive may be added.

Example 5

If one score is chosen at random from the following distribution, what is the probability that it is (a) 4 or less? (b) 14 or less?

X	f
0– 4	3
5– 9	4
10–14	7
15–19	2
20–24	9
	25

(a) $\text{Pr}(0\text{–}4) = \dfrac{3}{25} = .12.$

(b) $\text{Pr}(0\text{–}4 \text{ or } 5\text{–}9 \text{ or } 10\text{–}14) = \text{Pr}(0\text{–}4) + \text{Pr}(5\text{–}9) + \text{Pr}(10\text{–}14) = \dfrac{3}{25} + \dfrac{4}{25} + \dfrac{7}{25} = \dfrac{14}{25} = .56.$

5.5 AXIOMS OF PROBABILITY

All further results in probability theory can be derived from the following three axioms or assumptions:

√√ *(handwritten)*

1. $\Pr(S) = 1$ *P(SOMETHING WILL HAPPEN = 1* *(handwritten)*
2. $0 \leq \Pr(A) \leq 1$ (Read $\leq$ as "less than or equal to.")
3. If A and B are mutually exclusive events, then $\Pr(A \text{ or } B) = \Pr(A) + \Pr(B)$ — $P(A+B)$ *(handwritten)*

If these three axioms are not crystal clear to you, read this chapter again and study the examples carefully. Remember that if an event is certain to happen, then the probability of its happening is 1; and if an event is certain **not** to happen, then the probability of its happening is 0. The second axiom tells you that the answer to every probability problem will be a number between 0 and 1, inclusive. Check your work and find your mistake if you come up with an answer like 2.31 or $-.50$!

If A is an event in the sample space S, let $\overline{A}$ be the event "A does **not** occur." It follows immediately* from the axioms that:

$$\Pr(A) = 1 - \Pr(\overline{A})$$

If the probability that George Stone is seasick on a cruise to Bermuda is .05, then the probability that he is not seasick is .95. If the probability of rain today is .40, the probability of no rain today is .60.

Example 1   If two dice are tossed, what is the probability that the sum of the two numbers showing is 4 or more? (See Example 8, page 94, for Venn diagram.)

$\Pr(4 \text{ or more}) = 1 - \Pr(2 \text{ or } 3) = 1 - [\Pr(2) + \Pr(3)] = 1 - (1/36 + 2/36)$ $= 11/12$. Note that the events "the sum is 2" and "the sum is 3" are mutually exclusive, so $\Pr(2 \text{ or } 3) = \Pr(2) + \Pr(3)$.

√ 5.6   PR(A OR B) IF A AND B ARE NOT MUTUALLY EXCLUSIVE

Why can't $\Pr(A \text{ or } B) = \Pr(A) + \Pr(B)$ be used if A and B are not mutually exclusive? The reason is that in such a case A and B overlap in a Venn diagram, and the elements in the overlap are counted twice. Suppose, for example, that 1/3 of the 12,000 voters in Mendham are Republicans and 1/2 are women. If a voter is chosen at random and with replacement 600 times, we expect that $(1/3)(600) = 200$ will be Republicans and $(1/2)(600) = 300$ will be women. But the number in our sample who are Republicans OR women is not $200 + 300 = 500$. Those in the shaded area have been counted twice.

---

*Assume, from the definition of $\overline{A}$, that every element of S is in either A or $\overline{A}$, and that A and $\overline{A}$ are mutually exclusive.

$\Pr(S) = 1$ (Axiom 1)
$\Pr(A \text{ or } \overline{A}) = 1$
$\Pr(A) + \Pr(\overline{A}) = 1$ (Axiom 3)
$\Pr(A) = 1 - \Pr(\overline{A})$

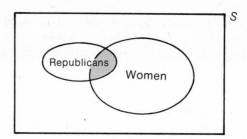

If we know that 1/2 of the Republicans are women, then we can work out the answer: 100 of the 200 Republicans in our sample are women and are counted twice, so the total of Republicans or women is 200 + 300 − 100 = 400, and Pr(Republican or woman) = 400/600 = .67. But note that Pr(Republican or woman) $= \dfrac{200 + 300 - 100}{600} = \dfrac{200}{600} + \dfrac{300}{600} - \dfrac{100}{600} =$ Pr(Republican) + Pr(woman) − Pr(Republican and woman).

This is true in general for any events A and B:

$$\text{Pr}(A \text{ or } B) = \text{Pr}(A) + \text{Pr}(B) - \text{Pr}(A \text{ and } B)$$

The formula you studied earlier for mutually exclusive events is a special case of this, since Pr(A and B) = 0 if A and B are mutually exclusive.

Example 1

If one card is drawn at random from a standard deck of cards, what is the probability that it is a king or a heart?

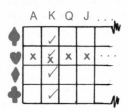

$$\text{Pr}(K) = \frac{4}{52}; \ \text{Pr}(\heartsuit) = \frac{13}{52}.$$

Pr(K and $\heartsuit$) = the probability the king of hearts is drawn

$$= \frac{1}{52}.$$

$$\text{Pr}(K \text{ or } \heartsuit) = \text{Pr}(K) + \text{Pr}(\heartsuit) - \text{Pr}(K \text{ and } \heartsuit)$$

$$= \frac{4}{52} + \frac{13}{52} - \frac{1}{52} = \frac{12}{52}$$

$$= .31.$$

Example 2   Of 200 seniors at Morris College, 98 are women, 34 are majoring in English, and 20 English majors are women. If one student is chosen at random from the senior class, what is the probability that the choice will be either an English major or a woman (or both).

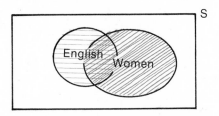

$$Pr(\text{English major or woman}) = Pr(\text{English major}) + Pr(\text{woman}) - Pr(\text{English major and woman})$$

$$= \frac{34}{200} + \frac{98}{200} - \frac{20}{200} = \frac{112}{200}$$

$$= .56.$$

5.7 EXERCISES

1. Which of the following are mutually exclusive?
    (a) George won and Mary won when they played tennis together.
    (b) George lost and Mary won when they played tennis together.
    (c) George won and Mary won when each played tennis with another player.
    (d) The average age of college freshmen is 18 or it is 19.
    (e) The Yankees won the World Series and Los Angeles won the National League Pennant.

2. Prints of 90 paintings are sold at a college book store. 40 are by Impressionists, 12 are by Renaissance artists, and the remainder are by modern artists. If 1 print is chosen at random, what is the probability it is of a painting (a) done by an Impressionist? (b) done by an Impressionist or a Renaissance artist? (c) not done by an Impressionist?

3. In a fashion show of evening dresses, 4 dresses are made of taffeta, 6 of organdy, 7 of silk, and 3 of cotton. What is the probability that one dress chosen at random is (a) made of organdy? (b) not made of organdy? (c) made of organdy or silk?

4. 104 houses are for sale. 52 have fireplaces, 13 have a red roof, and 8 of the houses with a red roof have a fireplace. What is the probability one of these houses has (a) a red roof? (b) a fireplace? (c) a red roof or a fireplace? (d) a red roof and a fireplace? (e) neither a red roof nor a fireplace?

5. Of 22 calculators displayed at Goody's Discount Store, 16 are made abroad and 6 in the United States. 5 of the United States-made and 14 of the foreign-made calculators find square roots. If 1 calculator is purchased at random, what is the probability that it (a) will find square roots? (b) was made in the United States? (c) will find a square root or is American-made?

6. You randomly choose 1 recipe from 6: A, B, C, D, E, F. Onions are used in A, B, and C, milk in A, D, F, and cheese in D, E, F. What is the probability that the recipe you choose uses (a) onions? (b) milk? (c) onions or cheese? (d) neither milk nor onions? (e) onions and milk? (f) onions and milk and cheese?

7. The probability it will be over 80°F in Phoenix, Arizona, on July 9 is .85. What is the probability it is 80° or less on July 9 in Phoenix?

8. If two dice are tossed, what is the probability that the sum of the two dice will be (a) 3? (b) 3 or less? (c) 13? (d) an even number? (e) an even number or 3? (f) an even number or 4?

9. A die and a coin are tossed. What is the probability of getting (a) a 3 and heads? (b) an even number and heads? (c) either a 3 and heads, or an even number and heads? (d) a 3 and heads, and an even number and heads?

10. If 40% of the households in Petsville have a cat, 55% have a dog, and 25% have a cat and a dog, (a) what percentage of the households have a cat or a dog? (b) what percentage have neither a cat nor a dog? (c) what percentage have a cat but no dog? (d) what percentage of those that have a dog have a cat too? (e) what percentage have no pets at all?

11. You will win $5 if you toss two dice and the sum is 7, or if you draw a card at random from a deck of 52 cards and it is a king or a queen. Which would you prefer to do? Why?

ANSWERS

1. (a), (d).

2. (a) $\frac{40}{90}$ = .44; (b) $\frac{40 + 12}{90}$ = .58; (c) 1 − .44 = .56.

3. (a) $\frac{6}{20}$ = .30; (b) 1 − .3 = .7; (c) $\frac{13}{20}$ = .65.

4. (a) $\frac{13}{104}$ = .13; (b) $\frac{52}{104}$ = .50; (c) .13 + .50 − $\frac{8}{104}$ = .55; (d) $\frac{8}{104}$ = .08; (e) 1 − .55 = .45.

5. (a) $\frac{19}{22}$ = .86; (b) $\frac{6}{22}$ = .27; (c) $\frac{19}{22}$ + $\frac{6}{22}$ − $\frac{5}{22}$ = .91 or $\frac{14 + 6}{22}$ = .91.

6. (a) $\frac{3}{6}$ = .50; (b) $\frac{3}{6}$ = .50; (c) 1.00; (d) $\frac{2}{6}$ = .33; (e) $\frac{1}{6}$ = .17; (f) 0.

7. 1 − .85 = .15.

8. (a) $\frac{2}{36}$ = .06; (b) $\frac{3}{36}$ = .08; (c) 0; (d) $\frac{18}{36}$ = .50; (e) .50 + .06 = .56; (f) .50 (4 is an even number).

9. (a) $\frac{1}{12}$ = .08; (b) $\frac{1}{4}$ = .25; (c) .08 + .25 = .33; (d) 0.

10. (a) 40% + 55% − 25% = 70%; (b) 100 − 70 = 30%; (c) 40% − 25% = 15%; (d) $\frac{25}{55}$(100) = 45%; (e) Can't tell; no information about alligators is given.

11. Toss two dice. Pr(7 on two dice) = $\frac{6}{36}$ = .17, > Pr(K or Q) = $\frac{8}{52}$ = .15.

5.8   INDEPENDENT AND DEPENDENT EVENTS, PR(A AND B)

If a coin is flipped twice, the outcome of the second flip doesn't depend on whether the first one comes up heads or tails. If two dice are tossed, the number on the second die is independent of the number that comes up on the first die. Suppose a ball is chosen at random from a bag containing four red and six white balls, its color is observed, and then the ball is replaced in the bag and the bag is well shaken. Now a second ball is drawn from the bag. What is the probability that the second ball is red? The answer (4) does not depend on the first choice; since that ball was replaced, the two events are independent: the outcome of one does not affect the outcome of the other.

   If, however, the first ball is **not** replaced before the second ball is drawn, what is the probability that the second is red? If the first ball drawn was red, there now remain 9 balls in the bag, of which 3 are red. Pr(the second is red) = 3/9. But if the first ball was white, then 4 of the 9 balls left in the bag are red and Pr(the second is red) = 4/9. The result of the second draw depends on what happened on the first; we say the two events are **dependent.**

> A and B are **independent** events if the occurrence of event A has no effect on the probability of occurrence of event B and vice versa. If random draws are made with replacement, the draws are independent.
>
> A and B are **dependent** events if the occurrence of one event does affect the probability of occurrence of the other. If random draws are made without replacement, the draws are dependent.

Example 1   Candidates A and B each have speakers' lists of 10 men and 8 women. One representative is to be chosen at random from each list for a debate on the candidates' policies. The events C = {candidate A's speaker is a woman} and D = {candidate B's speaker is a man} are independent, since a man is neither more nor less likely to be chosen for candidate B if a woman is chosen for candidate A; the selection was specified to be "at random." (Is this example realistic in today's politics?)

Example 2   Two cards are drawn at random, one after the other but without replacement, from a pack of 52 cards. The events $A = \{$the first card is a heart$\}$ and $B = \{$the second card is a heart$\}$ are dependent; if the first card is a heart, then there are 51 choices for the second card, of which 12 are hearts, and the probability that the second card is a heart is 12/51; but if the first card is not a heart, then there are 13 hearts left among the 51 from which the second card is chosen, and the probability of B is 13/51.

> **Notation:** the symbol $\Pr(B/A)$ means "the probability of event B, given that event A occurs."

In Example 1 above (a woman is chosen for candidate A and a man for candidate B from speakers' lists of 10 men and 8 women), $\Pr(D/C) = \Pr($a man speaking for candidate B given that a woman is to speak for candidate A) = 10/18. Note that in this case $\Pr(D/C) = \Pr(D)$.

In Example 2 above (2 cards drawn from a deck of 52), $\Pr(B/A) = \Pr($the second card is a heart given that the first is a heart) = 12/51.

Whenever you wish to find the probability that events A and B both happen, you must first ask (and answer) the question "Are A and B independent events or dependent events?"

> If A and B are independent, $\Pr(A \text{ and } B) = \Pr(A)\Pr(B)$.
> If A and B are dependent, $\Pr(A \text{ and } B) = \Pr(A)\Pr(B/A)$.

Example 3   Representatives for a debate are chosen at random from separate speakers' list of 10 men and 8 women for candidates A and B. What is the probability that candidate A's speaker is a woman and candidate B's is a man?
$C = \{$Candidate A's speaker is a woman$\}$, $D = \{$Candidate B's speaker is a man$\}$, $\Pr(C) = 8/18$, $\Pr(D) = 10/18$. C and D are independent; $\Pr(C \text{ and } D) = (8/18)(10/18) = 20/81 = .25$.

Example 4   Two cards are drawn, one after the other but without replacement, from a standard deck of 52 cards. What is the probability that both cards are hearts?
$A = \{$the first card is a heart$\}$, $B = \{$the second card is a heart$\}$. A and B are dependent; $\Pr(A \text{ and } B) = \Pr(A)\Pr(B/A) = (13/52)(12/51) = 1/17 = .06$.

Example 5   A card is drawn from a standard deck of 52 cards, its suit is observed, and the card is then replaced before a second one is drawn. What is the probability that both are hearts?
Here the two events are clearly independent; the probability of each is $13/52 = 1/4$, and $\Pr(2 \text{ hearts are drawn}) = (1/4)(1/4) = 1/16$. Note: An alternative (and common) way of asking the same questions is, "If two cards are drawn with

replacement from a standard deck of 52 cards, what is the probability that both are hearts?''

Example 6

If two cards are drawn from a standard deck of 52 cards, what is the probability of getting 1 heart and 1 spade?

The small trap: "two cards are drawn" presents to you the same problems as "one card is drawn and not replaced; then a second card is drawn." The larger trap: "getting one heart and one spade" does not describe order. The first card may be a heart and the second a spade or the first a spade and the second a heart. (If you insist on drawing two cards at once, think of both hands reaching out at the same time; you may have a heart in the left and a spade in the right, or vice versa.)

A tree diagram is helpful here:

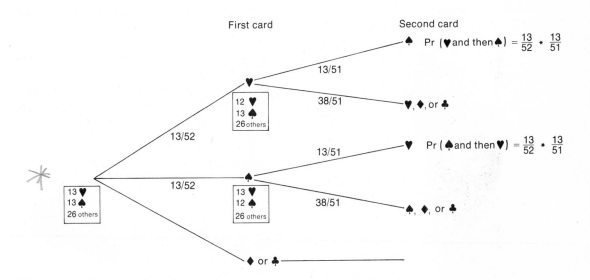

$$\text{Pr(1 heart and 1 spade)} = \frac{13}{52}\left(\frac{13}{51}\right) + \frac{13}{52}\left(\frac{13}{51}\right) = 2\left(\frac{13}{204}\right) = \frac{13}{102}$$

Sometimes it's easier to find a solution without drawing a tree or a Venn diagram. See if you can follow this reasoning:

Pr(1 heart and 1 spade) = Pr[(first card is a heart and second is a spade) or (first card is a spade and second a heart)]

= Pr(first card is a heart and second is a spade) + Pr(first card is a spade and second a heart)

= Pr(first card is a heart) · Pr(second is a spade, given that the first is a heart) + Pr(first card is a spade) · Pr(second is a heart, given that the first is a spade)

$$= \frac{13}{52}\left(\frac{13}{51}\right) + \frac{13}{52}\left(\frac{13}{51}\right)$$

$$= 2\left(\frac{13}{204}\right)$$

$$= \frac{13}{102} = .13$$

Mutually exclusive events allow use of $\Pr(A \text{ or } B) = \Pr(A) + \Pr(B)$.

Note: Life is too short to write out each problem in such detail. See if you can follow this translation: $\Pr(1\heartsuit \text{ and } 1\spadesuit) = \Pr(\heartsuit_1\spadesuit_2 \text{ or } \spadesuit_1\heartsuit_2) = \Pr(\heartsuit_1\spadesuit_2) + \Pr(\spadesuit_1\heartsuit_2) = \Pr(\heartsuit_1) \cdot \Pr(\spadesuit_2/\heartsuit_1) + \Pr(\spadesuit_1) \cdot \Pr(\heartsuit_2/\spadesuit_1) = (13/52)(13/51) + (13/52)(13/51) = 13/102$.

Example 7

What is the probability of heads exactly 3 times if a coin is flipped 4 times?

Here the notation $\Pr(HHTH)$ will be used as shorthand for "the probability of heads on the first, second, and fourth tosses, but tails on the third toss." This is even more abbreviated than the last example, since even subscripts are omitted.

$$\Pr(3H \text{ in } 4 \text{ tosses}) = \Pr(HHHT \text{ or } HHTH \text{ or } HTHH \text{ or } THHH) \tag{1}$$

$$= \Pr(HHHT) + \Pr(HHTH) + \Pr(HTHH) + \Pr(THHH) \tag{2}$$

$$= \Pr(H)\Pr(H)\Pr(H)\Pr(T) + \Pr(H)\Pr(H)\Pr(T)\Pr(H) + \Pr(H)\Pr(T)\Pr(H)\Pr(H) + \Pr(T)\Pr(H)\Pr(H)\Pr(H) \tag{3}$$

$$= \left(\frac{1}{2}\right)^4 + \left(\frac{1}{2}\right)^4 + \left(\frac{1}{2}\right)^4 + \left(\frac{1}{2}\right)^4 \tag{4}$$

$$= 4\left(\frac{1}{2}\right)^4$$

$$= .25$$

Explanations

(1) These are 4 different events, each consisting of heads 3 times and tails once.
(2) The 4 events in (1) are mutually exclusive; **or** is translated into **+**.
(3) Each of the events in (2) is made up of 4 independent events, so **and** is translated into multiplication.
(4) $\Pr(H) = 1/2$, $\Pr(T) = 1/2$ (assuming the coin is fair).

Do not confuse "mutually exclusive" and "independent." Suppose George and Mary are both playing tennis on Saturday at 3 P.M. Let G be the event George wins his match, and let M be the event Mary wins hers. If the two are playing in different matches, G and M are independent events but are not mutually exclusive; if they are playing against each other, G and M are mutually exclusive events which are not independent.

If you are finding $\Pr(A \text{ or } B)$, ask "Are A and B mutually exclusive?" If they are, **add** $\Pr(A)$ and $\Pr(B)$.

If you are finding $\Pr(A \text{ and } B)$, ask "Are A and B independent?" If they are, **multiply** $\Pr(A)$ and $\Pr(B)$.

5.9 EXERCISES

1. Jim's chance of winning a heat in a race is .60. If he wins his heat, he takes part in a run-off and his probability of winning that is .20. What is the probability he wins the gold medal?

2. The probability that Mr. Fenton, an ambitious lawyer, is elected to the legislature is .60. If he gets elected, the probability that he will get put on the Judiciary Committee is .70. What is the probability that he will be on the Judiciary Committee?

3. In the past the Seafarer's Union has petitioned for election 1,200 times and gotten it 980 times; it has won the election for workers' representative 70% of the time, and it has won a contract without a strike 480 times during 696 times it has acted as representative of workers. What is the probability it will get an election, win it, and then get a contract without a strike?

4. If a die is tossed 3 times, what is the probability of 6 coming up at least once?

5. In a herd of sheep, 70% have black wool and 30% have white wool; 40% are males and 60% are females.
   (a) How small can the percentage of black-wooled female sheep be?
   (b) How large can the percentage of black-wooled female sheep be?
   (c) What is the percentage of black-wooled female sheep if sex and color of wool are independent?

6. If 10% of the sales in a men's clothing store are raincoats and 60% of the raincoats sold have linings, (a) what percentage of all the sales are raincoats without linings? (b) what is the probability that the next sale, if at random, will not be a raincoat with a lining?

7. 60% of the movies shown at the Garbage Theater are X-rated and 20% are R-rated. If half of the X-rated, three quarters of the R-rated, and one quarter of the other movies are made in Hollywood, and the rest are foreign, find the probability that the movie showing on a night chosen at random will be (a) a Hollywood movie, (b) a movie that is foreign and X-rated, (c) neither made in Hollywood nor R-rated.

8. Scores in a population are 1, 1, 2, 3, 8, 9, 12, 13, 19, 27, 43, 87.
   (a) What is the probability that a random sample of 2 scores without replacement consists of the scores (i) 2 and 3? (ii) 1 and 3?
   (b) What is the probability that a random sample of 3 scores without replacement consists of the scores 1, 1, and 2?
   (c) Is the random sample described in (b) a representative sample?

   If $\Pr(A) \neq 0$, then $\Pr(A \text{ and } B) = \Pr(A) \cdot \Pr(B/A)$ can be solved for $\Pr(B/A)$:

   $$\Pr(B/A) = \frac{\Pr(A \text{ and } B)}{\Pr(A)}$$

   This equation is useful in some of the remaining exercises.

9. If the probability that a paper is well organized is .70 and the probability that it is well organized and well written is .50, what is the probability that it is well written, given that it is well organized?

10. If the probability that a student passes calculus is .90, and the probability that he passes calculus and statistics is .90, what is the probability that he passes statistics, given that he passes calculus?

11. What is the probability of getting heads 5 times in a row in 5 tosses of a fair coin?

12. If $\Pr(A) = .4$, $\Pr(B) = .7$, and $\Pr(B/A) = .25$, what is (a) $\Pr(A \text{ and } B)$? (b) $\Pr(A/B)$?

ANSWERS

1. $.60(.20) = .12.$

2. $.60(.70) = .42.$

3. $\dfrac{980}{1200}\left(\dfrac{7}{10}\right)\left(\dfrac{480}{696}\right) = .39.$

4. $1 - \Pr(\text{no 6's}) = 1 - \left(\dfrac{5}{6}\right)^3 = .42.$

5. (a) $70\% - 40\% = 30\%$; (b) 60%; (c) $.60(.70) = .42$; 42%.

6. (a) 40% of raincoats have no linings. $.10(.40) = .04$; 4% are raincoats without linings. (b) $1 - .10(.60) = 1 - .06 = .94.$

7. (a) Pr(X-rated and Hollywood, or R-rated and Hollywood, or other rating and Hollywood) $= \dfrac{6}{10}\left(\dfrac{1}{2}\right) + \dfrac{2}{10}\left(\dfrac{3}{4}\right) + \dfrac{2}{10}\left(\dfrac{1}{4}\right) = \dfrac{5}{10} = .50.$

(b) Pr(foreign and X-rated) = Pr(X-rated $\cdot$ Pr(foreign/X-rated) $= \left(\dfrac{6}{10}\right)\left(\dfrac{1}{2}\right) = .3.$

(c) Pr(X-rated and foreign, or other rating and foreign) $= \dfrac{1}{2}\left(\dfrac{6}{10}\right) + \dfrac{1}{4}\left(\dfrac{2}{10}\right) = .35.$

8. (a) (i) $2\left(\dfrac{1}{12}\right)\left(\dfrac{1}{11}\right) = \dfrac{1}{66} = .015$; (ii) $2\left(\dfrac{2}{12}\right)\left(\dfrac{1}{11}\right) = .03.$

(b) $3\left(\dfrac{2}{12}\right)\left(\dfrac{1}{11}\right)\left(\dfrac{1}{10}\right) = .0045.$

(c) Apparently not, but I'm not sure I know what a representative sample from this population is.

9. $\dfrac{.50}{.70} = .71.$

10. $\dfrac{.90}{.90} = 1.00.$

11. $\left(\dfrac{1}{2}\right)^5 = \dfrac{1}{32} = .03.$

12. (a) $\Pr(A) \cdot \Pr(B/A) = .4(.25) = .10;$

(b) $\dfrac{\Pr(A \text{ and } B)}{\Pr(B)} = \dfrac{.10}{.70} = .14.$

5.10   VOCABULARY AND SYMBOLS

probability   Venn diagram
chosen at random   mutually exclusive events
chosen with replacement $\Pr(A \text{ or } B)$
equally likely events $\bar{A}$
outcome   independent events
event   dependent events
sample space $\Pr(A/B)$
tree diagram $\Pr(A \text{ and } B)$

5.11   REVIEW EXERCISES

(Use a Venn diagram or a tree wherever it is helpful.)

1.   20 freshmen and 30 students from other classes live on the third floor of Baldwin Hall. One student is chosen, at random and with replacement, 500 times.

(a) How many times do you expect that freshman Bill Carroll is chosen?
(b) How many times do you expect a freshman to be chosen?
(c) What proportion of the time is (i) Bill Carroll, (ii) a freshman, (iii) a student from some other class chosen?
(d) If one student is chosen at random, what is the probability that the person chosen is (i) Bill Carroll? (ii) a freshman? (iii) a student from some other class?

2.   Assume that the probabilities that a student's score on an SAT test is over 600, 500–599, 400–499, or under 400 are .16, .34, .34, and .16, respectively. What is the probability that his score is (a) more than 400? (b) either less than 400 or between 500 and 599?

3.   Assume that 70% of crimes occur at night, and that 40% of crimes at night and 20% of daytime crimes are violent. If the record of one crime is chosen at random from the police files of 1,000 crimes on which these percentages are based, what is the probability that it is the record of a violent crime?

4.   If one card is drawn at random from a well shuffled standard (bridge) deck of 52 cards, what is the probability that it is (a) an ace? (b) an ace, a 4 of diamonds, or a 7 of hearts? (c) a 2 or a 5? (d) a club or a spade? (e) not a 3? (f) a 10 or a 9 or a club?

5. A ball is drawn at random from a bag containing 5 red, 10 black, and 10 white balls. What is the probability that it is (a) red? (b) red or white? (c) not red? (d) neither red nor white?

6. Find the probability of a 3 turning up at least once when a die is tossed twice.

7. If a fair coin is tossed 3 times, find the probability of (a) no heads; (b) exactly one head; (c) at least one head. Use a tree diagram.

8. There are three children in a family. Find the probability that (a) all are girls; (b) one is a boy; (c) at least one is a boy. Assume the probability of a boy at any given birth is 1/2.

9. A man has 6 keys, only one of which fits the door of his house. If he tries the keys one at a time, choosing at random from those that have not yet been tried, what is the probability that he opens the door of his house (a) on the first try? (b) on the third try? (c) on the fifth try?

10. During the fraternity rush, $\Delta\Phi\Omega$ puts a flyer in half the campus mail boxes, choosing them at random, and APO puts a flyer in every third mail box. What is the probability that a randomly selected student will get (a) both flyers in his mail box? (b) neither flyer? (c) at least one flyer?

11. Acme Computer Company manufactures 10,000 units each week. Each unit passes through three inspection stations, A, B, and C, before being shipped out. 2% are rejected at station $A;$ of those remaining, 5% are rejected at station B, and finally approximately 1% of the rest are rejected at station C. What is the probability that a randomly chosen unit will pass all three inspections?

12. Fifty-two men and 48 women climb Mount Washington, and 14 of them get blisters; of those with blisters, 6 are women. If one member of the group is chosen at random, what is the probability that the person chosen either has blisters or is a woman (or both)?

13. A doctor discovers that the probability is .60 that patients with symptom A have tuberculosis, and that the probability is .50 that those with symptom B have this disease. What is the probability that those with symptom A or B have tuberculosis?

14. If three dice are tossed, what is the probability that their sum is (a) 3? (b) 4? (c) greater than 4?

15. (a) One of the digits 1, 2, 3 is chosen at random. What is the probability it is an even number?
   (b) Three of the digits 1, 2, 3 are chosen at random, with replacement. What is the probability that (i) none of the three digits is even? (ii) exactly one is even? (iii) exactly two are even? (iv) all three are even?

16. A cigarette box contains 10 cigarettes with filters and 5 without filters. If 2 cigarettes are chosen at random from the box, what is the probability that (a) both have filters? (b) at least one has a filter?

17.   Four tennis players are each to play a set against each of the other three. A handicapper assigns probabilities that a player will win at least two sets as follows:

Player	A	B	C	D
Probability	.7	.8	.2	.3

Can the sum of the probabilities be more than 1? Explain.

18.   A child tries to find his way through the maze below. He chooses at random whenever a choice of paths must be made, and never retraces his steps except when returning down a blind alley. If he makes no mistakes, he can find his way to the center in 40 seconds; each mistake causes a delay of 8 seconds. What is the probability that he gets to the center in 70 seconds or less?

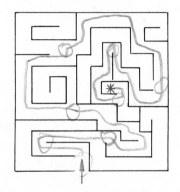

19.   A die is tossed 4 times. What is the probability that the first time a 5 comes up is on the fourth toss?

20.   On a multiple choice exam there are 10 questions and 5 choices for each answer. If a student knows the answer to 5 questions and randomly guesses at the answers to the other 5, what is the probability he passes the exam? He needs a grade of 70 or more to pass.

21.   Mr. and Mrs. Able live in Newark, New Jersey. The probability that Mr. Able will go to New York on December 31 is .2; the probability that Mrs. Able will go is .1; the probability that Mr. Able will go, given that Mrs. Able goes, is .3. What is the probability that
    (a) both Mr. and Mrs. Able will go to New York on December 31?
    (b) at least one of them will go to New York on December 31?
    (c) neither Mr. Able nor Mrs. Able will go to New York on December 31?

22.   If 3 different pieces of fruits are drawn at random from a bowl containing 4 apples, 5 pears, and 1 peach, what is the probability that
    (a) the peach is selected?
    (b) 2 or more apples are selected?
    (c) a peach and 2 apples are selected?

23. A bag contains 4 red and 6 white balls. Balls are drawn at random; WR means with replacement, WOR means without replacement. Fill in the four columns below.

	2 balls drawn		3 balls drawn	
	WR	WOR	WR	WOR
Pr(none are white)				
Pr(exactly one is white)				
Pr(exactly two are white)				
Pr(exactly three are white)				

24. A die is rolled twice. What is the probability that the sum of the two throws is (a) 11? (b) 11 or more? (c) 11 if the first die comes up with a 5? (d) 11 if at least one of the two throws comes up 5?

25. If 1 spark plug in your 6-cylinder car is not working, what is the probability that the defective spark plug is replaced (a) if 1 spark plug, chosen at random, is replaced with a new one? (b) if 2 spark plugs, chosen at random, are replaced with new ones?

26. If a bill must be passed by the House, Senate, and President to become law, and the probabilities of these three actions for a particular bill if presented to that body or person are .40, .30, and .20, respectively, what is the probability of that particular bill becoming law?

27. In a simplified form of the game of Bingo, balls numbered 0 to 9 are drawn at random and players cover the corresponding numbers on their cards, the winner being the first to cover five numbers in any row or diagonal. If you need the numbers 3, 5, 9, what is the probability that you will complete the row (a) after 3 more calls? (b) after 4 more calls? (Compute to 3 decimal places.)

28. In a hat there are 2 coins. One is "fair" $[\Pr(H) = \Pr(T) = 1/2]$, but the other one has two heads $[\Pr(H) = 1, \Pr(T) = 0]$. One coin is chosen at random from the hat and flipped 6 times. The first 5 times it comes up heads. What is the probability of heads on the sixth toss? Hints: If $\Pr(A) \neq 0$, $\Pr(B/A) = \Pr(A \text{ and } B)/\Pr(A)$; Pr(heads on first 5 tosses) = Pr([fair coin and heads on first 5 tosses] or [two-headed coin and heads on first 5 tosses]). Let A be the event $\{$heads on the first 5 tosses$\}$, B be the event $\{$heads on the sixth toss$\}$.

ANSWERS

1. (a) 10; (b) 200; (c) (i) $\frac{10}{500}$, (ii) $\frac{200}{500}$, (iii) $\frac{300}{500}$; (d) (i) .02, (ii) .40, (iii) .60.

3. Of every 1,000 crimes, 700 are at night and 280 of these are violent, while 300 are in the daytime and 60 of these are violent; $280 + 60 = 340$ crimes are violent, Pr(the record is of a violent crime) $= \frac{340}{1000} = .34$.

5.   (a) $\dfrac{5}{25} = \dfrac{1}{5}$; (b) $\dfrac{15}{25} = \dfrac{3}{5}$; (c) $1 - \dfrac{1}{5} = \dfrac{4}{5}$ or $\dfrac{10}{25} + \dfrac{10}{25} = \dfrac{20}{25} = \dfrac{4}{5}$; (d) $\dfrac{10}{25} = \dfrac{2}{5}$.

7.   (See page 98 for the tree diagram.) (a) $\dfrac{1}{8}$; (b) $\dfrac{3}{8}$; (c) $1 - \dfrac{1}{8} = \dfrac{7}{8}$.

9.   (a) $\dfrac{1}{6}$; (b) Pr(FFS) $= \dfrac{5}{6}\left(\dfrac{4}{5}\right)\left(\dfrac{1}{4}\right) = \dfrac{1}{6}$; (c) $\dfrac{5}{6}\left(\dfrac{4}{5}\right)\left(\dfrac{3}{4}\right)\left(\dfrac{2}{3}\right)\left(\dfrac{1}{2}\right) = \dfrac{1}{6}$.

11. Of 10,000 units, 9,800 pass inspection at station A; of these, 9,310 pass station B, and 9,217 pass station C also. Pr(passing all three) is $\dfrac{9,217}{10,000} = .92$ or Pr(passing station A, and passing station B given that station A has been passed, and passing station C given that stations A and B have been passed) $= (1 - .02)(1 - .05)(1 - .01) = .98\,(.95)(.99) = .92$.

13. No answer. Did you try Pr(A or B) = Pr(A) + Pr(B) = .60 + .50 = 1.10?! Probability can never be > 1. Not enough information—such as Pr(A and B)—is given.

15. (a) Pr(even) $= \dfrac{1}{3}$. (b) (i) $\left(\dfrac{2}{3}\right)^3 = \dfrac{8}{27}$; (ii) $3\left(\dfrac{1}{3}\right)^1\left(\dfrac{2}{3}\right)^2 = \dfrac{4}{9}$; (iii) $3\left(\dfrac{1}{3}\right)^2\left(\dfrac{2}{3}\right) = \dfrac{2}{9}$;

(iv) $\left(\dfrac{1}{3}\right)^3 = \dfrac{1}{27}$.

17. Yes. Read "Pr(A w B)" as "the probability that player A wins his set with player B." Then Pr(X w Y) + Pr(Y w X) must equal 1 for each pair of players. The answer is easier to understand if players A and B are tennis stars and players C and D are beginners. Then the assigned probabilities of winning at least 2 sets are 1, 1, 0, 0, respectively. The following assignment of probabilities to individual matches may seem unlikely, but would result in the handicapper's predictions:

	A	B	C	D
Pr(A w -)		.5	.4	1
Pr(B w -)	.5		1	.6
Pr(C w -)	.6	0		.3
Pr(D w -)	0	.4	.7	

19. Pr(no 5 on the first 3 tosses and then 5 on the fourth toss) $= \left(\dfrac{5}{6}\right)^3\left(\dfrac{1}{6}\right) = .10$.

21. (a) Pr(Mr. and Mrs.) = Pr(Mrs.)Pr(Mr./Mrs.) = .1(.3) = .03.
     (b) Pr(Mr. or Mrs.) = Pr(Mr.) + Pr(Mrs.) − Pr(Mr. and Mrs.) = .2 + .1 − .03 = .27.
     (c) Pr(neither Mr. nor Mrs.) = 1 − Pr(Mr. or Mrs.) = 1 − .27 = .73.

23.

Probabilities

Number of white	2 balls drawn		3 balls drawn	
	WR	WOR*	WR	WOR
0	$(.4)^2 = .16$	$\dfrac{4}{10}\left(\dfrac{3}{9}\right) = .13$	$(.4)^3 = .06$	$\dfrac{4}{10}\left(\dfrac{3}{9}\right)\left(\dfrac{2}{8}\right) = .03$
1	$2(.4)(.6) = .48$	$2\left(\dfrac{4}{10}\right)\left(\dfrac{6}{9}\right) = .53$	$3(.4)^2\,(.6) = .29$	$3\left(\dfrac{4}{10}\right)\left(\dfrac{3}{9}\right)\left(\dfrac{6}{8}\right) = .30$
2	$(.6)^2 = .36$	$\dfrac{6}{10}\left(\dfrac{5}{9}\right) = .33$	$3(.4)(.6)^2 = .43$	$3\left(\dfrac{4}{10}\right)\left(\dfrac{6}{9}\right)\left(\dfrac{5}{8}\right) = .50$
3	0	0	$(.6)^3 = .22$	$\dfrac{6}{10}\left(\dfrac{5}{9}\right)\left(\dfrac{4}{8}\right) = .17$

*Because of round-off, the sum of these probabilities does not equal 1.00.

25. (a) $\dfrac{1}{6} = .17$; (b) $1 - \dfrac{5}{6}\left(\dfrac{4}{5}\right) = \dfrac{1}{3}$ or $\dfrac{1}{6}\left(\dfrac{5}{5}\right) + \dfrac{5}{6}\left(\dfrac{1}{5}\right) = \dfrac{1}{3}$.

27. (a) $.3(.2)(.1) = .006$. (b) To win on 4 calls, the last one must give you bingo. A further complication is that if you get a 3, say, on the first call, you can get 3 again on the second or third calls and still win in 4; but if you do not get a 3, 7, or 9 on the first call then there can be no repetitions in the 4 calls.

$$\text{Pr}(3, 5, 9 \text{ [in any order] on calls } 2, 3, 4) = .7(.3)(.2)(.1) = .0042$$
$$\text{Pr}(3, 5, 9 \text{ [in any order] on calls } 1, 3, 4) = .3(.8)(.2)(.1) = .0048$$
$$\text{Pr}(3, 5, 9 \text{ [in any order] on calls } 1, 2, 4) = .3(.2)(.9)(.1) = \underline{.0048}$$
$$.0138$$

$\text{Pr}(\text{winning on 4 calls}) = .014.$

6
DISCRETE PROBABILITY DISTRIBUTIONS

In this chapter you are introduced to probability distributions. The bionomial and normal distributions are particular kinds of probability distributions. The binomial distribution, taken up in this chapter, is important in itself—you will see it applied several times in later chapters—but it also heads you toward the normal distribution, the subject of the following chapter. And an understanding of the normal distribution is basic to much of the rest of the book.

6.1   RANDOM VARIABLES

Sometimes the "points" or elements of a sample space are numbers. For example, the number of heads that come up when a coin is tossed four times is 0, 1, 2, 3, or 4. When a die is tossed once, a 1, 2, 3, 4, 5, or 6 comes up. If you are to find the probability that a woman aged 20 has a certain height, you might measure the heights of many women to the nearest inch, and the elements of your sample space would be numbers such as 60 or 61. But in other cases, the element of a sample space is a category and not a number. If you are finding the probability that a randomly chosen senior is a man or a woman, your sample space has 2 points or elements: men and women (or male and female). Can we assign numbers to these categories so that the sample space consists of numbers instead of categories? Certainly. One way is to assign 0 to a man, 1 to a woman. Then if there are 2,500 men and 2,000 women in the senior class, Pr(man) = 2,500/4,500 = .56 becomes $P(0) = 2{,}500/4{,}500 = .56$.

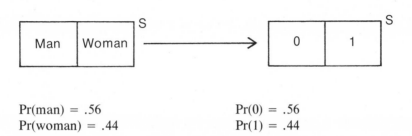

Pr(man) = .56   Pr(0) = .56
Pr(woman) = .44   Pr(1) = .44

Do you remember studying functions in high-school algebra? We are setting up a function or correspondence between the elements of a sample space and the real numbers. In statistics, this function is called a **random variable.**

Example 1

Set up a random variable for each of the following sample spaces:

(a) Toss of 1 coin: $S = \{H, T\}$
(b) Republicans, Democrats, and Socialists in Pawtucket: $S = \{R, D, S\}$.
(c) Toss of 1 die: $S = \{1, 2, 3, 4, 5, 6\}$.

(a) $H \rightarrow 0, T \rightarrow 1$ (or $H \rightarrow 10, T \rightarrow 20$)
(b) $R \rightarrow 1, D \rightarrow 2, S \rightarrow 3$ (or $R \rightarrow 2, D \rightarrow 1, S \rightarrow 0$).
(c) $1 \rightarrow 1, 2 \rightarrow 2, 3 \rightarrow 3, 4 \rightarrow 4, 5 \rightarrow 5, 6 \rightarrow 6$.

Notice that two ways of doing this are shown in (a) and (b), where the original elements were not numbers, and there are infinite possibilities. But when the elements of the sample space are already numbers, then the easiest way is to assign to each number its own value.

The same number may be assigned to many elements of a sample space. When two dice are tossed, for example, the elements in the sample space are (1, 1), (1, 2) . . . , (1, 6), (2, 1) . . . , (2, 6), . . . , (6, 6). (See Example 8, page 93.) We might assign to each of these the sum on the two dice; then (1, 3), (2, 2), and (3, 1) would all be assigned the number 4. Another way to do it would be to assign the number 1 if the sum is odd, and the number 2 if the sum is even; then 1 would be assigned to (4, 1), (5, 2), (1, 6) and others, and 2 would be assigned to (2, 2), or (6, 2), for example.

But the important thing to know is not that a random variable is a function; the important thing is that the elements of the sample space are numbers.

> For a random variable, the elements of the sample space are numbers.

Why is this done? It started when mathematicians wanted to multiply an element of a sample space by the probability assigned to it. 1/3(man) + 2/3(woman) didn't make sense, but 1/3(0) + 2/3(1) did. We'll come back to this idea later.

A random variable can take on a countable number of different values ("countable" means "finite or just as many as there are integers") or it can be continuous; see the discussion of discrete and continuous variables on page 41. In the first case, the random variable is called discrete.

Example 2

In each of the following, is the random variable discrete or continuous?

(a) Number of baskets made by the Indiana State basketball team each year.
(b) Time to run a 100-yard dash.
(c) Time to run a 100-yard dash, measured to the nearest tenth of a second.
(d) Height of a child.
(e) Height of a child, measured to the nearest quarter of an inch.

Discrete: (a), (c), (e). Continuous: (b), (d).

6.2 DISCRETE PROBABILITY DISTRIBUTIONS

With each value of a random variable is associated a probability; if we show this association in a table or by a function, then we have a probability distribution.

> A probability distribution assigns a probability to each value of a discrete random variable.

In this chapter we'll look at probability distributions for discrete random variables; in the next chapter we'll look at a particular and very important probability distribution for a continuous random variable.

Example 1   Two coins are tossed. The probability distribution for the number of heads that come up is:

Number of heads (X)	Pr(X)
0	.25
1	.50
2	.25
	1.00

Example 2   A random variable X can take on values 1, 2, 3. If $\Pr(X) = X/6$, then we have a probability distribution. Instead of using the formula $\Pr(X) = X/6$, we could show the same information in a table.

X	Pr(X)
1	$\dfrac{1}{6}$
2	$\dfrac{2}{6}$
3	$\dfrac{3}{6}$
	$\dfrac{6}{6} = 1.00$

Note that in both these examples, the sum of the probabilities is 1. This is, of course, always the case, since $\Pr(S) = 1$: the sum of the probabilities assigned to all the outcomes in S must always be 1.

A probability distribution can be illustrated by a probability histogram. On the following page are the histograms for the examples above:

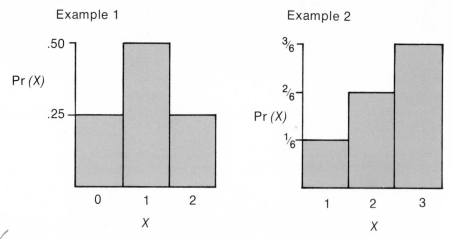

✓ Note that the probability of getting 0 or 1 head is represented by the area of the bars over 0 and 1: Pr(0 or 1) = 1/4 + 1/2 = 3/4, and the total area under each histogram is 1.

What is the relationship between a probability histogram and a relative frequency histogram? If you were to toss two fair coins 50 times, you might get no heads 14 times, 1 head 24 times, and 2 heads 12 times. The relative frequency histogram resembles that for Example 1 above:

Number of heads (X)	f	Relative frequency
0	14	.28
1	24	.48
2	12	.24
	50	1.00

The resemblance would get better as the number of times the two coins are tossed is increased. Note, however, that a probability distribution is given only for a random variable, while a relative frequency distribution can be given for any kind of data (including categorical data).

Example 3   Compare the probability histogram for the digits in a table of random numbers with the relative frequency histogram of the following count of the first digit in each number in Column 1 of Table 3 in Appendix C:

Digit (X):	0	1	2	3	4	5	6	7	8	9
Frequency	7	12	9	8	11	11	8	10	15	9
Relative frequency	.07	.12	.09	.08	.11	.11	.08	.10	.15	.09
Pr(X)	.10	.10	.10	.10	.10	.10	.10	.10	.10	.10

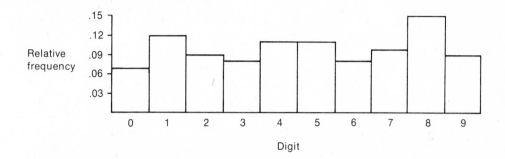

A frequency distribution has a mean and a standard deviation. Does a probability distribution? The answer is "yes" and the formulas are:

$$\mu = \Sigma X \, \Pr(X). \qquad \sigma = \sqrt{\Sigma(X - \mu)^2 \, \Pr(X)}.$$

These are your old friends $\mu = \dfrac{\Sigma Xf}{N}$ and $\sigma = \sqrt{\dfrac{\Sigma(X - \mu)^2 f}{N}}$ (see pages 59 and 70), but now f is replaced by $\Pr(X)$ and $\Sigma f = N$ is replaced by $\Sigma \Pr(X) = 1$, so the denominator seems to disappear. And the formulas make sense because X is a random variable, so both X and $\Pr(X)$ are numbers which can be multiplied and added.

Example 4   After much experimentation, the distribution shown in the box below has been determined for the probability that a rat will need X trials to solve a certain maze within 60 seconds. (a) Find the mean and standard deviation of this distribution. (b) What is the probability that a rat will need 3 trials or less before solving the maze in the allotted time?

X	$\Pr(X)$	$X \Pr(X)$	$X - \mu$	$(X - \mu)^2$	$(X - \mu)^2 \Pr(X)$
1	.20	.20	−1.5	2.25	.450
2	.40	.80	−.5	.25	.100
3	.20	.60	.5	.25	.050
4	.10	.40	1.5	2.25	.225
5	.10	.50	2.5	6.25	.625
		2.50			1.450

(a) $\mu = \Sigma X \Pr(X) = 2.5$

$\sigma^2 = \Sigma(X - \mu)^2 \cdot \Pr(X) = 1.450$

$\sigma = \sqrt{1.450} = 1.20$

(b) $\Pr(X = 1, 2, \text{ or } 3)$

$= .20 + .40 + .20 = .80$

6.3 EXERCISES

1. (a) Which of the following are random variables?
   (i) $S = \{\text{red, green, blue}\}$
   (ii) $S = \{1, 2, 3, \ldots, 11\}$
   (iii) $S = \{\text{real numbers} > 0\}$
   (iv) $S = \{\text{species of birds}\}$
   (b) Assign numerical values so that those in (a) which are not already random variables become so.

2. Which of the following random variables are discrete and which are continuous?
   (a) $S = \{\text{integers between 0 and 480}\}$.
   (b) $S = \{\text{real numbers between 0 and 480}\}$.
   (c) Boiling point of isopropyl alcohol mixed with water.
   (d) Error in repeated measurement of 10 cc. of water.
   (e) Number of tomatoes grown in a 10 by 10 foot plot.
   (f) Weight of tomatoes grown in a 10 by 10 foot plot.
   (g) Weight of tomatoes grown in a 10 by 10 foot plot, measured to the nearest ounce.

3. Which of the following can be probability distributions? If not, explain why not.
   (a) $S = \{1, 2, 3\}$; $\Pr(1) = .4$, $\Pr(2) = .4$, $\Pr(3) = .2$.
   (b) $S = \{1, 2, 3\}$; $\Pr(1) = .5$, $\Pr(2) = .4$, $\Pr(3) = .2$.
   (c) $S = \{R, G, B\}$; $\Pr(R) = .3$, $\Pr(G) = .4$, $\Pr(B) = .3$.
   (d) $S = \{1, 2, 3\}$; $\Pr(1) = .5$, $\Pr(2) = -.1$, $\Pr(3) = .6$.

4. Which of the following are probability distributions? If not, explain why not. $S = \{0, 1, 2, 3, 4\}$
   (a) $\Pr(X) = \dfrac{X}{10}$
   (b) $\Pr(X) = \dfrac{2X - 3}{5}$
   (c) $\Pr(X) = \dfrac{2X + 2}{30}$

5. Find the probability distribution for the number of Republicans that are chosen when 2 people are chosen at random and without replacement from a group of 6 Republicans and 4 Democrats.

6. Find the probability distribution for the number of hearts drawn when 3 cards are randomly drawn (all at the same time) from a standard deck of cards.

7. Draw probability histograms to illustrate the probability distribution in (a) Exercise 4(a), (b) Exercise 5, (c) Exercise 6.

8. What is the difference between a probability distribution and a relative frequency distribution?

9. Find the mean and the standard deviation of the probability distribution in (a) Exercise 4(a), (b) Exercise 5, (c) Exercise 6.

10. What is (a) the mean and (b) the standard deviation of each of the following probability distributions?

(i) X	Pr(X)	(ii) X	Pr(X)	(iii) X	Pr(X)
0	.062	0	.729	0	.002
1	.250	1	.243	1	.026
2	.375	2	.027	2	.154
3	.250	3	.001	3	.410
4	.062			4	.410

ANSWERS

1. (a) (ii), (iii)

(b) (i) Red → 0, green → 1, blue → 2; (iv) Assign the number 4 to each species. (There are, of course, very many other answers in (b).)

2. Discrete: (a), (e), (g).

3. (a); (b) is not because the sum of the probabilities is not 1.0; in (c) there is no random variable; in (d), Pr(2) is less than 0.

4. (a) and (c). (b) is not because Pr(0) and Pr(1) are less than 0.

5. $S = \{0, 1, 2\}$. Pr(0) $= \dfrac{4}{10}\left(\dfrac{3}{9}\right) = .13$, Pr(1) $= 2\left(\dfrac{6}{10}\right)\left(\dfrac{4}{9}\right) = .53$, Pr(2) $= \dfrac{6}{10}\left(\dfrac{5}{9}\right) = .33$.

6. $S = \{0, 1, 2, 3\}$. Pr(0) $= \dfrac{39}{52}\left(\dfrac{38}{51}\right)\left(\dfrac{37}{50}\right) = .41$, Pr(1) $= 3\left(\dfrac{13}{52}\right)\left(\dfrac{39}{51}\right)\left(\dfrac{38}{50}\right) = .44$,

Pr(2) $= 3\left(\dfrac{13}{52}\right)\left(\dfrac{12}{51}\right)\left(\dfrac{39}{50}\right) = .14$, Pr(3) $= \dfrac{13}{52}\left(\dfrac{12}{51}\right)\left(\dfrac{11}{50}\right) = .01$.

7.

(a)

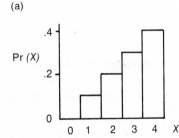

(b)

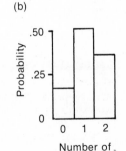

Number of Republicans

(c)

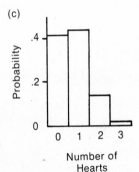

Number of Hearts

8. A probability distribution is given only for a random variable; a relative frequency distribution may be given for categorical data in which the categories are not numbers. A relative frequency distribution may be given for a small number of cases; a probability distribution is either a theoretical model or is based (presumably) on a relatively large number of cases.

9. (a) $\mu = 0(0) + 1(.1) + 2(.2) + 3(.3) + 4(.4) = 3.0; \sigma^2 = (-3)^2(0) + (-2)^2(.1) + (-1)^2(.2) + 0^2(.3) + 1^2(.4) = 1.0; \sigma = 1.0$.
   (b) $\mu = 1.2, \sigma = .65$; (c) $\mu = .75, \sigma = .54$.

10. (a) $\mu = 2.0, \sigma = 1.0$; (b) $\mu = .30, \sigma = .52$; (c) $\mu = 3.2, \sigma = .8$.

6.4 ASSUMPTIONS FOR A BINOMIAL DISTRIBUTION

We shall continue with the study of probability distributions and the corresponding histograms, but shall limit our study for the moment to problems which satisfy the following assumptions:

1. The same experiment is carried out n times (n trials are made).
2. Each trial has two possible outcomes. Let us call these two outcomes "success" and "failure." (A successful outcome doesn't imply a good one, nor failure a bad outcome. If a trial consists in throwing one die, success may mean "a 2 comes up"; then failure would mean "1, 3, 4, 5, or 6 comes up.") Let P be the probability of success in one trial, then $1 - P$ is the probability of failure.
3. The result of each trial is independent of the result of any other trial.

We shall call these the **binomial assumptions**; the reason for the word **binomial** is that there are two choices, success or failure, for the outcome of each trial. Do you remember the binomial theorem for the expansion of $(x + y)^n$? The probabilities that we shall soon develop are identical with the coefficients in this expansion if $x = 1 - P$ and $y = P$.

If a fair coin is tossed three times, the binomial assumptions are satisfied—$n = 3$; "success" is interpreted, say, as getting a head on any one toss, so $P = 1/2$, and whether you get heads or tails on one toss doesn't affect the outcome of any other toss.

Now consider a doctor testing a new drug on 35 patients with a certain disease. He wants to compare the new drug with one now used, and he knows that this relieves the symptoms of 70% of patients who use it. Then $P = .70$, $1 - P = .30$, $n = 35$.

On the other hand, suppose that four cards are drawn, without replacement, from a deck of 52 cards, and suppose "success" means drawing a heart. Here the binomial assumptions are not satisfied; the probability of a heart on the first draw is 13/52, but on the second draw it is 12/51 if a heart was chosen on the first draw, and 13/51 if a club, diamond, or spade was drawn first; the result of the second trial is not independent of the result of the first trial.

6.5 Pr(X SUCCESSES IN n TRIALS)

Now let's look at the probability of getting exactly 2 red balls if 4 balls are drawn with replacement from the bag of 4 red and 6 green balls. One way is to draw

RRGG; the probability of doing this is $\left(\frac{4}{10}\right)\left(\frac{4}{10}\right)\left(\frac{6}{10}\right)\left(\frac{6}{10}\right) = \left(\frac{4}{10}\right)^2\left(\frac{6}{10}\right)^2$ since the draws are independent. Another way is to draw RGGR; the probability is $\left(\frac{4}{10}\right)\left(\frac{6}{10}\right)\left(\frac{6}{10}\right)\left(\frac{4}{10}\right) = \left(\frac{4}{10}\right)^2\left(\frac{6}{10}\right)^2$ —exactly the same as for RRGG. In fact any pattern of 2 R's and 2 G's will have the same probability. So all we have to do is count the number of ways we can fill 4 slots with 2 R's and 2 G's and multiply this number by $\left(\frac{4}{10}\right)^2\left(\frac{6}{10}\right)^2$ to find the probability of exactly 2 reds in 4 draws.

Mathematicians have found that the number of ways of filling 4 slots with 2 R's and 2 G's is $\dfrac{(4)(3)(2)(1)}{(2)(1)(2)(1)} = 6$. (Be sure you cancel all you can before multiplying.) So the probability of exactly 2 red balls in 4 draws is $6\left(\frac{4}{10}\right)^2\left(\frac{6}{10}\right)^2 = .35$.

We can improve the appearance of the formula by using factorial notation: $3(2)(1) = 3!$ (read "3 factorial"), $5(4)(3)(2)(1) = 5!$

> **Factorial notation:** $n! = n(n-1)(n-2)\ldots(2)(1)$
>
> **Definition:** $0! = 1$

Then the number of ways of filling 4 slots with 2 R's and 2 G's is $\dfrac{4!}{2!\,2!}$, and the probability of getting 2 red balls in 4 draws with replacement is $\dfrac{4!}{2!\,2!}\left(\frac{4}{10}\right)^2\left(\frac{6}{10}\right)^2 = .35$.

> If the binomial assumptions are satisfied, the probability of exactly X successes in n trials is
>
> $$\frac{n!}{X!\,(n-X)!}\,P^X\,(1-P)^{n-X}.$$
>
> This probability distribution is called the binomial distribution.

There is a different binomial distribution for each different pair of n and X values. X can't be greater than n, of course.

Example 1 If 4 cards are drawn at random and with replacement from a standard deck, what is the probability that exactly 1 is a spade?

The binomial assumptions are satisfied. $n = 4$, $P = \dfrac{1}{4}$, $X = 1$. Pr(exactly 1 is a spade) $= \dfrac{4!}{1!\,3!}\left(\dfrac{1}{4}\right)^1\left(\dfrac{3}{4}\right)^3 = \dfrac{54}{128} = .42$.

Table 2, Appendix C, gives values of

$$\frac{n!}{P^X\,(n-X)!}\,P^X\,(1-P)^{n-X}$$

for n from 2 to 20 for various values of P.

Example 2 Find $\dfrac{15!}{4!\,11!}\,P^4\,(1-P)^{11}$ if $P = .60$.

Here $n = 15$, $X = 4$. Find 15 in the n column of Table 2, Appendix C, and then go down that section in the table until you find 4 in the X column. Now look across the row to the $P = .60$ column; you will find the number .007. Therefore, $\dfrac{15!}{4!\,11!}\,(.60)^4(.40)^{11} = .007$.

Example 3 A fair coin is flipped three times. (a) What is the probability of heads exactly 0, 1, 2, 3 times? (b) What is the probability of at most 1 head in 3 flips? (c) What is the probability of at least 1 head?
 (a) $n = 3$, $P = .50$; let X = number of heads. Use Table 2, Appendix C.

X	$Pr(X)$
0	.125
1	.375
2	.375
3	.125

 (b) $Pr(X = 0 \text{ or } 1) = Pr(X = 0) + Pr(X = 1) = .125 + .375 = .50$.
 (c) $Pr(X = 1, 2, \text{ or } 3) = 1 - Pr(X = 0) = 1 - .125 = .875$.

Example 4 A family has 5 children. What is the probability (a) that at least 1 child is a boy? (b) that at least 2 children are boys? Assume the probability of the birth of a boy is 1/2 at each birth.
 $P = .5$; let X equal the number of boys, and use Table 2, Appendix C.
 (a) $Pr(X = 0) = .031$.
  $Pr(X = 1 \text{ or more}) = 1 - Pr(X = 0) = 1 - .03 = .97$.
 (b) $Pr(X = 1) = .156$; $Pr(X = 2 \text{ or more}) = 1 - Pr(X = 0 \text{ or } 1) = 1 - (.031 + .156) = .81$; or $Pr(X = 2 \text{ or more}) = .312 + .312 + .156 + .031 = .81$. But

don't just blast your way through to an answer; see if you can find the most graceful way.

6.6 THE MEAN AND STANDARD DEVIATION OF THE BINOMIAL DISTRIBUTION

It is particularly easy to find the mean and standard deviation for a binomial distribution. They are given by the following formulas:

$$\mu = nP, \ \sigma = \sqrt{nP(1 - P)}$$

The formula for the mean can be intuitively found: if a fair coin is flipped 100 times, you would expect heads to come up 50 times on the average. But here $n = 100$, $P = 1/2$, $nP = 100 \, (1/2) = 50$. If a die is tossed 60 times, you would expect, on the average, 3 would come up 10 times. $n = 60$, $P = 1/6$, $nP = 60 \, (1/6) = 10$.

Example 1 Find the mean and standard deviation of a binomial distribution with $n = 3$, $P = .5$.

$$\mu = nP = 3(.5) = 1.5$$

$$\sigma = \sqrt{nP(1 - P)} = \sqrt{3(.5)(.5)} = .87$$

Example 2 Balls are taken twice, with replacement, from a bag containing 4 red and 6 green balls. Find the mean and standard deviation for the probability distribution of the number of red balls which are chosen.

$$P = .4, n = 2$$

$$\mu = nP = 2 \, (.4) = .8$$

$$\sigma = \sqrt{nP(1 - P)} = \sqrt{2(.4)(1 - .4)} = \sqrt{.48} = .7$$

6.7 APPLICATIONS OF THE BINOMIAL DISTRIBUTION

Knowledge of the binomial distribution will help you understand the normal distribution to be taken up in the next chapter, but some applications are of immediate interest.

Example 1 A doctor studying cures for leukemia knows that 30% of patients will be alive 8 years after onset of the disease with the present treatment. He tries a new drug on 10 leukemia patients, chosen at random, and finds that 7 are still alive 8 years after onset. Does it seem likely that the new treatment is much preferable to the present treatment?

The question is first rephrased in terms of probabilities: What is the probability that at least 7 out of 10 leukemia patients will be alive after 8 years of the present treatment? $P = .3$; Pr(at least 7 are alive = Pr(7 or 8 or 9 or 10 are alive) $= .009 + .0010 + .000 + .000 = .01$. There is approximately a 1% probability that at least 7 out of 10 patients will live 8 years under the present treatment, so the new drug looks pretty promising. The doctor should certainly continue his work with it, and try the new treatment on more patients.

Example 2   If 1% of the watches manufactured by the Mainspring Company are defective, what is the probability that none is defective in a random shipment of 4 watches?

Are the binomial assumptions satisfied? Strictly, they are not. As each succeeding watch is chosen from the shipment, the probability it is defective depends on whether those already chosen were defective. But if the shipment is a small proportion of the number manufactured by the company, then the binomial distribution works very well as a model of what happens, even though the binomial assumptions are not satisfied to the letter.

$$n = 4, P = .01, X = 0, \text{ so Pr(none is defective)} = .96.$$

Example 3   A machine shop foreman has found that, in the past, 1% of the bolts produced by a machine are defective. He tests a sample of 600 and finds that 8 are defective. What is the probability that 8 or more will be defective in a sample of 600 if the 1% figure still holds?

$P = .01, 1 - P = .99$; Pr(8 or more defective out of 600) $= 1 - $ Pr(7 or less are defective). But $n = 600$, and Table 2, Appendix C, does not contain binomial probabilities for such large n. Computing the answer does not seem to be a lovely prospect. But in the next chapter we shall learn how to approximate it easily using the normal distribution, so don't stay up all night carrying out such computations.

6.8   EXERCISES

1.   What are the binomial assumptions?

2.   (a) 3! = ? (b) 5! = ? (c) 0! = ?

3.   (a) $\dfrac{5!}{3! \, 2!} = $? (b) $\dfrac{8!}{8! \, 0!} = $? (c) $\dfrac{7!}{6! \, 1!} = $? (d) $\dfrac{120!}{119! \, 1!} = $?

4.   Find the value of $\dfrac{n!}{X! \, (n - X)!} \, P^X \, (1 - P)^{n-X}$ if (a) $n = 4, P = \dfrac{1}{2}, X = 3$; (b) $n = 10, P = .3, X = 2$; (c) $n = 19, P = .8, X = 17$; (d) $n = 4, P = \dfrac{1}{3}, X = 2$.

5.   What is the probability that a fair die will come up 6 exactly once when it is tossed 3 times?

6.   What is the probability that (a) exactly 1, (b) 1 or 0, (c) more than 1 woman will be included in a group of 5 people chosen at random from a group of 1,200 men and 800 women?

7. Assume the probability is .60 that a student can remember what he or she was doing on Monday of last week at 7 P.M. If 14 students are chosen at random, what is the probability that 11 or more will remember what they were doing at that time?

8. Assume that 30 of the 300 children in third grade in Georgetown watch TV 1 hour a day or less, and 270 watch TV more than 1 hour. Make up a question which can be answered by using the binomial formula.

9. Which of the following is (a) not a probability distribution? (b) not a binomial distribution? (c) For each binomial distribution, what is P? (d) Draw the histogram for each probability distribution.

(i) X	Pr(X)	(ii) X	Pr(X)	(iii) X	Pr(X)	(iv) X	Pr(X)
0	.410	0	.100	0	.100	0	.026
1	.410	1	.200	1	.200	1	.154
2	.154	2	.100	2	.400	2	.346
3	.026	3	.200	3	.200	3	.346
4	.002	4	.100	4	.100	4	.130

10. 25% of Democrats voting in the primary election voted for Oshkosh for Senator. In a random sample of 4 Democrats who voted in the primary, what is the probability that (a) exactly 3, (b) 3 or more voted for Oshkosh?

11. If 2 balls are drawn at random and without replacement from a bag containing 4 red and 6 green balls, what is the probability that (a) both are red, (b) one is red and one is green?

12. What is (a) the mean and (ii) the standard deviation of a binomial distribution with (a) $n = 9$, $P = .2$; (b) $n = 400$, $P = .2$?

13. Suppose an examination given to students after 2 years of medical school is designed so that 80% of students will pass it. Of 18 randomly chosen medical students who take this examination, what is the probability that (a) exactly 16, (b) 16 or more, (c) 10 or less will pass it? What is the mean number in a randomly chosen group of 18 who will pass it? (e) What is the standard deviation?

14. If 5% of food processors sold at Goody's Discount Store are defective, what is the probability that (a) exactly 2, (b) 2 or less are defective in a random sample of 20 sales of food processors there? (c) What is the mean number of defective food processors in 20 sales? (d) What is the standard deviation?

ANSWERS

1. (a) There are n trials of the same experiment; (b) There are 2 possible outcomes of each experiment; (c) Trials are independent.

2. (a) 6; (b)120; (c) 1.

3. (a) 10; (b) 1; (c) 7; (d) 120.

4. (a) .250; (b) .233; (c) .154; (d) $\dfrac{4!}{2!\,2!}\left(\dfrac{1}{3}\right)^2\left(\dfrac{2}{3}\right)^2 = .296$. Table 2, Appendix C cannot be used for (d) since $P = .33$ is not included there.

5. $\dfrac{3!}{1!\,2!}\left(\dfrac{1}{6}\right)^1\left(\dfrac{5}{6}\right)^2 = 3\left(\dfrac{25}{216}\right) = .35$.

6. $P = .40$, $n = 5$; (a) .26; (b) $.078 + .259 = .34$; (c) $1 - .34 = .66$.

7. $P = .60$, $n = 14$, $.085 + .032 + .007 + .001 = .125$.

8. What is the probability that exactly 2 in a random sample of 9 children in third grade in Georgetown watch TV 1 hour or less per day?

9. (a) (ii) is not, since the sum of the probabilities does not equal 1; (b) Neither (ii) nor (iii). (ii) is symmetric, so if it were binomial P must equal .5 but these are not the correct probabilities for $P = .5$, $n = 4$; (c) (i) $P = .20$, (iv) $P = .60$.

(d)

(i)

(iii)

(iv)

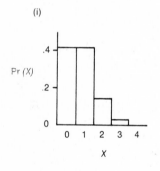

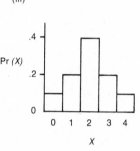

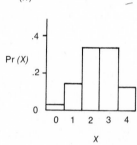

10. $P = .25$, $n = 4$. (a) $\Pr(X = 3) = \dfrac{4!}{3!\,1!}\left(\dfrac{1}{4}\right)^3\left(\dfrac{3}{4}\right) = \dfrac{3}{64} = .047$ (or .05).

(b) $.047 + \dfrac{4!}{4!\,0!}\left(\dfrac{1}{4}\right)^4 = .047 + .004 = .051$ (or .05).

11. (Not binomial) (a) $\dfrac{4}{10}\left(\dfrac{3}{9}\right) = .13$; (b) $\dfrac{4}{10}\left(\dfrac{6}{9}\right) + \dfrac{6}{10}\left(\dfrac{4}{9}\right) = .53$.

12. (a) (i) $9(.2) = 1.8$, $\sqrt{9(.2)(.8)} = 1.2$; (b) (i) $400(.2) = 80$, (ii) $\sqrt{400(.2)(.8)} = 8$.

13. $P = .80$, $n = 18$. (a) .172; (b) $.172 + .081 + .018 = .27$; (c) $.012 + .003 + .001 = .02$; (d) $18(.80) = 14.4$; (e) $\sqrt{18(.8)(.2)} = 1.7$.

14. $P = .05$, $n = 20$; (a) .189; (b) $.189 + .377 + .358 = .92$; (c) $20(.05) = 1.0$; (d) $\sqrt{20(.05)(.95)} = .97$.

6.9 VOCABULARY AND SYMBOLS

probability distribution
factorial
random variable $\qquad n!$
binomial distribution
$$\dfrac{n!}{X!\,(n - X)!}\,P^X\,(1 - P)^{n-X}$$
binomial coefficients

6.10 REVIEW EXERCISES

1. Which of the following are random variables, if each describes outcomes of an experiment or points in a sample space?
   (a) Freshmen, sophomore, junior, senior classes.
   (b) Income less than $10,000, between $10,000 and $15,000, or over $15,000.
   (c) IQ's of students varying from 103 to 157.
   (d) Sum of two dice is divisible by 3 or it is not.
   (e) Number of heads when 4 coins are tossed.

2. Describe a discrete probability distribution. How does it differ from a frequency distribution?

3. Draw probability histograms for the following probability distributions:

(a)	X	$\Pr(X)$	(b)	Y	$\Pr(Y)$	(c)	Z	$\Pr(Z)$
	0	.40		1	.1		0	.24
	1	.50		2	0		1	.10
	2	.10		3	.7		2	.20
				4	.2		3	.10
							4	.36

$$\mu = \Sigma X \Pr_X \qquad \sigma = \sqrt{\Sigma(X - \mu)^2 \cdot \Sigma}$$

4. Find (a) the mean and (b) the standard deviation for each of the probability distributions in Exercise 3.

5. Compute $\dfrac{n!}{X!\,(n-X)!}$ if (a) $n = 4$, $X = 2$; (b) $n = 7$, $X = 5$; (c) $n = 139$, $X = 139$; (d) $n = 500$, $X = 1$.

6. What is the probability of exactly three successes in 4 trials if $P = \dfrac{1}{4}$ and the binomial assumptions are satisfied?

7. In which of the following are the binomial assumptions satisfied? If not, why not? You wish to know the probability of
   (a) getting 3 queens when 3 cards are dealt from the top of a well-shuffled deck.
   (b) choosing exactly 3 women when a committee of 5 is chosen at random from a group of 14 women and 12 men.
   (c) getting a pair of kings when 4 cards are drawn with replacement from a standard deck.
   (d) getting a pair when 4 cards are drawn with replacement from a standard deck.
   (e) getting 2 or less defective cans of tuna fish in a shipment of 400 cans chosen at random from the output of Marine Canneries, Incorporated.

8. If one-third of skiers who break a bone have a broken leg, what is the probability that exactly 1 out of 5 skiers with broken bones has a broken leg?

9. If two-thirds of women who are divorced before age 30 marry again, what is the probability that (a) exactly 3, (b) 3 or more of 4 women who are divorced and are less than 30 will marry again?

10. If 90% of absences from work are claimed to be due to illness, what is the probability that 15 or less of 20 absences are due to illness?

11. Draw the probability histogram for a binomial distribution (a) with $P = .5$, $n = 6$; (b) with $P = .3$, $n = 5$; (c) with $P = .7$, $n = 0$.

12. What is the mean number of successes expected in 200 trials if the binomial assumptions are satisfied and the probability of success in any one trial is .4? What is the standard deviation of the probability distribution?

13. Assume that, when a child is born, the probability it is a girl is $\frac{1}{2}$ and that the sex of the child does not depend on the sex of an older sibling. (a) Find the probability distribution for the number of girls in a family with 4 children. (b) Find the mean and standard deviation of this distribution.

14. If 3 cards are dealt from a standard deck of 52 cards, (a) find the probability distribution for the number of 10's that are dealt; (b) find the mean of this probability distribution.

15. There are 10 questions in a multiple-choice quiz, with 5 answers to each. If answers are chosen at random, (a) what is the probability of 3 correct answers? (b) What is (i) the mean number of correct answers expected? (c) What is the standard deviation of the probability distribution?

16. A pair of dice is tossed 200 times. (a) What is the mean number of times the sum of the two dice is 7? (b) What is the most probable number of times the sum of the two dice is 7?

ANSWERS

1. (c) and (e).

3. (a)   (b)   (c)

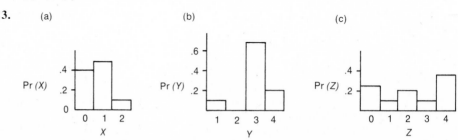

5. (a) 6; (b) 21; (c) 1; (d) 500.

7. (c), (e). In (a), the first "trial" is dealing 1 card of 52, the second is dealing 1 card of 51; different trials are not identical, nor are the results independent. In (b), choices are without replacement, so again trials are neither identical nor independent. In (d) there are more than 2 outcomes. Even in (e) the binomial assumptions are not strictly satisfied, since the cans are chosen without replacement. Presumably, however, 400 cans is such a small proportion of the canning company's production that the probability that a can is defective does not change appreciably.

9. (a) $\dfrac{4!}{3!\,1!}\left(\dfrac{2}{3}\right)^3\left(\dfrac{1}{3}\right) = \dfrac{32}{81} = .40$; (b) $\dfrac{32}{81} + \dfrac{4!}{4!\,0!}\left(\dfrac{2}{3}\right)^4 = \dfrac{32}{81} + \dfrac{16}{81} = \dfrac{48}{81} = .59.$

11.

(a) $P = .5, n = .6$ (b) $P = .3, n = 5$ (c) $P = .7, n = 9$

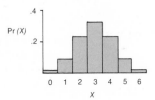 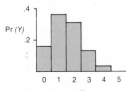 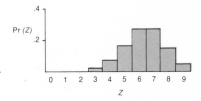

13. $P = .50$; $X = 0, 1, 2, 3, 4$; $\Pr(X) = .063, .250, .375, .250, .063.$ (b) $\mu = 4(.5) = 2$, $\sigma = \sqrt{4(.5)(.5)} = 1.0.$

14. The binomial assumptions are not satisfied.

(a)

X = Number of Hearts	$\Pr(X)$	$X\,\Pr(X)$
0	$\dfrac{39}{52}\left(\dfrac{38}{51}\right)\left(\dfrac{37}{50}\right) = .41$	.0
1	$3\left(\dfrac{13}{52}\right)\left(\dfrac{39}{51}\right)\left(\dfrac{38}{50}\right) = .43$	.43
2	$3\left(\dfrac{13}{52}\right)\left(\dfrac{12}{51}\right)\left(\dfrac{39}{50}\right) = .14$	.28
3	$\dfrac{13}{52}\left(\dfrac{12}{51}\right)\left(\dfrac{11}{50}\right) = .01$	.03

(b) $\mu = .75$ .75

15. $n = 10$, $P = \dfrac{1}{5}$ (a) $\dfrac{10!}{3!\,7!}\left(\dfrac{1}{5}\right)^3\left(\dfrac{4}{5}\right)^7 = .20$; (b) $\mu = \left(\dfrac{1}{5}\right)(10) = 2$; (c) $\sigma = \sqrt{\left(\dfrac{1}{5}\right)(10)\left(\dfrac{4}{5}\right)} = 1.3.$

7

NORMAL DISTRIBUTIONS

In the probability distributions of the last chapter, the random variable X could take on only a finite number of values. Now you will get better acquainted with continuous probability distributions, in which an infinite number of different scores are possible. The probability histogram used to illustrate a distribution with discrete values of X will become a smooth curve.

By far the most important of the continuous probability distributions are the normal distributions. There are many of these; you will first become familiar with one of them, called the standard normal distribution, whose mean is 0 and whose standard deviation is 1. Then you are reminded again of standard scores in which the unit of measurement is one standard deviation. Standard scores make familiarity with other normal distributions easy once you know the standard normal distribution.

In the remainder of the book you will face a great many formulas, but most of them will involve simply changing scores in a normal distribution into standard scores. If you thoroughly master the standard normal curve, use of the z table for probabilities in any normal distribution won't cause any trouble. Use of the normal distribution to approximate a binomial distribution will need some careful attention. Remember that the mean of a binomial distribution $= nP$ and its standard deviation is $\sqrt{nP(1 - P)}$.

7.1 HISTOGRAMS OF BINOMIAL DISTRIBUTIONS

Consider the binomial distributions for the number of heads which turn up when a fair coin is tossed 3, 9, or 81 times (remember that these are probability distributions).

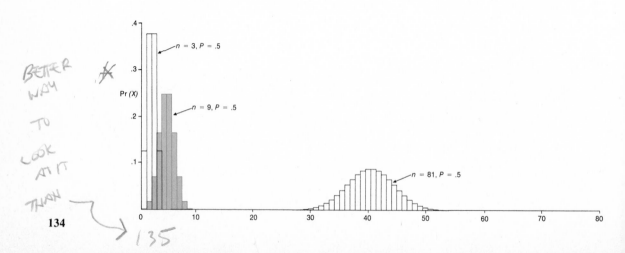

BETTER WAY TO LOOK AT IT THAN

134

135

As *n* increases, the histogram flattens out. So let's separate the 3 and change the scales of each so you can see better what's happening. Observe, in each case,

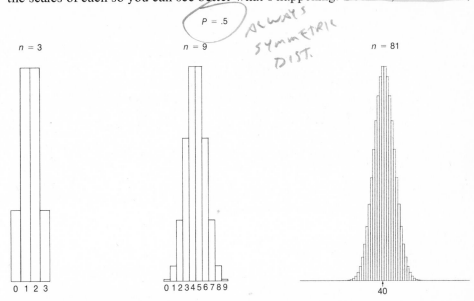

P = .5 *ALWAYS SYMMETRIC DIST.*

n = 3   n = 9   n = 81

0 1 2 3   0 1 2 3 4 5 6 7 8 9   40

that the histogram is symmetrical about its mode, that the area under each histogram is 1, and that the outline of the histogram gets closer to a bell-shaped curve as *n* gets larger.

Now look at the histograms for the binomial distribution with *P* = .3 and *n* = 3, 9, or 81. Again the scale has been changed so the histograms have the same

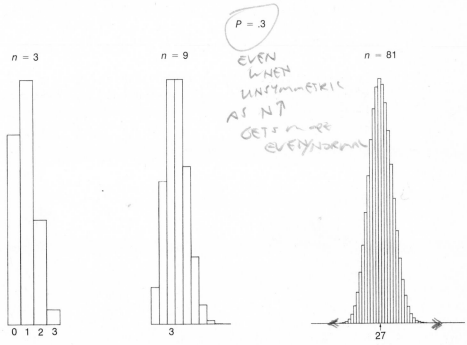

P = .3 *EVEN WHEN UNSYMMETRIC AS N↑ GETS more EVEN/NORMAL*

n = 3   n = 9   n = 81

0 1 2 3   3   27

height. Here the probability histograms are not symmetrical about the mode for small values of n. As n gets larger, however, the histogram becomes more bell-shaped (and more symmetrical about the mode).

7.2   NORMAL CURVES

One kind of continuous probability distribution, called the **normal distribution,** is fundamental in modern statistics; the curve representing it is called the **normal curve.** Historically, it was discovered in the eighteenth century in the study of errors: if a gun is shot at a target, and the horizontal distance from the point where each bullet hits the target to the center is measured, assume there is no bias; that is, the gun doesn't generally shoot a little to the left, but rather the shots average in the center of the target. But there are still random errors, and these cause the differences that form a normal curve. For this reason, a normal curve is often called an "error curve." Random fluctuations about a center occur frequently, and are often sufficiently well approximated by a normal curve. Another reason for using a normal curve is convenience. As we shall soon discover, it is

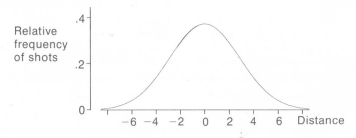

easy to find areas under the normal curve, and the area from a to b represents the probability that a score randomly chosen from the normal distribution is between a and b. The normal curve is easier to understand, however, as the symmetrical bell-shaped curve which a binomial distribution approaches as n gets very large. There is a whole family of normal curves, differing in mean or standard deviation. First, we shall look only at one member of the family—the **standard normal curve** with $\mu = 0$, $\sigma = 1$—and shall study the rest of the family in the following section.

7.3   THE STANDARD NORMAL CURVE

You certainly met simple equations such as $y = x + 1$ and $y = x^2$ in high school. The equation of the standard normal curve is rather more complicated; it is

$$y = \frac{1}{\sqrt{2\pi}} e^{-z^2/2}$$

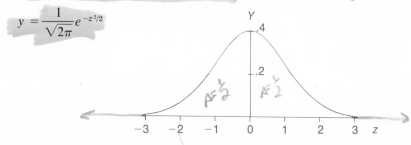

There is no need whatever to memorize this equation! It is included simply to reassure you that the standard normal curve is the graph of a particular equation. Here z is used for the independent variable instead of the x to which we are accustomed. This is done to distinguish the standard normal curve, which has μ = 0 and σ = 1 so that scores are in standard deviation units, from other normal curves. e is an irrational number which is usually first met in studying calculus; its approximate value is 2.7. Note the different scales on the z and y axes. This is a probability distribution and we shall be interested in areas rather than in particular y values, so the scale on the vertical axis will be omitted after this.

But even if you don't need to know the equation, you should become very familiar with certain characteristics of the standard normal curve:

1. It extends indefinitely in both directions. The curve gets very close to the horizontal axis without actually touching it as z gets either very large or very small. Fortunately, most of the area under the curve occurs between −3 and +3 and it won't be necessary to extend the tails very far.

2. The highest point of the curve is at 0; the mode of the standard normal distribution is 0.

3. The curve is symmetrical about 0: the height of the curve above the z axis is the same for 3 and −3, for 1.4 and −1.4, for z and −z. Therefore **the mean of the standard normal distribution is 0.**

4. A curve is "concave up" if it is part of a bowl that will hold water: $\cup$ or $\smallsmile$ or $\llcorner$; a curve is "concave down" if it is part of a bowl that will spill water: $\cap$ or $\smallfrown$ or $\ulcorner$. The standard normal curve changes from concave up to concave down at +1 and −1: This knowledge will help you in sketching the curve and in setting the scale on the horizontal axis.

POINT OF INFLECTION

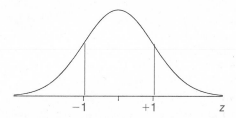

5. **The standard deviation of the standard normal distribution is 1.** A z score of 2.3 is, then, 2.3 standard deviations above the mean. You are already familiar with z scores as **scores in standard deviation units.**

6. The standard normal distribution is a probability distribution: the area under the whole curve is 1, and the probability that a score is between a and b equals the area under the normal curve between a and b.

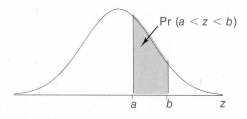

7.4 AREAS UNDER THE NORMAL CURVE

Areas under the normal curve are given in Table 4, Appendix C. The entries in this table give the area under the normal curve between the mean ($z = 0$) and the given positive value of z. The integral value and the first decimal place of z are given in the vertical column at the left; the second decimal place of z is given in the horizontal row at the top. **Always** sketch the area needed.

✓ **Example 1**   What is the probability that a z score lies between 0 and 1.32?

Look down the left-hand column of Table 4 to 1.3 and across the top row to .02. The intersection of the 1.3 row and the .02 column contains the number .4066. Rounding off to two decimal places, $\Pr(0 < z < 1.32) = .41$.

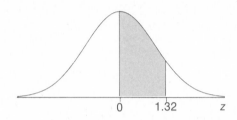

Notation: $A(z)$ will be used for the area under the standard normal curve between 0 and z when z is a positive number.

In Example 1, $A(z) = .41$.

✓ **Example 2** $\Pr(-2.03 < z < 0) = ?$

Negative values of z are not given in the table. However, the standard normal curve is symmetrical about $z = 0$, so $\Pr(-2.03 < z < 0) = \Pr(0 < z < 2.03) = .48$.

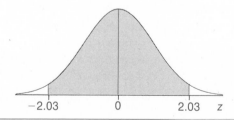

✓ **Example 3**   If a z score is chosen at random, what is the probability that it is at least 1.25 standard deviation units above the mean?

Since the mean is 0 and the standard deviation is 1, the question is a more interesting way of asking "$\Pr(z > 1.25) = ?$"

Table 4 gives areas starting at $z = 0$. Because of symmetry, and since the total area under the curve is 1, the area to the right of $z = 0$ is .50; the area between 0 and 1.25 can be found in the table. So $\Pr(z > 1.25) = .50 - \Pr(0 < z < 1.25) = .50 - .39 = .11$. See if you understand this graphical "arithmetic."

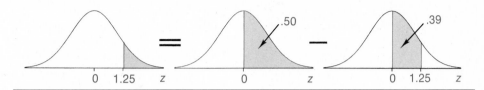

Example 4

$\Pr(z > -0.34) = ?$

By symmetry, the area between $z = -.34$ and $z = 0$ is the same as that between $z = 0$ and $z = +.34$. To this must be added the area of the right half of the curve.

Symbolically,

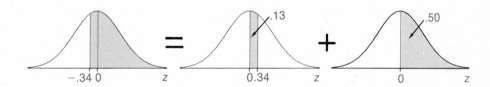

$$\Pr(z > -0.34) = \Pr(0 < z < .34) + \Pr(0 < z) = .13 + .50 = .63.$$

Example 5

If one score is chosen at random from a population which is normally distributed with mean 0 and standard deviation 1, what is the probability that the score is between 0.34 and 2.30 standard deviation units above the mean?

$\Pr(0.34 < z < 2.30) = ?$

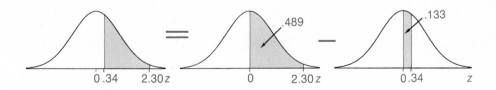

$$\Pr(0.34 < z < 2.30) = \Pr(0 < z < 2.30) - \Pr(0 < z < 0.34) = .489 - .133$$
$$= .36.$$

Example 6

What proportion of z scores is between -1.96 and 2.30?

This is equivalent to the question "$\Pr(-1.96 < z < 2.30) = ?$" $\Pr(-1.96 < z < 2.30) = \Pr(0 < z < 1.96) + \Pr(0 < z < 2.30) = .475 + .489 = .96.$

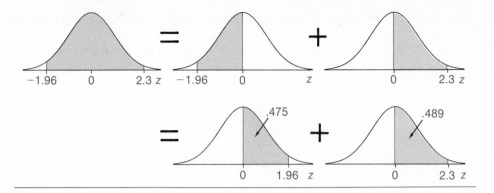

Compare Examples 5 and 6 carefully; note that probabilities are subtracted when boundary values of z have the same sign, but are added when the boundary values have opposite signs.

Always rely on drawing quick sketches of the normal curve and the area involved until you have done enough problems yourself so that you can see the sketch in your mind.

Sometimes you will be faced with the reverse problem: an area under the normal curve is given, and a value of z is to be determined. This still involves use of Table 4, but here a number in the body of the table determines a value of z on the edges.

Example 7 Find z if the area under the normal curve between 0 and z is .36 (if $A(z) = .36$).

In Table 4, the number closest to .36 in the body of the table is located to the right of the z value 1.0 and below the value .08; $z = 1.08$. If $z = -1.08$, the area between z and 0 is also .36, by symmetry.

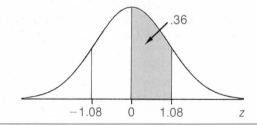

Example 8 Find the 98th percentile of z scores.

This is equivalent to "find z if the area to the left of z is .98." The area below $z = 0$ equals .50, so the area bounded by 0 and z is $.98 - .50 = .48$; using Table 4, $z = 2.05$. (Use the number closest to .48 in the table.)

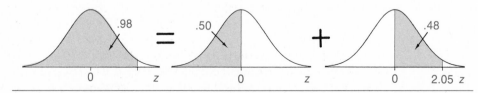

Example 9   What is z_0 if the probability is .70 that a randomly chosen z score is greater than z_0?

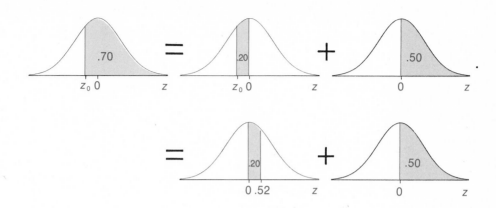

z_0 must be negative, since the area sought is greater than .50. An area of .20 corresponds to a z value of 0.52; the value sought, then, must be $z_0 = -0.52$.

Always remember that a normal distribution is a probability distribution, and therefore there is a direct correspondence between probabilities and areas under the curve.

7.5 PERCENTAGES OF SCORES WITHIN 1, 2, AND 3 STANDARD DEVIATIONS OF THE MEAN

$\Pr(0 < z < 1) = .341$, so $2(.341) = .682$; approximately 68% of scores **in a normal distribution** fall within 1 standard deviation of the mean.

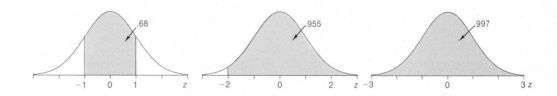

In like fashion it is determined that, **in a normal distribution,** approximately 95.5% of scores fall within 2 standard deviation units of the mean, and approximately 99.7% of scores fall within 3 standard deviations of the mean.

The words "in a normal distribution" are in bold type above to emphasize that the statements made are true only for this case. Many novices have the odd belief that, for any distribution, roughly 2/3, 95% and 99.7% of the scores fall within 1, 2, and 3 standard deviations, respectively, of the mean. This is errant nonsense, of course, as the following example shows.

Example 1 X: 1, 1, 1, 1, 1, 1, 1, 1, 1, 9; mean = 1.8, standard deviation = 2.4.* What percentage of scores falls within (a) 1, (b) 2, and (c) 3 standard deviations of the mean?

The answer to all these questions is 90%!

7.6 EXERCISES

In each exercise, make a quick sketch of the normal curve and the area involved.

1. Find the area of a standard normal distribution between
   (a) $z = 0$ and $z = 1.41$.
   (b) $z = -.6$ and $z = 0$.
   (c) $z = -1.23$ and $z = 0.53$.
   (d) $z = -1.23$ and $z = -0.53$.
   (e) $z = .46$ and $z = 2.31$.
   — (f) $z = -\infty$ and $z = 1.28$.

2. What is the probability that a z score chosen at random from a normal distribution is
   (a) greater than 1.65?
   (b) between 0.46 and 2.59?
   (c) between −0.46 and 2.59?
   (d) less than 0.46?
   (e) greater than −1.3?
   (f) less than −1.17?
   (g) within 1.96 standard deviation units of the mean?

3. Find z if $A(z) = .424$.

4. Find z if the probability that a score in a normal distribution falls between z and 1.42 is .20.

5. In a normal distribution, what percentage of the z scores falls between −.16 and +1.97?

6. If a set of measurements are normally distributed with $\mu = 0$ and $\sigma = 1$, what percentage of the measurements are
   (a) inside the range $\mu \pm 1.2\sigma$?
   (b) outside the range $\mu \pm 2\sigma$?

7. Find z if (a) 95% (b) 98% of normally distributed scores fall between $-z$ and $+z$.

8. Find (a) the 99th percentile, and (b) the 75th percentile of normally distributed z scores.

---

*Check these values of μ and σ.

ANSWERS

1. (a) .42
    (b) .23
    (c) .3907 + .2019 = .59

    (d) .3907 − .2019 = .19
    (e) .4986 − .1772 = .32
    (f) .5 + .3997 = .90.

2. (a) .5 − .451 = .05
    (b) .495 − .177 = .32
    (c) .495 + .177 = .67
    (d) .500 + .177 = .68

    (e) .500 + .403 = .90
    (f) .500 − .3790 = .12
    (g) 2(.475) = .95.

3. 1.43.

4. $\Pr(0 < z < 1.42) = .42$, and .42 + .20 = .62 which is greater than .5, so z must be less than 1.42. .42 − .20 = .22 = $A(z)$, so z = .59.

5. .0636 + .4756 = .5392; 53.9%.

6. (a) 2(.385) = .77; 77%; (b) 1 − 2(.477) = .05; 5%.

7. (a) $\dfrac{.95}{2}$ = .4750, z = 1.96; (b) $\dfrac{.98}{2}$ = .49, z = 2.33.

8. (a) $\Pr(z < z_1) = .99$, so $\Pr(0 < z < z_1) = .49$; $z_1 = 2.33$. (b) $\Pr(z < z_2) = .75$, so $\Pr(0 < z < z_2) = .25$; $z_2 = .67$.

7.7 NORMAL DISTRIBUTIONS IN GENERAL

There is a (double) family of normal distributions. Particular values of μ and σ give particular members of the family. Here are some examples:

The standard normal curve is one member of the family (with $\mu = 0$, $\sigma = 1$).

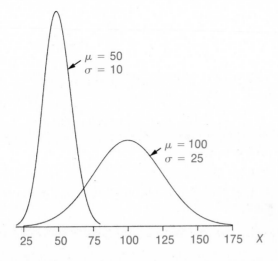

All members of the family have certain things in common; if you are friendly with the standard normal curve, none of these should surprise you:

1. They are all probability distributions—that is, the area between the curve and the horizontal axis is always equal to 1, and the area between the curve,

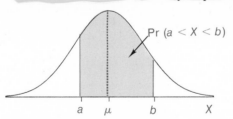

the X axis, and the lines $X = a$, $X = b$ represents the probability that an X score will fall between a and b.

2. All normal curves extend indefinitely, both to the right and to the left. Only a small portion about the mode is sketched, since the curve gets (and remains) very close to the horizontal axis on both sides of the mode.

3. Each normal curve is bell-shaped and is symmetrical about the mode, and the mode, median, and mean coincide.

4. Each normal curve changes from concave down ⌢ to concave up ⌣ at a point 1 standard deviation away from the mean. Thus, if $\mu = 50$, $\sigma = 20$, these "points of inflection" will be at 30 and 70. Also, check the figures at the beginning of this section.

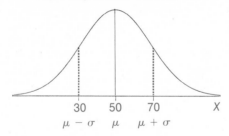

7.8 AREA UNDER ANY NORMAL CURVE

To find the area under a normal curve (with mean μ and standard deviation σ) between $X = a$ and $X = b$, find the z scores corresponding to a and b (call them z_1 and z_2), and then find the area under the standard normal curve between z_1 and z_2, using Table 4, Appendix C. Some examples will demonstrate how easily this is done.

Example 1 X scores are normally distributed with $\mu = 100$ g., $\sigma = 10$ g. What percentage of X scores will fall between (a) 100 and 110 g.? (b) between 80 and 105 g.?

(a) $z = \dfrac{X - 100}{10}$.

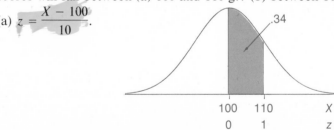

When $X = 100$, $z = \dfrac{100 - 100}{10} = 0$; when $X = 110$, $z = 1$. $\Pr(100 < X < 110) =$ $\Pr(0 < z < 1) = .341$; 34% of X scores will fall between 100 and 110 g.

   (b) When $X = 80$, $z = \dfrac{80 - 100}{10} = -2$; when $X = 105$, $z = \dfrac{105 - 100}{15} =$ $.5$. $\Pr(80 < X < 105) = \Pr(-2.0 < z < .5) = .477 + .192 = .669$; 66.9% of the scores are between 80 and 105 g.

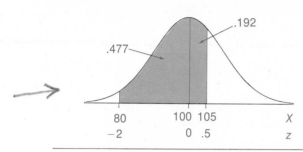

It will frequently be helpful to show on the horizontal axis both X scores and their corresponding z scores, as was done in the sketches above.

Example 2 X scores are normally distributed with $\mu = 50$ km., $\sigma = 5$ km. What is the probability that a score is (a) between 45 and 48 km.? (b) over 58 km.?

   Here $z = \dfrac{X - \mu}{\sigma}$ becomes $z = \dfrac{X - 50}{5}$. It is often convenient to arrange corresponding X and z scores in parallel columns.

X	z
45	-1
48	$-.4$
58	1.6

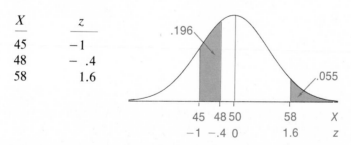

(a) $\Pr(45 < X < 48) = \Pr(-1 < z < -.4) = .341 - .155 = .19$.

(b) $\Pr(X > 58) = \Pr(z > 1.6) = .5 - .445 = .06$.

In Examples 1 and 2 it was assumed that X scores formed a continuous distribution. It is time to meet a complication which arises when they form a discrete distribution.

Example 3   The weights of 1,000 male students are found to be normally distributed with mean of 175 pounds and standard deviation of 12 pounds. How many of them weigh 186 pounds or less? (All measurements were made to the nearest pound.)

If the number of X scores is finite, then they must come from a discrete distribution. How can a discrete distribution be normally distributed, since a normal distribution is continuous? This "mistake" will be made frequently and casually in future pages. Of course an approximation is implied: It is assumed that if X scores are changed to standard scores, the distribution is sufficiently close to the standard normal distribution so that Table 4, Appendix C, may be used to determine probabilities. But whenever a discrete distribution is approximated by a continuous one, a **correction for continuity** must be used. Below is shown part of a histogram for discrete X scores with $\mu = 175$, $\sigma = 12$, and a normal approximation; the bar over 186 is magnified.

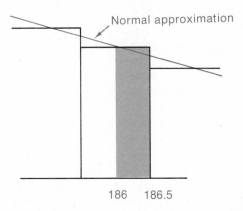

Since measurements are made to the nearest pound, a man who weighs in at 186 pounds weighs less than 186.5, and it is this latter figure which is used in computing the z score for weights of 186 pounds or less. If this is not done, the shaded area in the figure above is omitted.

$$z = \frac{X - 175}{12}; \Pr(X < 186.5) = \Pr(z < .96) = .500 + .332 = .832; .832(1,000)$$

$= 832$ of the students weigh 186 pounds or less.

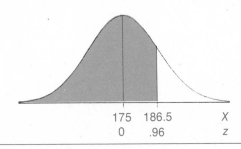

Example 4   The diameters of ball bearings are normally distributed with mean .5000 cm. and standard deviation .0010 cm. Of 10,000 ball bearings, measured to .0001 cm., how many will have diameters (a) between .5005 and .5020 cm.? (b) equal to .4992 cm.?

$$z = \frac{X - .5000}{.0010}$$

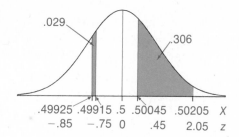

	X	z
(a)	.50045	.45
	.50205	2.05
(b)	.49915	−.85
	.49925	−.75

(a) Pr(.50045 < X < .50205) = Pr(.45 < z < 2.05) = .480 − .174 = .306; approximately 3,060 bearings will have diameters between .5005 and .5020 cm.

(b) Pr(X = .4992 cm.) = Pr(.49915 cm. < X < .49925 cm.) = Pr(−.85 < X < −.75) = .302 − .273 = .029; approximately 290 bearings will have diameters equal to .4992 cm.

Sometimes it is necessary to look up areas and find the corresponding X value as the following example demonstrates.

Example 5   On a statistics quiz the grades are normally distributed with a mean of 77 and a standard deviation of 7. If the top 10% of students get A's, what is the lowest grade to which an A will be assigned?

.50 − .10 = .40 = Pr(0 < z < 1.28)

$$z = \frac{X - 77}{7}$$

If z = 1.28, X = 7(1.28) + 77 = 85.96.

The lowest grade for an A is 86.

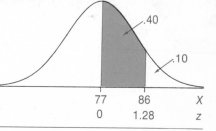

7.9   EXERCISES

1.   If one score is chosen at random from a normal distribution with μ = 60 inches, σ = 5 inches, what is the probability it is (a) between 60 and 63 inches?; (b) greater than 65 inches?; (c) between 62 and 65 inches?; (d) between 58 and 65 inches?; (e) less than 58 inches?; (f) within 2 standard deviation units of the mean?; (g) within 2 inches of the mean?

2.   In a normal distribution with mean of 16°C and standard deviation of 1.4°C, find X_0 if (a) 10% of the temperatures are larger than X_0; (b) 10% of the temperatures are smaller than X_0; (c) 90% of the temperatures are between $-X_0$ and $+X_0$; (d) 1% of the temperatures are larger than X_0.

3. 95% of the weights in a normal distribution with mean of 48.2 pounds and standard deviation of 4.1 pounds fall in an interval with μ as center. Find (a) the width of the interval in standard deviation units, (b) the width of the interval in pounds, (c) the largest weight which falls in the interval.

4. Assume that the average height of men drafted for the Army during the war in Vietnam was 69.8 inches, with a standard deviation of 3.0 inches; heights were normally distributed.

    (a) A draftee whose height was 70.3 inches was how many standard deviations above the mean in height?

    (b) 2% of draftees were over what height?

    (c) 90% of draftees were over what height?

    (d) What proportion of draftees were between 67 and 71.2 inches in height?

5. If SAT scores (which are measured to the nearest integer) are normally distributed with $\mu = 465$ and $\sigma = 100$, what is the probability that one randomly chosen SAT score is (a) over 700? (b) at least 500 but not more than 600? (c) 64% of SAT scores are over what number?

6. A manufacturer of plastic bottles believes the capacities of his pint bottle are normally distributed with $\mu = 1.04$ pints and $\sigma = .012$ pints. (a) What proportion of his bottles will not hold a full pint? (b) 90% of the pint bottles will hold more than what volume?

7. Three hundred women students have a mean height of 64.0 inches and standard deviation 2.0 inches. The 300 heights are normally distributed and are measured to the nearest inch.

    (a) How many of these students are 65 to 69 inches tall?

    (b) How many of them are 63 inches or less?

    (c) 30% of the students are below what height?

    (d) How many students have heights which differ from the mean by more than one standard deviation?

8. Five hundred students taking Psychology I have a mean grade for the course on all exams of 77, with a standard deviation of 10. Final grades are to be assigned "on the curve" so that 10% of the class gets A, 30% gets B, 50% gets C, and 10% gets no credit. Find the student averages for each of these grades.

9. A machine makes bolts whose mean diameter is .240 inches with standard deviation .005 inches. (a) If the diameters of 6,000 bolts produced by the machine are normally distributed, how many are more than .239 inches? (b) If the machine is adjusted so the mean diameter of the bolts is increased by .006 inches but the standard deviation is unchanged, what percent of the bolts it produces is over .239 inches in diameter?

10. (a) When is a "correction for continuity" needed when areas under a normal curve are being determined? (b) Why is it needed?

ANSWERS

1. $z = \dfrac{X - 60}{5}$. (a) $\Pr(60 < X < 63) = \Pr(0 < z < .6) = .23$; (b) $\Pr(z > 1.00) = .5 - .341 = .16$; (c) $\Pr(.4 < z < 1.00) = .341 - .155 = .19$; (d) $\Pr(-.4 < z < 1.00) = .155 + .341 = .50$; (e) $\Pr(z < -.40) = .500 - .155 = .35$; (f) $2(.4773) = .95$; (g) $\Pr(-.4 < z < .4) = 2(.155) = .31$.

2. $z = \dfrac{X - 16}{1.4}$. (a) 40% of z scores are between 0 and 1.28, so 10% of z scores are over 1.28; then 10% of X scores are over $1.28(1.4) + 16 = 17.8°$; (b) 10% of z scores are less than -1.28; the corresponding X_0 is $-1.28(1.4) + 16 = 14.2°$; (c) 90% of z scores are between -1.64 and $+1.64$, so 90% of X scores are between $16 - 1.64(1.4)$ and $16 + 1.64(1.4) = 13.7°$ and $18.3°$; (d) 49% of z scores are between -2.33 and 0, so 99% are above -2.33. $X_0 = -2.33(1.4) + 16 = 12.7°$.

3. (a) $2(1.96) = 3.92$; (b) $3.92(4.1) = 16.1$ pounds; (c) $48.2 + \dfrac{16.1}{2} = 56.3$ pounds.

4. $z = \dfrac{X - 69.8}{3.0}$. (a) $\dfrac{70.3 - 69.8}{3.0} = .17$; (b) $\Pr(0 < z < z_0) = .5 - .02 = .48$, $z_0 = 2.05$; the corresponding height is 76.0 inches; (c) If $\Pr(z > z_0) = .90$, then $\Pr(z_0 < z < 0) = .40$, so $z_0 = -1.28$, $X_0 = -1.28(3.0) + 69.8 = 66.0$ inches; (d) $\Pr(67 < X < 71.2) = \Pr(-.93 < z < .47) = .324 + .181 = .51$.

5. $z = \dfrac{X - 46.5}{100}$. (a) $\Pr(X > 700.5) = \Pr(z > 2.36) = .500 - .491 = .01$; (b) $\Pr(499.5 < X < 600.5) = \Pr(.35 < z < 1.35) = .412 - .137 = .28$; (c) $\Pr(z > -.36) = .64$; $X = -.36(100) + 465 = 429$.

6. $z = \dfrac{X - 1.04}{.012}$. (a) $\Pr(X < 1.00) = \Pr(z < -3.33) < .5 - .49952 = .05\%$; (b) $\Pr(z > -1.28) = .90$, $-1.28(.012) + 1.04 = 1.025$ pints.

7. $z = \dfrac{X - 64.0}{2.0}$. (a) $\Pr(64.5 < X < 69.5) = \Pr(.25 < z < 2.75) = .497 - .099 = .398$; $.398(300) = 119$ women; (b) $\Pr(X < 63.5) = \Pr(z < .25) = .5 - .099 = .599$; $.599(300) = 180$ women; (c) $A(z) = .50 - .30 = .20$, $z = -.52$, $X < 63$ inches. (d) $1 - 2(.341) = .318$; $.318(300) = 95$ women.

8. $A(z) = .40$, $z = 1.28$, so $X = 1.28(10) + 70 = 89.8$; students whose average is 89.8 or over get A. If $A(z) = .50 - (.10 + .30) = .10$, then $z = .25$ and $X = .25(10) + 77 = 79.5$; those with 79.5 to 89.7 get B. The lower boundary of the C grades is -1.28 z units ($A(z) = .40$ again); the comparable X score is 64.2, so a grade of C is given for an average of 64.2 to 79.5. Below 64.2 rates no credit.

9. (a) $z = \dfrac{X - .240}{.005}$. $\Pr(X > .2395) = \Pr(z > -.10) = .5 + .040 = .540$; approximately $.540(6,000) = 3,240$ bolts; (b) $z = \dfrac{X - .246}{.005}$. $\Pr(X > .2395) = \Pr(z > -1.30) = .5 + .403 = .903$; approximately 90%.

10. (a) When a discrete distribution is being approximated by a normal distribution; (b) because without it half of the bar in the histogram at an endpoint of the area being considered is omitted.

7.10 NORMAL APPROXIMATION TO THE BINOMIAL DISTRIBUTION

Good for Big n's

As the number of trials n gets larger and larger, a binomial distribution gets closer and closer to a normal distribution; this is how you were first introduced to normal distribution (see page 135). How large must n be if the approximation is to be useful? The answer is about 5 if $P = .5$, but n must be larger if P is not close to .5. So we shall adopt the following rule: a normal distribution with mean $\mu = nP$ and standard deviation $\sigma = \sqrt{nP(1 - P)}$ can be used to estimate binomial probabilities **if both nP and $n(1 - P)$ are 15 or more.** Remember to use the correction for continuity, since the binomial distribution is discrete and the normal distribution is continuous.

Example 1   Find the probability of getting at most 3 heads in 12 flips of a fair coin by using (a) the binomial distribution and Table 2, Appendix C; (b) the normal approximation to the binomial distribution.

(a) $n = 12$, $P = .5$

X	Pr(X)
0	.000+
1	.003
2	.016
3	.054
	.073

IF P = .5 CAN USE NORMAL APPROX FOR n ≥ 15

Pr(at most 3 heads) = .073.

(b) To apply the normal distribution to discrete data, it is necessary to treat the scores as though they were continuous. It follows that "at most 3 heads" should be treated as "less than 3.5 heads" for the normal distribution.

$\mu = nP = 12(1/2) = 6$. $Q = 1 - P = 1/2$.

$\sigma = \sqrt{nPQ} = \sqrt{12(1/2)(1/2)} = 1.73$

$z = \dfrac{X - 6}{1.73}$

If $X = 3.5$, $z = -2.50/1.73 = -1.45$

$Pr(X < 3.5) = Pr(z < -1.45) = .5 - .427 = .073$.

Example 2   10% of New Jersey residents had flu last year. In a random sample of 200 New Jersey residents, what is the probability that 19 or less had flu?

Since the population of New Jersey is over 7,000,000, the probability that the last person chosen had flu is approximately the same whether all in the sample had flu or all didn't have it; the binomial assumptions are approximately satisfied.

$\mu = nP = 200(.10) = 20$, $\sigma = \sqrt{nP(1 - P)} = \sqrt{200(.10)(.90)} = 4.24$

$$z = \frac{X - \mu}{\sigma} = \frac{19.5 - 20}{4.24} = -.12$$

$$\Pr(X < 19.5) = \Pr(z < -.12) = .5 - .048 = .45.$$

$$\mu = \frac{1}{6} \times 360 = 60.$$

$$\sigma^2 = 50. = (7.07)^2$$

7.11 EXERCISES

1. If a die is tossed 360 times, what is the probability that a 3 comes up (a) at least 60 but not more than 64 times? (b) less than 60 times?

2. 20% of the washing machines manufactured by the Wishwell Company need repair within 2 years. If 50 washing machines are randomly chosen from the production line, what is the probability that more than 12 need repair within 2 years?

3. 62% of freshmen entering Able State University apply for financial aid. In a random sample of 50 entering freshmen, what is the probability that exactly 33 will apply for financial aid?

4. If there is 7.1% unemployment, what is the probability that 11 or more in a random sample of 100 people in the labor force are unemployed?

5. If 50% of workers in manufacturing plants earn $6.04 or more per hour, what is the probability that 37 or less in a random sample of 80 workers in manufacturing plants make $6.04 or more per hour?

6. 10% of purchasers of soap powders for washing clothes favor Blue Suds. In a poll of 400 supermarket patrons, what is the probability that 46 or more prefer Blue Suds?

7. If 2% of long-term smokers get lung cancer, what is the probability that 18 to 26 (inclusive) in a random sample of 900 smokers get lung cancer?

8. 30% of the voters in a primary election in Tampa plan to vote for Wigsworth. In a random sample of 80 primary voters, what is the probability that at least 22 but not more than 26 will vote for Wigsworth?

9. 30% of the voters in a primary election in Tampa voted for Wigsworth. In a sample of 80 primary voters, what is the probability that at least 22 but not more than 26 voted for Wigsworth?

10. In the past, 10% of the students taking statistics have been absent from the first class in March because of illness. What is the probability that not more than 2 in a class of 30 will be absent from the first statistics class in March this year on account of illness?

11. After extensive tests, a seed company determines that 90% of its petunia seeds will germinate. It sells the seeds in packets of 100 and guarantees 85% germination. If the binomial assumptions are met, what is the probability that a particular packet will not meet the guarantee?

ANSWERS

1. $\mu = nP = 360\left(\frac{1}{6}\right) = 60$, $\sigma = \sqrt{360\left(\frac{1}{6}\right)\left(\frac{5}{6}\right)} = 7.07$, $z = \dfrac{X - 60}{7.07}$. (a) Pr$(59.5 < X < 64.5)$ = Pr$(-.07 < z < .64)$ = $.028 + .239 = .27$. (b) Pr$(X < 59.5)$ = Pr$(z < -.07)$ = $.47$.

2. $\mu = 50(.20) = 10$, $\sigma = \sqrt{50(.2)(.8)} = 2.83$, $z = \dfrac{X - 10}{2.83}$. Pr$(X > 12.5)$ = Pr$(z > .88)$ = $.5 - .311 = .19$. But nP is too small to trust the result.

3. $\mu = 50(.62) = 31$, $\sigma = \sqrt{50(.62)(.38)} = 3.43$, $z = \dfrac{X - 31}{3.43}$. Pr$(32.5 < X < 33.5)$ = Pr$(.44 < z < .73)$ = $.267 - .170 = .097$.

4. $\mu = 100(.071) = 7.1$, $\sigma = \sqrt{100(.071)(.929)} = 2.57$, $z = \dfrac{X - 7.1}{2.57}$. Pr$(X > 10.5)$ = Pr$(z > 1.32)$ = $.50 - .41 = .09$. Again, nP is too small to trust the result.

5. $\mu = 80(.50) = 40$, $\sigma = \sqrt{80(.50)(.50)} = 4.47$, $z = \dfrac{X - 40}{4.47}$. Pr$(X < 37.5)$ = Pr$(z < -.56)$ = $.50 - .21 = .29$.

6. $\mu = 400(.10) = 40$, $\sigma = \sqrt{400(.10)(.90)} = 6.0$; Pr$(X > 45.5)$ = Pr$(z > .92)$ = $.5 - .321 = .18$.

7. $\mu = 900(.02) = 18.0$; $\sigma = \sqrt{900(.02)(.98)} = 4.2$; Pr$(17.5 < X < 26.5)$ = Pr$(-.12 < z < 2.02)$ = $.048 + .478 = .53$.

8. $\mu = 80(.30) = 24.0$, $\sigma = \sqrt{80(.30)(.70)} = 4.1$; Pr$(21.5 < X < 26.5)$ = Pr$(-.61 < z < .61)$ = $2(.229) = .46$.

9. Either 0 or 1. Ask them; the number who voted for Wigsworth either is not or is between 22 and 26 (inclusive).

10. The binomial assumptions are not satisfied. One student who has flu or a cold, for example, may affect others. Outcomes are not independent.

11. $\mu = 100(.90) = 90$, $\sigma = \sqrt{100(.90)(.10)} = 3.0$; Pr$(X < 84.5)$ = $1 - $ Pr$(X > 84.5)$ = $1 - $ Pr$(z > -1.83)$ = $1 - .966 = .03$.

7.12 VOCABULARY AND SYMBOLS

normal distribution   concave up (down)
standard normal curve   A(z)

7.13 REVIEW EXERCISES

1. In a standard normal distribution,
   (a) What proportion of scores are (i) between $-.5$ and 1.3? (ii) more than 1.3? (iii) less than $-.5$?
   (b) Find z_0 if the probability is $.20$ that one score chosen at random is (i) between 0 and z_0, (ii) more than z_0, (iii) less than z_0.

2. 20,000 scores are normally distributed with $\mu = 0$, $\sigma = 1$. How many of them are (a) greater than 1? (b) between .50 and 1.53? What assumptions have you made?

3. X scores are normally distributed with $\mu = 30$ feet, $\sigma = 20$ feet. If one score is chosen at random, what is the probability that it is (a) between 60 and 80 feet? (b) less than 60 feet? (c) greater than 40 feet?

4. A department store uses fluorescent lights which have a mean life of 3,500 hours with a standard deviation of 600 hours; the lives are normally distributed. (a) If the lights are on 10 hours a day, 6 days a week, for 52 weeks, what proportion of the lights would need replacement? (b) After how many weeks would 10% of the lights need to be replaced?

5. Richard Smartt claims that he can distinguish wine A from three other brands of the same type. Over a period of a month he is given 30 trials, and is able to distinguish wine A 14 times. What is the probability that he would identify wine A 14 or more times if he were just guessing? Is his claim justified?

6. On the basis of tests and past experience with Model 101XE, a washing machine manufacturer decides its mean life in normal family use is 5.75 years with a standard deviation of 2 years. If the life of this model is normally distributed, (a) what guarantee should he offer if he is willing to repair only 1% of the machines sold during the guarantee period? (b) If he gives a 2-year guarantee, what percentage of machines will need repair before the guarantee period runs out?

7. Mendel discovered that when the seeds produced by tall F_1 peas were planted, 75% of the resulting plants were tall and 25% were dwarf. If 100 F_1 seeds are planted, what is the probability that at least 28 plants are dwarf?

8. On a 20-question true-false exam, what is the probability that a know-nothing student who answers each question by sheer guess will get 15 or more correct answers?

9. If a fair coin is tossed 100 times, what is the probability that the number of heads is

    (a) at least 40 and not more than 60?
    (b) at least 41 and not more than 59?
    (c) at least 43 and not more than 57?

10. A pair of fair dice is rolled 180 times. What is the probability that the same number comes up on both dice

    (a) at least 40 times?
    (b) between 30 and 40 times?
    (c) exactly 35 times?

11. A coffee machine is set to fill cups with 6 ounces of coffee, with a standard deviation of .40 ounces. If 7-ounce cups are used, what proportion of them will overflow?

12. The mean IQ of students admitted to St. James University is 118 with

a standard deviation of 5. If 400 students admitted have an IQ between 120 and 125, how many students were admitted? Assume their IQ's are normally distributed.

13. 5% of the washing machines manufactured by the Gilbert Washing Machine Company last year were defective. If 300 washing machines are chosen at random from the production line, what is the probability that 10 or less are defective?

ANSWERS

1. (a) (i) $.1915 + .4032 = .59$, (ii) $.5 - .403 = .10$, (iii) $.5 - .192 = .31$; (b) (i) .52, (ii) $A(z) = .30$, $z_0 = .84$, (iii) $-.84$.

3. $z = \dfrac{X - 30}{20}$.

X	z
60	1.5
80	2.5
40	.5

(a) $\Pr(60 < X < 80) = \Pr(1.5 < z < 2.5)$
$= .494 - .433 = .06$.

(b) $\Pr(X < 60) = \Pr(z < 1.5) = .5 + .433 = .93$.

(c) $\Pr(X > 40) = \Pr(z > .5) = .5 - \Pr(0 < z < .5)$
$= .5 - .192 = .31$.

5. $n = 30$, $P = \dfrac{1}{4}$, $\mu = nP = 7.5$, $\sigma = \sqrt{nPQ} = 2.37$, $z = \dfrac{13.5 - 7.5}{2.37} = 2.53$. Pr(identify 14 or more) $= .5 - .494 = .006$. The chance he is guessing is approximately 6 out of 1,000; his claim is probably justified.

7. $\Pr(\text{dwarf}) = P = \dfrac{1}{4}$, $n = 100$, $\mu = 25.0$, $\sigma = \sqrt{100\left(\dfrac{1}{4}\right)\left(\dfrac{3}{4}\right)} = 4.33$. If $X = 27.5$, $z = \dfrac{27.5 - 25.0}{4.33} = .58$. $\Pr(X > 27.5) = \Pr(z > .58) = .5 - .219 = .28$.

9. $n = 100$, $P = .5$, $\mu = nP = 50$, $\sigma = \sqrt{100(.5)(.5)} = 5$; $z = \dfrac{X - 50}{5}$.

X	z
39.5	-2.10
60.5	2.10
40.5	-1.90
59.5	1.90
42.5	-1.50
47.5	-1.50

(a) $\Pr(39.5 < X < 60.5) = \Pr(-2.10 < z < 2.10)$
$= 2(.482) = .96$.

(b) $\Pr(40.5 < X < 59.5) = \Pr(-1.90 < z < 1.90)$
$= 2(.471) = .94$.

(c) $\Pr(42.5 < X < 47.5) = \Pr(-1.50 < z < 1.50)$
$= 2(.433) = .87$.

11. $z = \dfrac{X - 6.0}{.40}$; $\Pr(X > 7.0) = \Pr(z > 2.5) = .5 - .4938 = .006$; .6% will overflow.

13. $\mu = 300(.05) = 15$, $\sigma = \sqrt{300(.05)(.95)} = 3.77$, $\Pr(X < 10.5) = \Pr(z < -1.19) = .5 - .383 = .12$.

8
CONFIDENCE INTERVALS FOR MEANS

Here at last you begin to do one of the most important things a statistician needs to do: to take a sample from the population, study the sample, and then make inferences (draw conclusions) about the population from which the sample was taken.

Part of this chapter is quite theoretical, but the sampling distribution of means must be thoroughly understood if you are to understand how inferences about the population mean are made. Then you finally get down to the nitty-gritty of estimating the population mean, knowing the mean of one random sample.

Two cases are discussed: (1) if σ is known, and (2) if σ is unknown but n is large (at least 30). If σ is unknown and n is smaller than 30, you will have to wait until Chapter 11 before estimating the population mean.

8.1 THE SAMPLING DISTRIBUTION OF MEANS

From a population of X scores with mean μ and standard deviation σ, take a sample of size n with replacement. Find the mean of this sample, $\overline{X}_1$ (the subscript 1 indicates that this is the mean of the first sample chosen). This number $\overline{X}_1$ is the first score in the sampling distribution of means for samples of size n. Now choose a different sample of size n, find its mean $\overline{X}_2$, and this mean is the second score in the new distribution. The collection of all the sample means for all samples forms a new distribution, called the **sampling distribution of means.**

Example 1

For a population of X scores: 0, 4, 8, find the sampling distribution of means for samples of size 2 taken with replacement.

Scores in Sample	$\overline{X}$
0, 0	0
0, 4	2
0, 8	4
4, 0	2
4, 4	4
4, 8	6
8, 0	4
8, 4	6
8, 8	8

The scores in the sampling distribution of means for samples of size 2 taken with replacement are shown in the box above. Note that the samples may contain the same numbers: first 0 and then 4 is a different sample from first 4 and then 0.

The sampling distribution of means is one of the most fundamental concepts of statistical inference, and it has remarkable properties. Since it is a frequency distribution, it has its own mean and standard deviation. Since the scores $(\overline{X})$ in the sampling distribution of means are themselves means (of individual samples), we shall use the notation $\sigma_{\overline{X}}$ (read "sigma sub X bar") for the standard deviation of the distribution: σ because it is a standard deviation, and the subscript $\overline{X}$ to distinguish this standard deviation from that of the population. One further complication: the standard deviation of the sampling distribution of means is called the **standard error of the mean.**

> **Notation:** $\sigma_{\overline{X}}$ = the standard error of the mean (= the standard deviation of the sampling distribution of means).

There are astonishing relations between the mean of the sampling distribution and μ and between $\sigma_{\overline{X}}$ and σ:

> 1. The mean of the sampling distribution = μ, the population mean.
>
> 2. $\sigma_{\overline{X}} = \dfrac{\sigma}{\sqrt{n}}$ *

 **Example 2**

(a) For the population of Example 1 (0, 4, 8) find the mean μ and the standard deviation σ.

(b) The scores in the sampling distribution of means for samples of size 2 were found to be 0, 2, 4, 2, 4, 6, 4, 6, 8. Find the mean of those scores, and find the standard error $\sigma_{\overline{X}}$.

(c) Does the mean of the sampling distribution = μ?

(d) Does $\sigma_{\overline{X}} = \dfrac{\sigma}{\sqrt{n}}$?

(a)

X	X^2
0	0
4	16
8	64
12	80

$$\mu = \frac{12}{3} = 4.0$$

$$\sigma = \sqrt{\frac{80 - 12^2/3}{3}} = \sqrt{\frac{32}{3}}$$

---

*If sampling is without replacement from a finite population of size N, then the mean of the sampling distribution is still μ, but it is not strictly true that $\sigma_{\overline{X}} = \dfrac{\sigma}{\sqrt{n}}$; rather, $\sigma_{\overline{X}} = \dfrac{\sigma}{\sqrt{n}}\sqrt{\dfrac{N-n}{N-1}}$. But this rather messy correction factor $\sqrt{\dfrac{N-n}{N-1}}$ is necessary in theory but seldom necessary in practice, since it is close to 1 unless the sample is an unusually large proportion of the population (over 5%, say).

$\sqrt{\dfrac{32}{3}}$ is left in this form instead of 3.3 so that its relation to the standard error of the mean will be easier to see.

      (b) $\bar{X}$: 0, 2, 4, 2, 4, 6, 4, 6, 8; $\Sigma\bar{X} = 36$
          $\bar{X}^2$: 0, 4, 16, 4, 16, 36, 16, 36, 64; $\Sigma\bar{X}^2 = 192$

The mean of the sampling distribution $= \dfrac{36}{9} = 4.0$; $\sigma_{\bar{X}} = \sqrt{\dfrac{192 - 36^2/9}{9}} =$

$\sqrt{\dfrac{192 - 144}{9}} = \sqrt{\dfrac{16}{3}}$. Notice that the denominator in the standard error is the number of scores in the sampling distribution (9), as it would be for finding σ, not one less than that number as in finding s.

      (c) Yes; $\mu = 4.0 =$ the mean of the sampling distribution of means.

      (d) Yes; $\sigma_{\bar{X}} = \sqrt{\dfrac{16}{3}}$ and $\dfrac{\sigma}{\sqrt{n}} = \sqrt{\dfrac{32/3}{2}} = \sqrt{\dfrac{16}{3}}$.

So far we have discussed the mean and standard error of the sampling distribution of means, but not the shape of the distribution. The incredible power of the sampling distribution of means depends upon the following theorem:

> **Central Limit Theorem:**
> 1. The sampling distribution of means is a normal distribution if the population is normally distributed.
> 2. Even if the population is not normally distributed, the sampling distribution of means is approximated by a normal distribution for large n. The more the population distribution differs from a normal distribution, the larger n must be for the sampling distribution to approximate a normal one. In practice, we shall assume the approximation is a workable one if n is 30 or more.

The proof will not be given, since it requires advanced mathematical techniques quite a bit beyond the prerequisites for this course. It is difficult to give examples that are persuasive. Would you happily find the means of all samples of size 30 from a population of 1,000, say? There would be 30^{1000}, a number bigger than 1 followed by 1400 zeroes, even for this small population. The only simple example is the trivial case $n = 1$ when the population is normally distributed; clearly the sampling distribution of means is normal too, since it is identical with the population. The example we took up earlier is hardly persuasive; the sampling distribution is not normal, since the population is not normal and n (=2) is too small. You will, alas, have to accept the theorem on faith.

The sampling distribution of means is of such importance that it is worth a summary.

1.   How is it formed? From the population, choose a sample of size n and find its mean $\bar{X}$. Do this for each different sample of size n; the set of all the different means is the sampling distribution of means. Note that each different value of n gives a different sampling distribution of means (even from the same population).

2. What about its mean and standard error? Its mean $= \mu$ and $\sigma_X = \dfrac{\sigma}{\sqrt{n}}$.

3. What about its shape? It is a normal distribution either if the population is normal or if the sample size n is at least 30.

APPROX. NORMAL

Example 3

A population has $\mu = 40$, $\sigma = 6$. What can you say about the sampling distribution of means for samples of size (a) 100, (b) 25?

(a) It is formed by finding the mean of every different sample of size 100. Its mean $= 40$, $\sigma_{\bar{X}} = \dfrac{6}{\sqrt{100}} = .60$. It is normal (even if the population is not) since *APPROX* $n = 100$, > 30.

(b) It is formed by taking the mean of every different sample of size 25. Its mean $= 40$ [as in (a)]; $\sigma_{\bar{X}} = \dfrac{6}{\sqrt{25}} = 1.2$. It is normal only if the population is normal. Nothing can be said about its shape if the population is not normal.

Example 4

A sampling distribution of means for samples of size 100 has mean equal to 300, standard error equal to 5, and is normally distributed. What can be said about the population from which it was taken?

$$\mu = 300; \; \sigma_{\bar{X}} = \dfrac{\sigma}{\sqrt{n}} \text{ so } 5 = \dfrac{\sigma}{\sqrt{100}}, \text{ and } \sigma = 50.$$

Nothing can be said about the shape of the population distribution. Since n is sufficiently large, the sampling distribution of means is normally distributed whether or not the population is normally distributed.

8.2 USES OF THE SAMPLING DISTRIBUTION OF MEANS

A psychologist may wish to measure the reaction time of 30 people to an electric shock and estimate the reaction time of all individuals to this shock. A manufacturer may find the mean life of 100 light bulbs and estimate the mean life of all light bulbs manufactured by a certain process. These are two examples of statistical inference which can be answered with the help of the sampling distribution of means and the Central Limit Theorem. For the moment, however, let us look at simpler problems about random samples when information about the population is given.

Example 1

A random sample of size 64 is chosen from a population of size 100,000 whose mean is 10 and whose standard deviation is 3. What is the probability that the mean of the sample is (a) between 9 and 11? (b) greater than 11?

The sampling distribution is approximately normal since $n > 30$. Its mean $= \mu = 10$.

$$\sigma_{\bar{X}} = \dfrac{\sigma}{\sqrt{n}} = \dfrac{3}{\sqrt{64}} = \dfrac{3}{8} = .375$$

Sketch the sampling distribution of means.

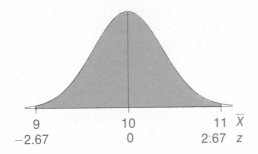

Note that the horizontal axis is labeled $\overline{X}$, since the distribution is made up of sample means.

(a) $\mathrm{Pr}(9 < \overline{X} < 11) = ?$ This is a familiar problem of finding areas under a normal curve. To transform to z scores, we always subtract the mean of the distribution and then divide by its standard deviation. In this case, the appropriate equation will be

$$z = \frac{\overline{X} - \mu}{\sigma_{\overline{X}}} = \frac{\overline{X} - \mu}{\dfrac{\sigma}{\sqrt{n}}}$$

$\sigma_{\overline{X}} = \dfrac{3}{\sqrt{64}} = .375.$ When $\overline{X} = 9$, $z = \dfrac{9 - 10}{.375} = -2.67$; when $\overline{X} = 11$, $z = \dfrac{11 - 10}{.375} = 2.67.$

$$\mathrm{Pr}(9 < \overline{X} < 11) = \mathrm{Pr}(-2.67 < z < 2.67) = 2(.496) = .99.$$

(b) $\mathrm{Pr}(\overline{X} > 11) = \mathrm{Pr}(z > 2.67) = .50 - .496 = .004$, rounding to three decimal places.

Example 2   All third-graders in New York City take the Peterson Interval Test; the mean score is 200 with a standard deviation of 20. What is the probability that a sample of 16 children chosen at random from the third grade in Browning School will have a mean score greater than 205?

There are two traps here; see if you can find both of them before you read on.

First, the sample is not a random sample from the specified population (all third-graders in New York City). Perhaps Browning School is a school for retarded children, and its third-graders do badly on all tests. All deductions made in this section about the probability of sample means, and all inferences that we shall make about the population mean after studying a sample in the next sections, are based on the premise that the sample is a random one. If this requirement is not satisfied, then the theory we are building up is not applicable.

The second trap is in the size of the sample: $n = 16$, < 30, so the sampling distribution is normal only if the population is normal. This fact is not explicitly stated; if only approximately true, then the sampling distribution of means is only approximately normal. Unfortunately, you have no way of judging the size of the error caused by assuming the population is normally distributed when it is not.

Example 3

A normal population has a mean of 50 and a standard deviation of 10. Ninety-five percent of random samples of size 25 have means in what interval? (Assume the interval has 50 as its midpoint.)

The area under the curve in the desired interval is .95, so the area to the right of $z = 0$ is .475. If $A(z) = .475$, then $z = 1.96$.

$$\mu = 50, \sigma = 10, n = 25, \sigma_{\overline{X}} = \frac{10}{\sqrt{25}} = 2$$

$$z = \frac{\overline{X} - \mu}{\sigma_{\overline{X}}} = \frac{\overline{X} - 50}{2}$$

$$\pm 1.96 = \frac{\overline{X} - 50}{2}.$$

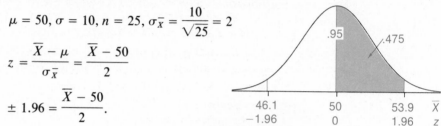

$\overline{X} = 50 \pm 3.92 = 53.92$ or 46.08; 95% of the sample means will be between 46.1 and 53.9.

8.3 EXERCISES

1. (a) Find the scores in the sampling distribution of means for samples of size 2 taken with replacement if the population consists of three scores: 0, 2, 4.
   (b) Find the mean and standard deviation of the population.
   (c) Find the mean of this sampling distribution of means and the standard error of the mean.
   (d) Compare the results of (b) and (c). Does the mean of the sampling distribution $= \mu$? Does $\sigma_{\overline{X}} = \frac{\sigma}{\sqrt{n}}$?

2. A population has a mean of 60 and a standard deviation of 4. Describe the sampling distribution of means for samples of size 64 taken with replacement (How is it formed? What is its mean? its standard error? Is it normally distributed?)

3. A population has a mean of 8.6 and a standard deviation of .20. Describe the sampling distribution of means for samples of size 25 taken with replacement (How is it formed? What is its mean? its standard error? Is it normally distributed?)

4. A sampling distribution of means for samples of size 16 has a mean of 27.0, a standard error of 4.3, and is normally distributed. What can you say about

the population from which it has been taken? (What is its mean? its standard deviation? Is it normally distributed?)

5. A sampling distribution of means for samples of size 100 taken with replacement has a mean of 364.1, a standard error of the mean equal to 18.2, and is normally distributed. What can you say about the population from which it has been taken? (What is its mean? its standard deviation? Is it normally distributed?)

6. (a) In a very large, normally distributed population with $\mu = 75$ hours and $\sigma = 10$ hours, 90% of scores fall between $75 - A$ and $75 + A$ hours. Find A.

   (b) In the sampling distribution of means for samples of size 25 chosen from this population, 90% of sample means fall between $75 - B$ and $75 + B$ hours. Find B.

   (c) Compare A and B. Comment.

7. (a) Assume that the heights of men aged 20 are normally distributed with $\mu = 69$ inches and $\sigma = 3$ inches. What proportion of all men aged 20 are between 68 and 70 inches tall?

   (b) What proportion of all samples of 25 men aged 20 have a mean height between 68 and 70 inches tall? [Consider the sampling distribution of means for samples of size 25 taken from the population described in (a).]

8. The mean weight of all women at Chanson University is 122 pounds, with a standard deviation of 9 pounds. If a random sample of 64 of these women is chosen, what is the probability that the mean weight of the women in the sample is (a) between 122 and 124 pounds? (b) over 124 pounds?

9. What happens to the standard error of the mean if the sample size is changed from 4 to (a) 16, (b) 64?

ANSWERS

1. (a) Samples: 0, 0; 0, 2; 0, 4; 2, 0; 2, 2; 2, 4; 4, 0; 4, 2; 4, 4.
   Sample means: 0, 1, 2, 1, 2, 3, 2, 3, 4.
   The sample means form the sampling distribution of means.

   (b) $\mu = \dfrac{6}{3} = 2.0, \sigma = \sqrt{\dfrac{20 - 6^2/3}{3}} = \sqrt{\dfrac{8}{3}}$

   (c) The mean of the sampling distribution $= \dfrac{18}{9} = 2.0, \sigma_{\bar{x}} = \sqrt{\dfrac{48 - 18^2/9}{9}} = \sqrt{\dfrac{4}{3}}.$

   (d) The mean of the sampling distribution $= \mu = 2.0, \sigma_{\bar{x}} = \dfrac{\sigma}{\sqrt{n}} = \dfrac{\sqrt{8/3}}{\sqrt{2}} = \sqrt{\dfrac{4}{3}}.$

2. It is formed by taking different samples of size 64 with replacement, and finding the mean of each sample. The means are the scores in the sampling distribution. Its mean $= 60. \, \sigma_{\bar{x}} = \dfrac{4}{\sqrt{64}} = .5.$ It is normally distributed (even if the population is not) since n is sufficiently large (> 30).

3. Find the mean of each different sample of size 25. The set of all these means forms the sampling distribution. Its mean $= 8.6, \sigma_{\bar{X}} = \dfrac{.20}{\sqrt{25}} = .04$. It is normally distributed only if the population is normally distributed, since n is less than 30.

4. $\mu = 27.0$. $\sigma_{\bar{X}} = \dfrac{\sigma}{\sqrt{n}}$ so $4.3 = \dfrac{\sigma}{\sqrt{16}}$ and therefore $\sigma = 4.3(4) = 17.2$. The population must be normally distributed since the sampling distribution is normal and n is small.

5. $\mu = 364.1$; $18.2 = \dfrac{\sigma}{\sqrt{100}}$ so $\sigma = 182$. There is no way of telling whether the population is normally distributed: the sampling distribution is normal since $n > 30$, whether or not the population is normal.

6. (a) If $\Pr(z_0 < z < z_0) = .90$, then $\Pr(0 < z < z_0) = .45$, so $z_0 = 1.64$. 1.64 standard deviation units $= 1.64(10) = 16.4$ hours $= A$.

   (b) $\sigma_{\bar{X}} = \dfrac{10}{\sqrt{25}} = 2.0$; 1.64 standard deviation units $= 1.64(2) = 3.3$ hours $= B$.

   (c) $B < A$. The means of all samples cluster about their mean more closely than do the individual scores in the population.

7. (a) $z = \dfrac{X - \mu}{\sigma} = \dfrac{X - 69}{3}$; $\Pr(68 < X < 70) = \Pr(-.33 < z < .33) = 2(.129) = .26$.

   (b) $z = \dfrac{\bar{X} - \mu}{\sigma_{\bar{X}}} = \dfrac{\bar{X} - 69}{3/\sqrt{25}}$; $\Pr(68 < \bar{X} < 70) = \Pr(-1.67 < z < 1.67) = 2(.453) = .91$.

   The proportion of sample means within 1 inch of their mean is more than 3 times as great as this proportion for individuals in the population.

8. (a) $\sigma_{\bar{X}} = \dfrac{9}{\sqrt{64}} = 1.13$; $z = \dfrac{\bar{X} - \mu}{\sigma_{\bar{X}}} = \dfrac{\bar{X} - 122}{1.13}$. $\Pr(122 < \bar{X} < 124) = \Pr(0 < z < 1.77) = .46$.

   (b) $.50 - .46 = .04$.

9. It becomes (a) $\dfrac{1}{2}$ or (b) $\dfrac{1}{4}$ as large as when $n = 4$.

8.4 STATISTICAL INFERENCE

You should now be quite comfortable with the statement, "the probability is .95 that the mean of a sample of 100 taken at random from a population with $\mu = 400$, $\sigma = 50$ will be between 391.2 and 409.8."

The main interest of the statistician, however, is in studying a sample and making inferences about the population from which the sample is taken. For

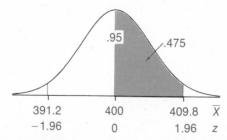

example, an economist wants to estimate the mean number of miles trucks are driven in a week, the proportion of carpenters with incomes over $10,000 per year, or the differences of mean wages of steelworkers and of men in automobile plants. A psychologist wants to know the mean reaction times of men to a certain stimulus, or the proportion of college students whose parents are not college-trained, or the differences of mean time to solve a maze for rats with and without vitamin A in their diets. A doctor studies the mean weight of children at birth, or the proportion of leukemia patients who live 5 years after treatment with a certain drug, or the difference in life expectancy of smokers and nonsmokers. A political scientist is interested in the mean waiting time for felony trials in New York City, or the proportion of voters who back Candidate A for senator, or the differences in mean age of those favoring a liberal and a conservative opposing each other for Congress.

Three problems are suggested in each case above; one involves a mean, one a proportion, and the third the differences of means. But in each case, full information about the population is not likely to be attainable, and the researcher must generalize after examining a sample taken from the population; he must make a statistical inference.

8.5   ESTIMATES OF THE POPULATION MEAN IF σ IS KNOWN   (read again)

Suppose that the mean of a random sample of size 36 is 100 pounds, and the standard deviation of the (very large) population is 24 pounds. An estimate of the population mean can be given in two ways. If a single number is given ("I estimate that μ = _____ pounds"), this is a **point estimate.** If a range of values is given ("I estimate that μ is between _____ and _____ pounds"), this is an **internal estimate.** *INTERVAL*

In the example at the beginning of this section, the best point estimate that can be made is that μ = 100 pounds. In general, the point estimate of μ is that it equals $\overline{X}$. We have as yet no idea of the accuracy of this estimate, and it is therefore useless to us. For example, two psychologists with two different samples may make different point estimates of a population mean, one of which may support a certain theory and the other deny it. We shall find that, as soon as we estimate the accuracy of a point estimate, we are really giving an interval estimate.

Since the sampling distribution of means is normal, we know that 95% of samples will have a mean whose z score is between -1.96 and $+1.96$, or, stated another way, 95% of sample means will be within 1.96 standard deviation units of the population mean.

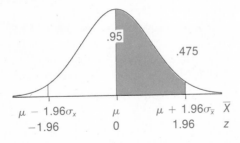

Assume that μ is unknown. It is known, however, that 95% of sample means will be within 1.96 $\sigma_{\bar{X}}$ score points of μ, since 1 standard deviation unit equals $\sigma_{\bar{X}}$ score points. In the example stated above, $\sigma = 24$ pounds, $n = 36$ pounds, and $\sigma_{\bar{X}} = \dfrac{24}{\sqrt{36}} = 4$ pounds; so 1.96 $\sigma_{\bar{X}} = 7.84$ pounds.

Now the researcher who chose the sample reasons this way: "Out of every 10,000 samples chosen from this population, about 9,500 of them will have a mean which is within 7.8 pounds of the population mean, and about 500 will have a mean which is more than 7.8 pounds away from the population mean. I don't know for certain which group my particular sample with a mean of 100 pounds is in, but I do know that if my sample is in the first group, then the population mean is within 7.8 pounds of 100—that is, between 92.2 and 107.8 pounds. I also know that if one sample is chosen at random, its probability of being in the first group is .95, and that the sample I am working with was chosen at random. **I am 95% confident that the population mean is between 92.2 and 107.8 pounds.**" (Note: if you understand every word of this paragraph, the rest is easy; the real kernel of this chapter is here.)

The range 92.2 to 107.8 pounds is called the **confidence interval;** 107.8 pounds is called the **upper confidence limit** and 92.2 pounds is called the **lower confidence limit** for μ. Here are three equivalent statements:

(1) I am 95% confident that the interval 92.2 to 107.8 pounds includes μ.

(2) A 95% confidence interval for μ is 92.2 to 107.8 pounds.

(3) An interval estimate for μ, at the 95% confidence level, is 92.2 to 107.8 pounds.

The confidence, expressed as a proportion, that the interval 92.2 to 107.8 pounds contains the unknown population mean is called the **confidence coefficient.** When this coefficient is .95, for example, if many different random samples are taken, and if the confidence interval for each is determined, then it is expected that 95% of these computed intervals will contain μ.

Example 1

The mean reading speed of a random sample of 81 adults is 325 words per minute. Find a 90% confidence interval for the mean reading speed of all adults if it is known that the standard deviation for all adults is 45 words per minute.

Cover the right-hand column below and see if you can answer the questions or fill in the blanks in the left-hand column.

1. If one z score is chosen at random from a standard normal distribution, the probability is .90 that it is between ___ and ___.

    1. -1.64 and $+1.64$.

2. Is the sampling distribution of means appropriate to this problem normal?

    2. Yes, because $n, = 81$, is sufficiently large.

3. What are the mean and standard error of the sampling distribution?

    3. Its mean is μ; its standard error is $\sigma_{\bar{X}} = \dfrac{\sigma}{n} = \dfrac{45}{\sqrt{81}} = 5$ words per minute.

4. 1 z or standard deviation unit is ___ words per minute, so 1.64 z units is ___ words per minute.

4. 5 words per minute; 1.64(5) = 8.2 words per minute.

5. The probability is .90 that one z score chosen at random is within 1.64 z units of 0. The probability is .90 that one $\overline{X}$ score chosen at random is within how many words per minute of μ?

5. 8.2 words per minute.

6. The probability is .90 that the interval $\overline{X}$ − ___ to $\overline{X}$ + ___ contains μ if a random sample with mean $\overline{X}$ is chosen.

6. $\overline{X}$ − 8.2 to $\overline{X}$ + 8.2.

7. The sample here was chosen at random with $\overline{X}$ = ___, so I have 90% confidence that the interval ___ to ___ words per minute includes μ.

7. $\overline{X}$ = 325; a 90% confidence interval for μ is 325 − 8.2 to 325 + 8.2 = 316.8 to 333.2 words per minute?

Note that, once a particular sample has been picked, its mean either is within 8.2 words per minute of μ or it is not, so the statement "the probability is .90 that 325 is within 8.2 words per minute of μ" doesn't make sense.

If you have thoroughly understood how the 90% confidence interval of 316.8 to 333.2 words per minute for μ was arrived at, then you should have no trouble understanding the following more formal solution of the same problem.

Example 2   Find a 90% confidence interval for n if σ = 45 and $\overline{X}$ = 325 words per minute in a random sample with n = 81.

90% confidence coefficient, so $z = \pm 1.64$; $\sigma = 45$, $n = 81$, $\sigma_{\overline{X}} = \dfrac{45}{\sqrt{81}} = 5.0$.

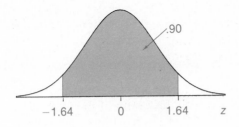

$$z = \frac{\overline{X} - \mu}{\sigma_{\overline{X}}}, \text{ so } \mu = \overline{X} \pm z\sigma_{\overline{X}} \qquad \text{(Note 1)}$$

$$= 325 \pm 1.64(5.0) = 325 \pm 8.2$$
$$= 316.8 \text{ or } 333.2 \text{ words per minute.} \qquad \text{(Note 2)}$$

A 90% confidence interval for μ is 316.8 to 333.2 words per minute.

Note 1. Strictly, $\mu = \overline{X} - z\sigma_{\overline{X}}$ and z can be either positive or negative; $\pm$ is used here to remind you to use both values of z.

Note 2. We have found the upper and lower confidence limits for μ.

> To find a confidence interval for μ, compute the confidence limits
> $$\overline{X} - z\sigma_{\overline{X}} \text{ and } \overline{X} + z\sigma_{\overline{X}}.$$

SAME   AS $\overline{X} \pm z \dfrac{\sigma}{\sqrt{n}}$

The most frequently used confidence levels are

Confidence level (%):	90	95	98	99
z:	± 1.64	± 1.96	± 2.33	± 2.58

Do not find a confidence interval for μ unless you are sure that the sampling distribution of means is (approximately) normal; this means that either the population must be normally distributed or the sample size n is sufficiently large (at least 30).

Now let's look at the relationship between the confidence coefficient and the width of the confidence interval.

Example 3   If $n = 81$, $\sigma = 45$, $\overline{X} = 325$, find a (a) 90%, (b) 95%, (c) 98%, (d) 99% confidence interval for μ.

$$\sigma_{\overline{X}} = \frac{45}{\sqrt{81}} = 5.0.$$

NUMBERS FROM $N(0,1)$ TABLE

Confidence Coefficient	z	$\overline{X}$	$\sigma_{\overline{X}}$	$z\sigma_{\overline{X}}$	Confidence Interval	Width of the Interval
.90	± 1.64	325	5	8.2	316.8–333.2	16.4
.95	± 1.96	325	5	9.8	315.2–334.8	19.6
.98	± 2.33	325	5	11.7	313.4–336.7	23.4
.99	± 2.58	325	5	12.9	312.1–337.9	25.8

Note that, as our degree of confidence that the given interval does include the population mean increases, the width of the interval increases.

Example 4   A random sample of 36 students chosen from a class of 7,200 has a mean weight of 140 pounds. It is known that the standard deviation of all students in the class is 5 pounds. Find a 90% confidence interval for the mean weight of all students in the class.

1. $\overline{X} = 140$.

2. $\sigma_{\bar{X}} = \dfrac{5}{\sqrt{36}} = .83$

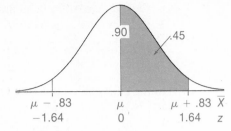

$\mu - .83$ μ $\mu + .83$ $\bar{X}$
-1.64 0 1.64 z

3. With a .90 confidence coefficient, $z = \pm 1.64$.

4. $\bar{X} \pm z\sigma_{\bar{X}} = 140 \pm 1.64(.83) = 138.6$ or 141.4.

A 90% confidence interval for the mean weight of all students in the class is 138.6 to 141.4 pounds.

I hope you have not just mechanically followed directions but have reasoned as you go along: "A 90% confidence interval is wanted. By using Table 4, I find $\Pr(-1.64 < z < 1.64) = .90$; then there is a .90 probability that a randomly chosen sample mean is within $1.64(.833) = 1.4$ pounds of μ. The sample I have here is random, and has a mean of 140 pounds. Therefore, I am 90% confident that the interval $140 - 1.4$ to $140 + 1.4$ (138.6 to 141.4) pounds includes μ."

The examples above may seem nonsensical in that σ is known—and if σ, then why not μ? It is true that, on the occasions you have determined σ, it has been easy (easier, even) to find μ. But there are indeed occasions of great practical interest in which σ can be estimated quite accurately but μ cannot, as the following example demonstrates.

Example 5   Each decade since 1900 the height of army recruits has been determined. Their mean height has gradually increased, but the standard deviation of heights of all recruits each decade has stayed constant at 3.0 inches. A random sample of 900 recruits are measured in 1980, and their mean height is found to be 70 inches. What is a 95% confidence interval for the mean height of all recruits in 1980?

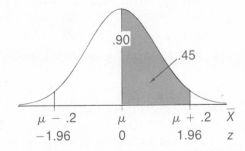

$\mu - .2$ μ $\mu + .2$ $\bar{X}$
-1.96 0 1.96 z

It is assumed that $\sigma = 3.0$ in 1980 as well.

$\bar{X} = 70, \ n = 900, \ \sigma_{\bar{X}} = \dfrac{3.0}{\sqrt{900}} = .10$; at a 95% confidence level,

$z = \pm 1.96$, and 1.96 standard deviation units $= 1.96(.10) = .2$ inches.
$\overline{X} \pm z\sigma_{\overline{X}} = 70 \pm .2 = 69.8$ or 70.2 inches.

A 95% confidence interval for μ in 1980 is 69.8 to 70.2 inches.

In the example with which we began this section ($\overline{X} = 100, n = 36, \sigma = 24$; see page 163), we said the point estimate of μ is 100, and the confidence limits are 100 ± 7.8 with 95% confidence. We can now tell how good a guess 100 is as a point estimate, just as an engineer prescribes the diameter of a shaft: .5000 inches $\pm$ .0025 inches—but the engineer hopes for 100% compliance with his prescription and the statistician does not. Without comments on its accuracy, a point estimate is pretty useless, but with an idea of its accuracy it becomes an interval estimate.

8.6 EXERCISES

1. (a) In a standard normal curve, 90% of z scores are between $-z_0$ and z_0. Find z_0.
   (b) Describe the sampling distribution of means for samples of size 400 taken from a population with (unknown) mean μ and standard deviation $\sigma = 16.0$ feet.
   (c) In this sampling distribution, 1.64 standard deviation units equals how many feet?
   (d) The probability is .90 that the mean of a random sample of size 400 from this population is within how many feet of μ?
   (e) One sample of size 400 is drawn at random from this population; its mean is 86.0 feet. You have 90% confidence that 86.0 feet is within how many feet of μ?
   (f) Find a 90% confidence interval for μ.

2. $\overline{X} = 60.0$ inches, $\sigma = 4.0$ inches; find a 95% confidence interval for μ if $n =$ (a) 64, (b) 100, (c) 400. (d) What happens to the width of a confidence interval as n increases?

3. $\overline{X} = 90.0, \sigma = 6.0, n = 100$. Find (a) 90%, (b) 95%, (c) 98% confidence interval. (d) What happens to the width of a confidence interval as the confidence coefficient increases?

4. Criticize and correct each of the following:
   (a) "The probability is .95 that μ falls between 91.2 and 96.4."
   (b) "The probability is .90 that the mean of a random sample of size 110 taken from this population is within 10 cm. of μ. My sample of size 110 was chosen at random. Therefore the probability is .90 that the mean of my sample is within 10 cm. of μ."

5. In the past, it has been found that the standard deviation of holes drilled by a certain drill press at the Upjack factory is .04 mm. The machine has been reset slightly, but there is no reason to believe the standard deviation has

changed. One hundred holes drilled on this drill press are chosen at random and found to have a mean diameter of 5.2 mm. Give an interval estimate for the mean diameter of all holes with this setting, with 90% confidence that your estimate includes the true mean.

6. Each of 200 investigators independently takes a random sample of size 40 from the same population, and each finds a 90% confidence interval for the population mean. How many of these confidence intervals do you expect will contain the true mean?

7. Can you find a 95% confidence interval for the mean of a population if the data is (a) categorical? (b) ranked? (c) metric?

8. Gollakata Jagannadham, in his doctoral research in geology at the University of North Carolina (1972), measured hundreds of zirconium silicate crystals. A sample of 209 zircon crystals from Suite B in the North Carolina mountains had a mean width/length ratio of 0.435 and a standard deviation of 0.126. Find a 99% confidence interval for the mean width/length ratio of all zircon crystals in Suite B.

9. The mean survival time of 91 laboratory rats after removal of the thyroid gland is 82 days with a standard deviation of 10 days. Find a 90% confidence interval for the mean survival time of all laboratory rats after thyroidectomy.

10. If a 95% confidence interval for μ is $98.00 to $102.00, which of the following is most likely to be a 99% confidence interval (based on the same sample)? (a) $97.37 to $102.00, (b) $97.37 to $102.63, (c) $98.00 to $102.00, (d) $98.40 to $101.60, (e) $98.40 to $102.60. Explain.

ANSWERS

1. (a) $z_0 = \pm 1.64$; (b) Its mean $= \mu$; $\sigma_{\bar{X}} = \dfrac{\sigma}{\sqrt{n}} = \dfrac{16.0}{\sqrt{400}} = .8$ feet; (c) $1.64(.8) = 1.31$

   feet; (d) 1.31 feet; (e) 1.31 feet; (f) $86.0 \pm 1.3 = 84.7$ or 87.3 feet. A 90% confidence interval for μ is 84.7 to 87.3 feet.

2. (a) $\sigma_{\bar{X}} = \dfrac{4.0}{\sqrt{64}} = .50$; $\bar{X} \pm z\sigma_{\bar{X}} = 60.0 \pm 1.96(.50) = 60.0 \pm .98 = 59.0$ or 61.0. A 95%

   confidence interval for μ is 59.0 to 61.0 inches if $n = 64$.

   (b) $\sigma_{\bar{X}} = \dfrac{4.0}{\sqrt{100}} = .40$; $\bar{X} \pm z\sigma_{\bar{X}} = 60.0 \pm 1.96(.40) = 60.0 \pm .78 = 59.2$ or 60.8. A

   95% confidence interval for μ is 59.2 to 60.8 inches if $n = 100$.

   (c) $\sigma_{\bar{X}} = \dfrac{4.0}{\sqrt{400}} = .20$; $60 \pm 1.96(.20) = 59.6$ or 60.4. A 95% confidence interval for

   μ is 59.6 to 60.4 inches if $n = 400$.

   (d) The interval gets smaller as n increases.

3. $\sigma_{\bar{X}} = \dfrac{6}{\sqrt{100}} = .60$. (a) $90.0 \pm 1.64(.60) = 89.0$ to 91.0 inches; (b) $90.0 \pm 1.96(.60) =$

   88.8 to 91.2 inches; (c) $90.0 \pm 2.33(.60) = 88.6$ to 91.4 inches; (d) the confidence interval becomes wider as the confidence coefficient increases.

4. (a) μ is fixed, so the probability is either 0 or 1. "I am 95% confident that the interval 91.2 to 96.4 inches includes μ." (b) Once you have drawn the sample, its mean either is or is not within 10 cm. of μ. "I have 90% confidence that the mean of my sample is within 10 cm. of μ."

5. $\sigma_{\bar{X}} = \dfrac{.04}{\sqrt{100}} = .004.\ 5.2 \pm 1.96(.004) = 5.19$ to 5.21 mm.

6. 90% of them = 180 of the intervals.

7. Only for metric data. (You can't find the mean of categorical or ranked data.)

8. $0.435 \pm 2.58 \left(\dfrac{0.126}{\sqrt{209}}\right) = .4125$ to $.4575$.

9. $82 \pm 1.64 \left(\dfrac{10}{\sqrt{91}}\right) = 80.3$ to 83.7 days.

10. (b). The interval is wider for 99% than for 95% so immediately the choice is narrowed to (a) or (b). The mid-point of the interval is the same, for both the 95% and 99% confidence intervals, and it is $100 for the 95% confidence interval, so (a) is impossible. (b) is correct.

8.7 INTERVAL ESTIMATES OF μ WHEN σ IS UNKNOWN AND $n \geqslant 30$

Our next aim is to find interval estimates for μ when σ is unknown. Can we use the sample standard deviation s and proceed merrily on our way? Sometimes the answer is "yes"; let us see when and why.

> **Definition:** A **parameter** is a numerical descriptive measure of a population. (μ and σ are examples of parameters.) A **statistic** is a numerical descriptive measure of a sample. ($\bar{X}$ and s are examples of statistics.)

To each sample statistic there corresponds a population parameter: to $\bar{X}$, μ; to s^2, σ^2; to s, σ; we shall soon deal with a sample proportion p and a population proportion P. We shall hope to use $\bar{X}$, s^2, s, p, etc., to estimate μ, σ^2, σ, P, etc. We have already found that the mean $\bar{X}$ of a sample can be used to estimate μ. This does not, of course, indicate that the mean of every sample will equal the population mean. But since the mean of the sampling distribution of means = μ, we do know that if we take all possible samples, the mean of their means will give us μ. Furthermore, a single sample taken at random has a mean which is neither consistently too low nor consistently too high. We say $\bar{X}$ is an unbiased estimator of μ.

Is s^2 an unbiased estimator of σ^2? The answer is "yes" because it was defined in such a way that it would be. (See the discussion on page 70.)

If σ is unknown *and n is at least 30*, $\sigma_{\bar{X}}$ is approximated by $\dfrac{s}{\sqrt{n}}$. If σ and s are both known, use $\dfrac{\sigma}{\sqrt{n}}$ in preference to $\dfrac{s}{\sqrt{n}}$. Why the restriction that the sample size be at least 30, and what do you do if σ is unknown but n is less than 30? Both these questions will be answered in Chapter 11.

$$\sigma_{\bar{X}} = \frac{s}{\sqrt{n}} \quad \text{if } \sigma \text{ is unknown and } n \geqslant 30$$

Example 1

Find a 90% confidence interval for the mean height of 2,200 students if a random sample of 100 students has a mean height of 67 inches and a standard deviation of 3.0 inches.

$\bar{X} = 67$, $s = 3$, $n = 100$, σ is unknown, so

$$\sigma_{\bar{X}} = \frac{s}{\sqrt{n}} = \frac{3}{\sqrt{100}} = .30$$

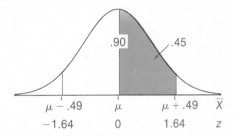

At a 90% confidence level, $z = \pm 1.64$.

1.64 standard deviation units $= z\sigma_{\bar{X}} = 1.64(.30) = .49$ inches.
$\bar{X} \pm z\sigma_{\bar{X}} = 67 \pm .49 = 66.5$ or 67.5 inches.

I am 90% confident that the mean height of all 2,200 students is between 66.5 and 67.5 inches.

Example 2

The mean weight of a sample of 36 Essex University students is 150 pounds, with a standard deviation of 15 pounds. It is known, from previous tests, that the standard deviation of the weights of all Essex students is 18 pounds. Find a 95% interval for the mean weight of all Essex students.

$\bar{X} = 150$, $n = 36$, $s = 15$, $\sigma = 18$.

When s and σ are both known, σ is used to determine $\sigma_{\bar{X}}$.

$$\sigma_{\bar{X}} = \frac{\sigma}{\sqrt{n}} = \frac{18}{6} = 3.0 \text{ pounds}$$

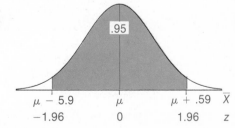

Confidence coefficient $= .95$, so $z = \pm 1.96$.

1.96 standard deviation units $= 1.96(3.0) = 5.9$ pounds.

$\bar{X} \pm z\sigma_{\bar{X}} = 150 \pm 5.9 = 144.1$ to 155.9 pounds.

A 95% confidence interval for the mean weight of all Essex students is 144.1 to 155.9 pounds.

Let me repeat the warning given before the last two examples: If σ is unknown, use $s/\sqrt{n}$ as an approximation of $\sigma_{\bar{x}}$ only if n is at least 30. If n is less than 30, a different technique will be used. Until you have studied this (in Chapter 11), you will have to increase the sample size if σ is unknown, so that the methods of this section can be used.

8.8 EXERCISES

1. A survey of 64 White Mountain College students reveals a mean weekly expenditure of $8.10 on books and records, with a standard deviation of $2.50. Make an interval estimate of the mean expenditure on books and records for all White Mountain College students, with a confidence level of 95%.

2. Thirty-six men in an army division are picked at random to test an obstacle course. Their mean time for the course is 36 seconds, with 4 seconds standard deviation. What is an interval estimate of the mean time for all troops in the division, with 90% confidence?

3. 3,600 randomly selected high school students are found to have watched a mean of 10.5 hours of television per week, with a standard deviation of 2 hours. Find a 98% confidence interval for the mean television viewing time of all high school students.

4. Find a 90% confidence interval for the age of 2,576 Central High School graduates if a random sample of 100 graduates indicates a mean age of 18.7 years with a standard deviation of .4 years.

5. Find a 98% confidence interval for radio ownership per family if a random sample of 900 families shows a mean 2.5 radios per family, with a standard deviation of .33.

6. A random selection of 49 students taking first-year typing are tested for speed; the mean speed is 26 words per minute, with a standard deviation of 5 words per minute. Find a 95% confidence interval for the mean speed of the class as a whole.

7. A random check of 81 of the 2,640 inmates at the State Prison indicates a mean of 2.1 previous terms per prisoner, with a standard deviation of .4. What is an interval estimate of the mean number of previous prison terms for all inmates, with 90% confidence?

8. A random survey of 25 Ph.D.'s in the social sciences shows a mean of 5.1 years of postgraduate study required for the degree, with a standard deviation of 1.2 years. Find a 90% confidence interval for the years required for a graduate degree in social sciences.

9. A safety consultant tests 36 bumper guards, chosen at random, on Chevrolets by bumping into a concrete wall at 5 miles per hour. The mean cost

of repairs is $120, with a standard deviation of $15. With what confidence can he claim that the mean cost of repairs for all Chevrolet cars under this test would be between $115 and $125?

10.   The mean weekly take-home pay of 10 wage earners in Michigan is $280 with a standard deviation of $60. Find a 95% confidence interval for the mean weekly take-home pay of all American workers.

ANSWERS

1. $\sigma_{\bar{x}} = \dfrac{2.50}{\sqrt{64}} = .31; \ 8.10 \pm 1.96(.31) = \$7.49 \text{ to } \$8.71.$

2. $\sigma_{\bar{x}} = \dfrac{4}{\sqrt{36}} = .667; \ 36 \pm 1.64(.667) = 34.9 \text{ to } 37.1 \text{ seconds.}$

3. $10.5 \pm 2.33 \left(\dfrac{2}{\sqrt{3600}} \right) = 10.5 \pm .08 = 10.4 \text{ to } 10.6 \text{ hours.}$

4. $18.7 \pm 1.64 \left(\dfrac{.4}{\sqrt{100}} \right) = 18.7 \pm .07 = 18.63 \text{ to } 18.77 \text{ years.}$

5. $2.5 \pm 2.33 \left(\dfrac{.33}{\sqrt{900}} \right) = 2.5 \pm .03 = 2.47 \text{ to } 2.53 \text{ radios.}$

6. $26 \pm 1.96 \left(\dfrac{5}{\sqrt{49}} \right) = 26 \pm 1.4 = 24.6 \text{ to } 27.4 \text{ words per minute.}$

7. $2.1 \pm 1.64 \left(\dfrac{.4}{\sqrt{81}} \right) = 2.1 \pm .07 = 2.03 \text{ to } 2.17 \text{ previous terms per prisoner.}$

8.   Beware! You don't know σ and n is less than 30. Wait until you have studied Chapter 11. (Also, the population is probably skewed to the right.)

9. $120 \pm z \left(\dfrac{15}{\sqrt{36}} \right) = 115$ or $125; \ 120 - 2.50z = 115,$ so $z = 2.00.$ Check: $120 +$ $z(2.50) = 125$, so $z = 2.00$. $\Pr(-2.00 < z < 2.00) = .955$, $115 to $125 would be a 95.4% confidence interval.

10.   You can't draw an inference about all American workers if the sample was restricted to Michigan. Also, n is too small (note that σ is unknown) and take-home pay of Michigan wage earners is probably skewed to the right rather than normally distributed.

8.9 VOCABULARY AND NOTATION

sampling distribution of means	upper confidence limit
standard error of the mean	lower confidence limit
Central Limit Theorem	confidence coefficient
point estimate	$\sigma_{\bar{x}}$
interval estimate	statistic
confidence interval	parameter

8.10 REVIEW EXERCISES

1. A population has 4 scores: 0, 4, 8, 16.
   (a) Find the sampling distribution of means for samples of size 2 taken from this population. (You will get 16 scores.)
   (b) Find the mean and the standard error of this sampling distribution.
   (c) Find μ and σ, using the population scores.
   (d) Compare the results of (b) and (c).

2. A population has a mean of 216 mm. and a standard deviation of 28 mm. Describe the sampling distribution for samples of size 49 for this population: Of what is it composed? What is its mean? its standard error? Is it normal?

3. The sampling distribution of means for samples of size 16 has a mean of 123.0, a standard error of the mean equal to 3.0, and is normally distributed. What is (a) μ, (b) σ, (c) the shape of the population?

4. What is the effect on the width of a confidence interval if the sample size is changed from (a) 25 to 100? (b) 25 to 2,500?

5. What is the probability that a random sample of 36 kittens aged 3 weeks will have a mean weight between 80.7 and 83.6 g., if the mean weight of all 3-week-old kittens is 82 g. with a standard deviation of 8 g.?

6. What assumptions must be satisfied about (a) the population, (b) the sample, if you are to find a confidence interval for μ?

7. Find a 90% confidence interval for the height of 1,200 students if it is found that a random sample of 49 students has a mean height of 67 inches, and it is assumed that the standard deviation for all students is 3 inches as it has been in the past.

8. A psychologist discovers that the mean reaction time of 36 rats to an 18-volt electric shock is .45 second with a standard deviation of .06 second. Find a 90% confidence interval for the mean reaction time of all rats of the same strain to an 18-volt shock.

9. A political scientist discovers that the mean waiting time between arrest and trial for 81 men chosen at random from those tried for felonies in New York City is 300 days, with a standard deviation of 30 days. He is 98% confident that the mean interval between arrest and trial for felony cases in New York City has what range?

10. A college housing officer wants to estimate the mean rental of rooms in the area. He randomly selects 30 of the 300 available rooms and finds a mean monthly rental of $120 per room, with a standard deviation of $10. Find the limits for the mean monthly rental per room of all 300 rooms, with 95% confidence.

11. A random check of 36 factory assemblers shows that the mean number of units completed per worker per hour is 16.2, with a standard deviation of 2.1. What is a 95% confidence interval for the mean number of units completed per hour by all 700 assemblers in the plant?

12. A national news magazine has 2,500,000 subscribers. A survey of 1,600 random subscribers shows a mean annual income of $18,100 with a standard deviation of $1,000. Find a 98% confidence level for the mean income of all subscribers.

13. A survey of 49 of the 250 newly appointed district attorneys in the country shows a mean of 13.7 years of legal experience prior to appointment, with a standard deviation of 2.8 years. Find a 90% confidence interval for the mean years of prior legal experience of all the newly appointed district attorneys.

14. A check of 100 random automobile accidents resulting in death reveals a mean driver's age of 24.5 years, with a standard deviation of 3.5 years. Find an interval estimate of the mean age of all drivers involved in fatal accidents, at a 95% confidence interval.

15. There are 2,150 supermarkets in the metropolitan area. A random sample of 64 supermarkets indicates a mean of 35 cases of baby food sold per week, with a standard deviation of 6 cases. Find a 95% confidence interval for the mean number of cases of baby food sold per week at all supermarkets in the metropolitan area.

16. An automobile manufacturer tests a random sample of 81 new Warlock cars and finds a mean of 20.5 miles per gallon in road tests, with a standard deviation of 2.7 miles per gallon. What is a 95% confidence interval for mileage per gallon of all new cars of that model?

17. A study of 50 households of four people each living in Central City, chosen at random, shows a mean expenditure of $96 a week for food, with a standard deviation of $3. Make an interval estimate of the average weekly cost of food for all households of four people in Central City, with 96% confidence.

18. XYZ tires are known to have a mean life of 20,000 miles, with a standard deviation of 2,000 miles. The manufacturer modifies the tires; he hopes he has increased the mean life but thinks the standard deviation will not be changed. Tests of 60 randomly chosen modified tires give a mean life of 21,500 miles, with a standard deviation of 2,100 miles. Find an interval estimate of the mean life of the modified tires, at a 90% confidence level.

ANSWERS

1. (a) Samples: 0, 0; 0, 4; 0, 8; 0, 16; 4, 0; 4, 4; 4, 8; 4, 16; 8, 0; 8, 4; 8, 8; 8, 16; 16, 0; 16, 4; 16, 8; 16, 16. $\bar{X}$ (= scores in the sampling distribution): 0, 2, 4, 8, 2, 4, 6, 10, 4, 6, 8, 12, 8, 10, 12, 16. (b) $\Sigma\bar{X} = 112$, $\Sigma\bar{X}^2 = 1,064$; mean of the sampling distri-

   bution $= \dfrac{112}{16} = 7.0$;

   $\sigma_{\bar{X}} = \sqrt{\dfrac{1,064 - 112^2/16}{16}} = 4.18.$

   (c) $\mu = \dfrac{28}{4} = 7.0$; $\sigma = \sqrt{\dfrac{336 - 28^2/4}{4}} = 5.92.$

(d) Mean of the sampling distribution $= \mu = 7.0$; $\dfrac{\sigma}{\sqrt{n}} = \dfrac{5.92}{\sqrt{2}} = 4.18 = \sigma_{\bar{x}}$.

3. (a) $\mu = 123.0$; (b) $\sigma_{\bar{x}} = 3.0 = \dfrac{\sigma}{\sqrt{16}}$, so $\sigma = 12.0$; (c) the population must be normally distributed since the sampling distribution is normal even though n is small.

5. $\mu = 82$, $\sigma = 8$, $n = 36$; $\sigma_{\bar{x}} = \dfrac{8}{\sqrt{36}} = 1.33$. $z = \dfrac{\overline{X} - \mu}{\sigma_{\bar{x}}} = \dfrac{\overline{X} - 82}{1.33}$. $\Pr(80.7 < \overline{X} < 83.6) = \Pr(-.98 < z < 1.20) = .337 + .385 = .72$.

7. $67 \pm 1.64 \left(\dfrac{3}{\sqrt{49}}\right) = 67 \pm .70$; 66.3 to 67.7 inches.

9. $300 \pm 2.33 \left(\dfrac{30}{\sqrt{81}}\right)$; 292.2 to 307.8 days.

11. $16.2 \pm 1.96 \left(\dfrac{2.1}{\sqrt{36}}\right)$; 15.5 to 16.9 units completed per hour by each worker.

13. $13.7 \pm 1.64 \left(\dfrac{2.8}{\sqrt{49}}\right)$; 13.0 to 14.4 years.

15. $35 \pm 1.96 \left(\dfrac{6}{\sqrt{64}}\right)$; 33.5 to 36.5 cases.

17. $96 \pm 2.05 \left(\dfrac{3}{\sqrt{50}}\right)$; $95.13 to $96.87 per week.

9
INTERVAL ESTIMATES OF PROPORTIONS, DIFFERENCE OF MEANS, AND DIFFERENCE OF PROPORTIONS; SAMPLE SIZE

You learned how to find an interval estimate for the population mean after studying the sampling distribution of means. Now you will learn how to make three other kinds of interval estimates—but each one will again be preceded by study of a sampling distribution. There is some new notation: P for proportion in a population, p for proportion in a sample, σ_p for standard error of proportions (= standard deviation of the sampling distribution of proportions), and so on. But once you get used to notation, look for the similarities between *all* the sampling distributions you study. Since the sampling distributions you will use in this chapter are always normal distributions, you will in every case be using standard scores to find areas under the normal curve. The formula for standard scores will change its appearance but not its essential character as the names of the scores, mean, and standard deviation change. Once you realize this, confidence intervals for proportions or for differences of means or of proportions will be no more difficult than for means in Chapter 8.

Sampling Distribution of:	Scores in Sampling Distribution	Mean	Standard Error	Standard Scores	Confidence Limits
Means	$\bar{X}$	μ	$\sigma_{\bar{X}}$	$z = \dfrac{\bar{X} - \mu}{\sigma_{\bar{X}}}$	$\bar{X} \pm z\sigma_{\bar{X}}$
Proportions	p	P	σ_p	$z = \dfrac{p - P}{\sigma_p}$	$p \pm z\sigma_p$
Difference of means	$\bar{X} - \bar{Y}$	$\mu_X - \mu_Y$	$\sigma_{(\bar{X}-\bar{Y})}$	$z = \dfrac{(\bar{X} - \bar{Y}) - (\mu_X - \mu_Y)}{\sigma_{(\bar{X}-\bar{Y})}}$	$(\bar{X} - \bar{Y}) \pm z\sigma_{(\bar{X}-\bar{Y})}$
Difference of proportions	$p_X - p_Y$	$P_X - P_Y$	$\sigma_{(p_X-p_Y)}$	$z = \dfrac{(p_X - p_Y) - (P_X - P_Y)}{\sigma_{(p_X-p_Y)}}$	$(p_X - p_Y) \pm z\sigma_{(p_X-p_Y)}$

9.1 THE SAMPLING DISTRIBUTION OF PROPORTIONS

Our first aim is to find point and interval estimates of a population proportion from a sample proportion. A typical problem is the following: A poll is conducted to find how many voters will vote for candidate A for President. A random sample is taken of 1,000 voters, and it is found that 520 plan to vote for candidate A. Estimate the percentage of the population that will vote for candidate A, at a 95% confidence level.

It should be intuitively apparent that a point estimate is $\frac{520}{1000} = 52\%$. This is the best single guess at population proportion—but, again, pretty useless without some idea of how good a guess it is. Before finding an interval estimate we must look at the sampling distribution of proportions.

Again we take all possible samples of given size n and find the proportion in each sample, thus forming the distribution of sample proportions—known, commonly but perversely, as the sampling distribution of proportions. See if these sketches help you understand the process; only two of the many scores in the sampling distribution are shown.

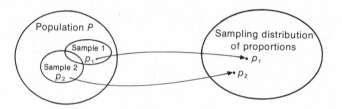

Some new notation is involved, but we won't need any new Greek letters.

Notation: P = proportion of "successes" or "Do's" in a population (a parameter).
$Q = 1 - P$ = proportion of "failures" or "Don'ts" in a population.
p = proportion of "successes" or "Do's" in a sample.
$q = 1 - p$
σ_p = standard deviation of the sampling distribution of proportions.
  = the standard error of proportions.
n = size of the sample.

The population represents categorical data (you are counting, for example, the number of voters who will vote for A for President), but the scores in the sampling distribution are proportions—numbers between 0 and 1—and this set of numbers has a mean and standard deviation. The sampling distribution of proportions has the following characteristics:

1. Its mean $= P$, the proportion in the population.

2. $\sigma_p = \sqrt{\dfrac{PQ}{n}}$

3. (Central Limit Theorem for proportions:) It is approximately a normal distri-

bution for large n, and the approximation gets better as n is increased. Care is needed here since the proportion P is always between 0 and 1. If $P \approx 1/2$, the sampling distribution is approximately normal if $n > 20$. But if P is close to 0 or 1, the sampling distribution will tend to have a longer tail to the right (or left), and n must be increased to get an approximation to a normal curve. In practice, let's assume that both nP and nQ are at least 15.

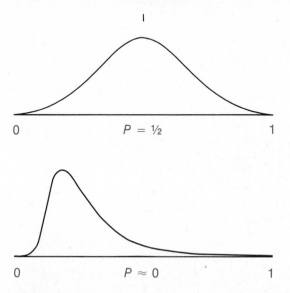

The mean and standard error formulas are familiar friends in a new guise. Suppose the random variable $X = 1$ is assigned to each "Do" in the population and $X = 0$ to each "Don't," and that there are a "Do's" in a population of size N:

X	f	Xf	X^2f
1	a	a	a
0	$N - a$	0	0
Sum:	N	a	a

$$\mu = \frac{a}{N} = P$$

$$\sigma = \sqrt{\frac{a - a^2/N}{N}} = \sqrt{\left(\frac{a}{N}\right) - \left(\frac{a}{N}\right)^2} = \sqrt{P - P^2} = \sqrt{P(1 - P)} = \sqrt{PQ},$$

so

$$\frac{\sigma}{\sqrt{n}} = \sqrt{\frac{PQ}{n}}$$

help!!

Example 1   In the town of Peacock there are three registered voters: A, B, and C. A will vote for M for governor; B and C will not vote for M. Consider the sampling distribution of proportions of samples of size 2 taken with replacement. Let P be the proportion who will vote for M.

$$P = \frac{1}{3},\ Q = \frac{2}{3},\ n = 2.$$

Sampling
Distribution
of Proportions

Sample	p
A, A	1
A, B	$\frac{1}{2}$
A, C	$\frac{1}{2}$
B, A	$\frac{1}{2}$
B, B	0
B, C	0
C, A	$\frac{1}{2}$
C, B	0
C, C	0

p	f	fp	fp^2
1	1	1	1
$\frac{1}{2}$	4	2	1
0	4	0	0
	9	3	2

$$\text{Mean} = \frac{3}{9} = \frac{1}{3}$$

$$\sigma_P = \sqrt{\frac{2 - 3^2/9}{9}} = \sqrt{\frac{1}{9}} = \frac{1}{3}$$

(a) Mean of the sampling distribution of proportions $= P\left(= \frac{1}{3}\right)$.

(b) $\sqrt{\dfrac{PQ}{n}} = \sqrt{\dfrac{1/3(2/3)}{2}} = \sqrt{\dfrac{1}{9}} = \dfrac{1}{3}$, so $\sigma_p = \sqrt{\dfrac{PQ}{n}}$.

(c) The sampling distribution of proportions is not normal here; nP, which equals $\frac{2}{3}$, is too small.

9.2 INTERVAL ESTIMATES OF PROPORTIONS

With the help of a normal curve sketch, you should now be able to solve problems such as this:

Example 1   If $P = .4$, the probability is .95 that a sample of size 40 will have a proportion p between what two values?

$$P = .4,\ Q = .6,\ n = 40.$$

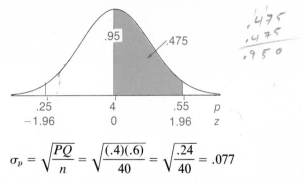

.475
.475
.950

$$\sigma_p = \sqrt{\frac{PQ}{n}} = \sqrt{\frac{(.4)(.6)}{40}} = \sqrt{\frac{.24}{40}} = .077$$

Probability $= .95$, so $A(z) = .475$ and $z = \pm 1.96$.

As always, the transformation from any scores to standard or z scores is made by subtracting the mean of the distribution (P) and dividing by its standard deviation (σ_p).

$$z = \frac{p - P}{\sigma_p}$$

$$\pm 1.96 = \frac{p - .4}{.077}$$

$$p = .4 \pm (1.96)(.077) = .4 \pm .15 = .25 \text{ or } .55.$$

There is a $.95$ probability that the sample proportion will be between $.25$ and $.55$.

A more interesting problem, however, is this one:

Example 2

A poll is conducted to find out how many voters will vote for candidate A for President. A random sample is taken of 1,000 voters, and it is found that 520 plan to vote for Candidate A. Find a 95% confidence interval for the proportion that will vote for candidate A.

We start out with the same kind of reasoning as for confidence intervals of means: 95% of the scores of any normal distribution are within 1.96 standard deviation units of the mean of the distribution. The sampling distribution of proportions is approximately a normal distribution ($np = 1,000(.52) > 15$, so is $nq = 1,000(.48)$), so there is a $.95$ probability that a random sample has a proportion which is within 1.96 standard deviation units or 1.96 σ_p score units of the unknown P.

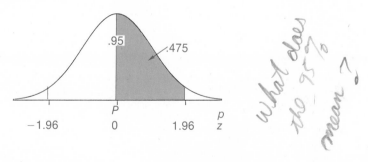

It is in finding the value of σ_p that we run into trouble. The standard error of the mean did not depend on μ, but the standard error of proportion does depend on P: $\sigma_p = \sqrt{\dfrac{PQ}{n}}$. **Fortunately, $\sqrt{\dfrac{pq}{n}}$ can be used as an approximation to σ_p when P is unknown.**

In the problem we are solving

$$\sigma_p = \sqrt{\frac{.52(.48)}{1,000}} = \sqrt{.000250} = .0158.$$

1.96 standard deviation units = 1.96(.0158) = .031 proportion units. The probability is .95 that the proportion in a random sample of size 1,000 from this population is within .031 of P. This sample was chosen at random, so I am 95% confident that this p (= .52) is within .031 of P. I am 95% confident that P is between .52 − .031 and .52 + .031, or between 48.9% and 55.1%.

Once you have thoroughly understood how the confidence limits are arrived at, you can use a short cut. We have essentially solved $z = \dfrac{p - P}{\sigma_p}$ for P:

$$P = p \pm z\sigma_p$$

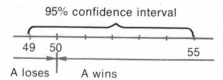

Confidence limits for P are given by

$$p - z\sigma_p \text{ and } p + z\sigma_p$$

where the value of z is determined by the confidence coefficient.

Note that if there are only two candidates for President, the above result does not predict the winner, since the 50% needed to win is inside the confidence interval.

95% confidence interval

49 50   55

A loses   A wins

What can be done? Try the same problem if the sample size is 4,000, and again 52% plan to vote for candidate A.

Answer

If n = 4,000, $\sigma_p = \sqrt{.52(.48)/4,000} = .0079$, so $P = .52 \pm 1.96(.0079) = .52 \pm .015$; a 95% confidence interval for P is 50.5% to 53.5%; you have 95% confidence that candidate A will win the election.

Example 3

Thirty thousand people turn up to see a football game. You discover that 420 of the first 900 people to arrive are women. Find a 95% confidence interval for the percentage of women at the game.

OUCH! Find the trap before you read on.

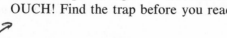

Is this a representative sample? Everything we have done earlier is predicated on random samples. Maybe women always come late to football games. Don't misuse statistics! *Fuck you! I'll misuse it if I wish!*

Example 4   Work out Example 3, but with a random sample of 900 people of whom 420 are women.

1. $p = \dfrac{420}{900}, q = \dfrac{480}{900}, n = 900.$

2. $\sigma_p = \sqrt{\dfrac{pq}{n}} = \sqrt{\dfrac{(420/900)(480/900)}{900}} = .0166.$

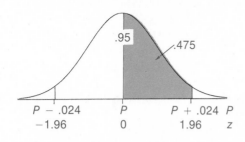

3. At a 95% confidence level, $z = \pm 1.96.$

4. $p \pm z\sigma_p = \dfrac{420}{900} \pm 1.96(.0166) = .467 \pm .0325 = .434$ or $.500.$ A 95% confidence interval for the percentage of women in the audience is 43 to 50.

9.3 EXERCISES

1. A population consists of 4 people; A and B who drive VW's and C and D who drive Cadillacs. Let P equal the proportion which drives VW's.
   (a) Find the sampling distribution of proportions for samples of size 2 from this population. (There will be 16 scores in it.)
   (b) Find the mean of this sampling distribution and find P, and show that they are equal.
   (c) Compute the standard error of proportions σ_p and show that it equals $\sqrt{\dfrac{PQ}{n}}$

2. A sampling distribution of proportions for samples of size 400 has a mean of .67, a standard error of .024, and is normally distributed. What can you say about the population from which it is taken?

3. In Congressional District 6, 54% of all voters will vote for Abring for representative. What can you say about the sampling distribution of proportions for samples of size 64 taken from all voters in this district? (How is it formed? What is its mean? its standard error? Is it a normal distribution?)

4. If 80% of all college freshmen take a course in English composition, what is the probability that, in a random sample of 100 college freshmen, 75% to 85% study English composition?

5. A sociologist is studying the proportion of college students neither of whose parents attended college, and would like to estimate this proportion with 95% confidence. He discovers that, in a random sample of 4,000 college students, the parents of 1,600 did not attend college. Help him with his estimate.

6. A random sample of 200 Madison wage earners shows that 80 have an income over $15,000. Find a 90% confidence interval for the percentage of all Madison wage earners who have incomes over $15,000.

7. (a) Postcards were sent to 10,000,000 voters asking their choices for President. One-fifth of them answered; 56% of those who returned the postcard reported they would vote for Mr. L. Find a 95% confidence interval for the proportion of all voters who would vote for Mr. L.

(b) In the election that followed, only 34% voted for Mr. L. What is the probability that, in a random sample of 2,000,000, 56% or more will vote for Mr. L?

(c) Can you explain the contradiction between (a) and (b)?

8. (a) In finding a confidence interval for a proportion, do you use categorical, ranked, or metric data?

(b) What assumption(s) must you make about (i) the population, (ii) the sample, if you are to find a confidence interval for the proportion in a population?

9. In field trials of the Salk vaccine for polio in 1955, 33 of 200,745 children injected with vaccine contracted paralytic polio. Find a 99% confidence interval for the percentage of all children injected with Salk vaccine who would get paralytic polio. $\sigma_p = .0000286$.

10. Of 40 self-referred patients treated by a 1-hour hypnosis session to kick the smoking habit, 32 stopped smoking, at least for the moment.* Find a 90% confidence interval for the proportion of all smokers who quit after electing this type of treatment.

11. In studying the randomly chosen records of 412 cars which have failed inspection at a New Jersey Motor Vehicles Inspection station, it is found that 200 have failed the emissions control test. Find a 90% confidence interval for the proportion of all failures in New Jersey car inspections that are due to the emissions control test.

12. A door-to-door cosmetics saleswoman spot checks 640 of her records for the past year, and finds that she had one sale in every 20 of her calls. Find a 95% confidence interval for the proportion of all her calls in which she made a sale.

---

*Berkowitz, B., Ross-Townsend, A., and Kohberger, R. "Hypnotic Treatment of Smoking: The Single-Treatment Method Revisited," *American Journal of Psychiatry, 136,* 1979, pp. 83–85.

13. In one branch of the National American Bank, 36% of 4,132 checking account customers have savings accounts as well. Find a 90% confidence interval for the proportion of checking account customers in all 36 branches of the National American Bank who also have savings accounts.

ANSWERS

1. (a) Samples: AA, AB, AC, AD, BA, BB, BC, BD, CA, CB, CC, CD, DA, DB, DC, DD. $p(=$ scores in the sampling distribution): 1, 1, .5, .5, 1, 1, .5, .5, .5, .5, 0, 0, .5, .5, 0, 0.

   (b) Mean $= \dfrac{8}{16} = .5$, $P = \dfrac{2}{4} = .50$.

   (c) $\sigma_p = \sqrt{\dfrac{6 - 8^2/16}{16}} = \sqrt{\dfrac{1}{8}} = .35$; $\sqrt{\dfrac{PQ}{n}} = \sqrt{\dfrac{.50(.50)}{2}} = \sqrt{\dfrac{1}{8}} = .35$.

2. Only that $P = .67$

3. It is formed by finding the proportion p in every sample of size 64 in this population; all these p's are the scores in the sampling distribution of proportions. Its mean is .54, its standard error $= \sigma_p = \sqrt{\dfrac{.54(.46)}{64}} = .062$, and it is normally distributed since both nP and nQ are over 15.

4. $P = .80$, $Q = .20$, $n = 100$, $\sigma_p = \sqrt{\dfrac{.80(.20)}{100}} = .040$, $z = \dfrac{p - P}{\sigma_p} = \dfrac{p - .80}{.04}$. $\Pr(.75 < p < .85) = \Pr(-1.25 < z < 1.25) = 2(.394) = .79$.

5. $p = \dfrac{1,600}{4,000} = .40$, $\sigma_p = \sqrt{\dfrac{.40(.60)}{4,000}} = .0077$; $.40 \pm 1.96(.0077) = .385$ or .415. A 95% confidence interval for the proportion of all college students neither of whose parents attended college is .385 to .415.

6. $p = \dfrac{80}{200} = .40$, $\sigma_p = \sqrt{\dfrac{.40(.60)}{200}} = .0346$; $.40 \pm 1.64(.0346) = .343$ or .457. A 90% confidence interval is 34.3% to 45.7%.

7. (a) These figures are (roughly) those obtained in the *Literary Digest* poll for Landon vs. Roosevelt in 1936. If the sample were random:

   $$\pm 1.96 = \dfrac{.56 - P}{\sqrt{\dfrac{(.56)(.44)}{2,000,000}}}; \ 55.93 \text{ to } 56.07\%.$$

   (b) $z = \dfrac{.56 - .34}{\sqrt{\dfrac{(.34)(.66)}{2,000,000}}} = 657$. The probability is incredibly small.

   (c) The sample was not a random one, of course. Ballots were sent to those who owned an automobile or had a telephone during a great depression.

8. (a) Categorical data. It can be used with metric data, but never is: why bother to take measurements if you only care about proportions?

   (b) (i) None. (It doesn't make sense to say categorical data should be normally distributed!) (ii) The sample must be random, and n must be large enough so that the sampling distribution of proportions is normal: nP and $nQ \geqslant 15$.

9. $.000164 \pm 2.58(.0000286)$; $.000090$ to $.000238$.

10. $p = .80$, $\sigma_p = \sqrt{\dfrac{.8(.2)}{40}} = .063$; $.80 \pm 1.64(.063) = .696$ to $.904$. But note that $nq = 40(.20) = 8$, < 15, so the results are very dubious.

11. $n = 412$, $p = \dfrac{200}{412}$, $q = 1 - .485 = .515$, $\sigma_p = \sqrt{\dfrac{.485(.515)}{412}} = .0246$; 90% confidence, so $z = \pm 1.64$. $P = .485 \pm 1.64(.0246) = .445$ or $.525$. I am 90% confident that the interval $.445$ to $.525$ includes the proportion of all cars failing inspection in New Jersey which fail the emissions control test.

12. $n = 640$, $p = \dfrac{1}{20} = .05$, $q = .95$, $\sigma_p = \sqrt{\dfrac{.05(.95)}{640}} = (.0086)$; 95% confidence, so $z = \pm 1.96$. $.05 \pm 1.96(.0086) = .033$ or $.067$. A 95% confidence interval for the probability of making a sale is $.03$ to $.07$.

13. This is not a random sample; perhaps wealthier customers do their banking at the branch from which the sample was taken. Don't rely on the hangman symbol to keep you alert to the misuse of statistics.

9.4 THE SAMPLING DISTRIBUTION OF DIFFERENCES OF MEANS

Consider two different populations (one consists of the weights of all men over age 17 in the United States, the other of the weights of all women over 17 in the United States, for example). The first population (X) has mean μ_X and standard deviation σ_X, the second (Y) has mean μ_Y and standard deviation σ_Y. (Do not confuse σ_X and $\sigma_{\bar{X}}$; in the former, the subscript X distinguishes one population (X scores) from another (Y scores), while in the latter the subscript indicates the standard error of the sampling distribution of means.) From the first population take a sample of size n_X and compute its mean $\bar{X}$; from the second population take independently a sample of size n_Y and compute $\bar{Y}$; then determine $\bar{X} - \bar{Y}$. Do this for all possible pairs of samples that can be chosen independently from the two populations. The differences, $\bar{X} - \bar{Y}$, are a new set of scores which form the **sampling distribution of differences of means.**

The characteristics of the sampling distribution of differences of means are

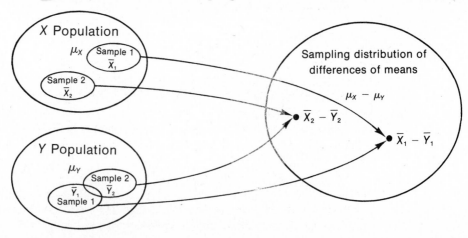

1. The mean of the sampling distribution of differences of means equals the difference of the population means:

$$\text{Mean} = \mu_X - \mu_Y$$

2. The standard deviation of the sampling distribution of differences of means, also called the standard error of differences of means, is denoted by $\sigma_{(\bar{X}-\bar{Y})}$.

$$\sigma_{(\bar{X}-\bar{Y})} = \sqrt{\sigma_{\bar{X}}^2 + \sigma_{\bar{Y}}^2}$$

where $\sigma_{\bar{X}}$ is the standard error of the mean of the first population and $\sigma_{\bar{Y}}$ is the standard error of the mean of the second population.

Note the **plus** sign between $\sigma_{\bar{X}}^2$ and $\sigma_{\bar{Y}}^2$, even though we are dealing with **differences** of means. This shouldn't surprise you. If two sets of scores have different means, their differences will have more spread than either set alone. Perhaps the following diagram will be suggestive:

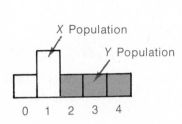

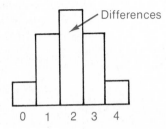

Since you need $\sigma_{\bar{X}}^2 = \dfrac{\sigma_X^2}{n}$, don't waste time taking a square root and finding σ_X; you will just have to square again.

3. (Central Limit Theorem for differences of means:) The sampling distribution is normal if both populations are normal, and is approximately normal if the samples are large enough (even if the populations aren't normal). The more skewed the populations, the larger the samples need to be. In practice, we'll assume that the sampling distribution of differences of means is normal if both n_X and n_Y are $\geqslant 30$.

9.5   USE OF THE SAMPLING DISTRIBUTION OF DIFFERENCES OF MEANS

In using this sampling distribution, it must be emphasized very strongly that the two samples must be picked independently. If a random sample of 36 students is tested for reaction time before and after drinking an ounce of alcohol, you have two sets of scores, but you do not have two independent samples. Techniques for making inferences from two sets of related or paired scores will be studied in Chapters 11, 14, 15, and 16.

Another limitation on the use of this sampling distribution of means is that n_X and n_Y must both be at least 30 if s_X^2 and s_Y^2 are used in finding $\sigma_{\bar{X}}^2$ and $\sigma_{\bar{Y}}^2$. What if this requirement is not met? A different method will be used, using a distribution which is not normal. Wait until Chapter 11.

Example 1 The mean IQ of 1,200 Defoe College students is 122 with a standard deviation of 6, while 2,000 students from Daniel College have a mean IQ of 118 with a standard deviation of 5. What is the probability that the mean IQ of a random sample of 36 Defoe students will be at least 6 points higher than the mean IQ of a random sample of 49 Daniel students? Try working this out before you read on.

$$\text{Defoe: } \mu_X = 122, \; \sigma_X = 6, \; n_X = 36, \; \sigma_{\bar{X}}{}^2 = \frac{6^2}{36} = 1$$

$$\text{Daniel: } \mu_Y = 118, \; \sigma_Y = 5, \; n_Y = 49, \; \sigma_{\bar{Y}}{}^2 = \frac{5^2}{49} = \frac{25}{49}$$

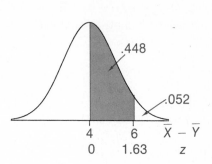

$$\mu_X - \mu_Y = 122 - 118 = 4$$

$$\sigma_{(\bar{X}-\bar{Y})} = \sqrt{1 + \frac{25}{49}} = 1.23$$

$$\bar{X} - \bar{Y} = 6$$

$$z = \frac{(\bar{X} - \bar{Y}) - (\mu_X - \mu_Y)}{\sigma_{(\bar{X}-\bar{Y})}} = \frac{6 - 4}{1.23} = 1.63$$

$$\Pr(\bar{X} - \bar{Y} > 6) = \Pr(z > 1.63) = .5 - \Pr(0 < z < 1.63) = .5 - .448 = .052.$$

The more interesting problem, as before, puts the emphasis the other way around: if the means of two samples, each chosen independently and at random from a different population, are known, and if the standard deviations, either of the populations or of the samples, are known, what statement can be made about the differences of the means of the populations from which the samples are taken?

Example 2 A psychologists tests in a maze 36 rats whose diet contains no Vitamin A, and finds their mean time to "solve" the maze is 4.0 minutes with a standard deviation of .50 minutes; then he tests in the same maze 50 rats whose diet is rich in Vitamin A, and finds a mean time of 3.2 minutes with a standard deviation of .40 minutes. At a 95% confidence level what can he say about the difference in mean times for solving the maze of rats with Vitamin A free and Vitamin A rich diets?

Population X (rats with no Vitamin A):

$$\bar{X} = 4.0, \; n_X = 36, \; \sigma_X \text{ is unknown}, \; s_X = .50, \; \sigma_{\bar{X}}^2 = \frac{s_X^2}{n_X} = \frac{.50^2}{36} = .0069.$$

Population Y (rats with Vitamin A):

$$\bar{Y} = 3.2, \; n_Y = 50, \; s_Y = .40, \; \sigma_{\bar{Y}}^2 = \frac{s_Y^2}{n_Y} = \frac{.40^2}{50} = .0032.$$

$$\bar{X} - \bar{Y} = 4.0 - 3.2 = .8.$$

$$\sigma_{(\bar{X}-\bar{Y})} = \sqrt{\sigma_{\bar{X}}^2 + \sigma_{\bar{Y}}^2} = \sqrt{.0069 + .0032} = .1007.$$

At a 95% confidence level, $z = \pm 1.96$.

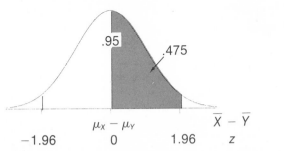

delp!!

1.96 z or standard deviation units $= 1.96(.1007) = .197$ minutes. The probability is .95 that the difference in means of two random and independent samples is within 1.96 z units or .197 minutes of $\mu_X - \mu_Y$. These samples are random and independent, and the difference in their means is .8 minutes. Therefore, I am 95% confident that the interval from $.8 - .197$ to $.8 + .197 = .60$ to 1.00 minutes includes $\mu_X - \mu_Y$.

A 95% confidence interval for the difference in mean times of rats without and with Vitamin A in their diets is .60 to 1.00 minute (36 to 60 seconds).

A formula for the confidence limits is found by solving

$$z = \frac{(\bar{X} - \bar{Y}) - (\mu_X - \mu_Y)}{\sigma_{(\bar{X}-\bar{Y})}} \text{ for } \mu_X - \mu_Y;$$

Confidence limits for the difference of means are

$$(\bar{X} - \bar{Y}) - z\sigma_{(\bar{X}-\bar{Y})} \quad \text{and} \quad (\bar{X} - \bar{Y}) + z\sigma_{(\bar{X}-\bar{Y})}$$

where z is determined by the confidence coefficient.

In the last example, $(\bar{X} - \bar{Y}) \pm z\sigma_{(\bar{X}-\bar{Y})} = .8 \pm 1.96(.1007) = .8 \pm .197 = .60$ or 1.00 minutes (36 to 60 seconds).

Example 3   If a random sample of 50 nonsmokers have a mean life of 76 years with a standard deviation of 8 years, and a random sample of 65 smokers live 68 years with a standard deviation of 9 years, find a 95% confidence interval for the difference of mean lifetime of nonsmokers and smokers.

Population X (nonsmokers):

$$n_X = 50, \bar{X} = 76, s_X = 8, \sigma_{\bar{X}}^2 = \frac{8^2}{50} = 1.28 \text{ years}$$

Population Y (smokers):

$$n_Y = 65, \bar{Y} = 68, s_Y = 9, \sigma_{\bar{Y}}^2 = \frac{9^2}{65} = 1.25 \text{ years}$$

$$\bar{X} - \bar{Y} = 76 - 68 = 8 \text{ years}$$
$$\sigma_{(\bar{X}-\bar{Y})} = \sqrt{1.28 + 1.25} = 1.59 \text{ years}$$

At a 95% confidence level, $z = \pm 1.96$.
1.96 standard deviation units $= 1.96(1.59) = 3.12$ years.
$(\bar{X} - \bar{Y}) \pm z\sigma_{(\bar{X}-\bar{Y})} = 8 \pm 3.1 = 4.9$ or 11.1 years.

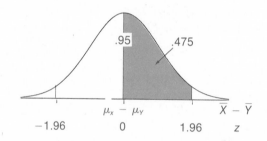

The upper confidence limit is 11.1; the lower confidence limit is 4.9; a 95% confidence interval is 4.9 to 11.1 years. On the basis of these two samples, I am 95% confident that the mean lifetime of nonsmokers is 4.8 to 11.2 years longer than the mean lifetime of smokers.

9.6 SAMPLING DISTRIBUTION OF DIFFERENCE OF PROPORTIONS

Consider two different populations again (voters in New York City and those in Chicago, for example). The first population has a proportion P_X who fit in a certain category (New York City voters who will vote for candidate A for President) and a proportion $Q_X = 1 - P_X$ who do not fit in that category (will not vote for candidate A for President); the same proportions in the second population are P_Y and Q_Y. Take a sample of size n_X from population X (all voters in New York City) and find the proportion p_X who fit a given category (will vote for candidate A for President). Then take **independently** a sample of size n_Y from population Y (all voters in Chicago) and find the proportion p_Y who fit the given category. Find the difference $p_X - p_Y$, and you have **one** score in the sampling distribution of differences of proportions. To find all the scores, go through the same process with all possible combinations of samples of sizes n_X and n_Y, respectively, chosen independently from the two populations.

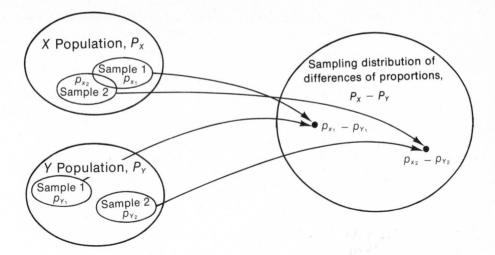

None of this should surprise you, nor should the following results:

1. The mean of the sampling distribution of difference of proportions equals the difference in the proportions P_X and P_Y in the X and Y populations:

Mean of this sampling distribution $= P_X - P_Y$.

2. $\sigma_{(p_X-p_Y)}$, the standard deviation of this new distribution (also called the **standard error of differences of proportions**), equals $\sqrt{\sigma_{p_X}^2 + \sigma_{p_Y}^2}$, where σ_{p_X} is the standard error of the proportion of the X population:

$$\sigma_{p_X} = \sqrt{\frac{P_X Q_X}{n_X}}; \text{ similarly, } \sigma_{p_Y} = \sqrt{\frac{P_Y Q_Y}{n_Y}}.$$

Note that:

$$\sqrt{\frac{P_X Q_X}{n_X}}^2 + \sqrt{\frac{P_Y Q_Y}{n_Y}}^2 = \frac{P_X Q_X}{n_X} + \frac{P_Y Q_Y}{n_Y}$$

$\mu = P_x - P_y$

$6\,(p_x - p_y) = \sqrt{\dfrac{P_x Q_x}{n_x} + \dfrac{P_y Q_y}{n_y}}$

so don't bother to find square roots since you are about to square. Therefore,

$$\sigma_{(p_X-p_Y)} = \sqrt{\frac{P_X Q_X}{n_X} + \frac{P_Y Q_Y}{n_Y}}$$

Again note the + sign under the square root, even though this is a distribution of differences.

3. (Central Limit Theorem for Differences of Proportions:) The sampling distribution of differences of proportions is approximately normal if the samples are large enough, and the approximation gets better as the sample sizes increase. Bigger samples are needed if P_X and P_Y are close to 0 or 1. In practice, we shall assume the sampling distribution is normal if $n_X P_X, n_X Q_X,$ $n_Y P_Y$ and $n_Y Q_Y$ are all ≥ 15.

Example 1   In New Jersey, 54% of voters will vote for A for President, in Arizona only 47% will vote for A. Describe the sampling distribution of differences of proportions for samples of 400 voters from New Jersey and 900 from Arizona: How is it formed? What is its mean? What is the standard error of differences of proportions? Is it normal?

Take a sample of 400 voters from New Jersey, and find the proportion p_X who will vote for A; then take a sample of 900 voters in Arizona and find the proportion p_Y in that sample who will vote for A. $p_X - p_Y$ is the first score in the sampling distribution difference of proportions. Continue to choose independently all of the different pairs of samples, one from New Jersey of size 400 and one from Arizona of size 900, and find the difference of the proportions in each pair. The set of all these differences form the sampling distribution of difference of proportions for samples of size 400 and 900, respectively. The mean of this sampling distribution $= P_X - P_Y = .54 - .47 = .07$. Its standard error $= \sigma_{(p_X - p_Y)} =$
$$\sqrt{\frac{.54(.46)}{400} + \frac{.47(.53)}{900}} = .0300.$$ It is normal since 400(.54), 400(.46), 900(.47), and 900(.53) are all ≥ 15.

Now let's see how this sampling distribution is used to find confidence limits for the difference of proportions in two populations.

Example 2

Each of two groups consists of 100 patients who have leukemia. A new drug is given to the first group but not to the second (the control group). It is found that in the first group 75 people have remission for 2 years; but only 60 in the second group. Find 95% confidence limits for the difference in the proportion of all patients with leukemia who have remission for 2 years.

$$p_X = .75,\ q_X = .25,\ n_X = 100,\ \sigma_{p_X}^2 = \frac{p_X q_X}{n_X} = \frac{.75(.25)}{100} = .001875$$

$$p_Y = .60,\ q_Y = .40,\ n_Y = 100,\ \sigma_{p_Y}^2 = \frac{p_Y q_Y}{n_Y} = \frac{.60(.40)}{100} = .002400$$

$$\sigma_{(p_X - p_Y)} = \sqrt{\sigma_{p_X}^2 + \sigma_{p_Y}^2} = \sqrt{.001875 + .00240} = .065$$

At a 95% confidence level, $z = \pm 1.96$.

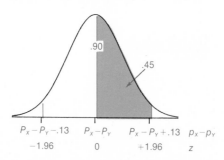

1.96 z units $= 1.96(.065) = .13$ difference-in-proportion units. The probability is .95 that the difference in the proportions in two randomly chosen, independent samples will be within .13 of the difference in the population proportions. These samples are random and independent, and their difference is $.75 - .60 = .15$, so I am 95% confident that .15 is within .13 of $P_X - P_Y$; a 95% confidence interval for the difference in the proportion with 2-year remission is $.15 - .13$ to $.15 + .13 = .02$ to .28.

As before, we have essentially solved the formula

$$z = \frac{(p_X - p_Y) - (P_X - P_Y)}{\sigma_{(p_X - p_Y)}} \text{ for } P_X - P_Y;$$

Confidence limits for $P_X - P_Y$:

$$(p_X - p_Y) \pm z\sigma_{(p_X - p_Y)}$$

where z is determined by the confidence coefficient and

$$\sigma_{(p_X - p_Y)} = \sqrt{\frac{p_X q_X}{n_X} + \frac{p_Y q_Y}{n_Y}}$$

In the example above, $(p_X - p_Y) \pm z\sigma_{(p_X - p_Y)} = (.75 - .60) \pm 1.96(.065) = .15 \pm .13 = .02$ or .28.

Example 3   The chairman of the Sociology Department is worried about grade inflation in his department. In a random sample of 100 students in sociology courses this year, 75 received A or B; in a sample of 120 grades in sociology 5 years ago, 66 were A or B. Find a 90% confidence interval for the difference in the percentage of A and B grades now and 5 years ago.

$$p_X = \frac{75}{100} = .75, \, n_X = 100, \, p_Y = \frac{66}{120} = .55, \, n_Y = 120.$$

$$\sigma_{(p_X - p_Y)} = \sqrt{\frac{.75(.25)}{100} + \frac{.55(.45)}{120}} = .0627$$

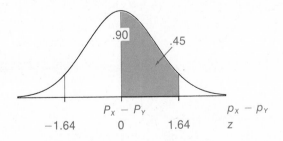

Confidence coefficient = .90, so $z = \pm 1.64$.

$$(p_X - p_Y) \pm z\sigma_p = .20 \pm 1.64(.0627) = .097 \text{ or } .303.$$

A 90% confidence interval for the grade inflation in the past 5 years is 9.7% to 30.3%.

9.7 SUMMARY OF FORMULAS FOR STATISTICAL INFERENCE

Inference About	z and Standard Error Formula	Confidence Limits
Mean		
If σ is known	$z = \dfrac{\bar{X} - \mu}{\sigma_{\bar{X}}}, \; \sigma_{\bar{X}} = \dfrac{\sigma}{\sqrt{n}}$	$\bar{X} \pm z\sigma_{\bar{X}}$
If σ is unknown	$z = \dfrac{\bar{X} - \mu}{\sigma_{\bar{X}}}, \; \sigma_{\bar{X}} = \dfrac{s}{\sqrt{n}}$	$X \pm z\sigma_{\bar{X}}$
Proportion		
If P is known	$z = \dfrac{p - P}{\sigma_p}, \; \sigma_p = \sqrt{\dfrac{PQ}{n}}$	
If P is unknown	$z = \dfrac{p - P}{\sigma_p}, \; \sigma_p = \sqrt{\dfrac{pq}{n}}$	$p \pm z\sigma_p$
Difference of means	$z = \dfrac{(\bar{X} - \bar{Y}) - (\mu_X - \mu_Y)}{\sigma_{(\bar{X}-\bar{Y})}}, \; \sigma_{(\bar{X}-\bar{Y})} = \sqrt{\sigma_{\bar{X}}^2 + \sigma_{\bar{Y}}^2}$	$(\bar{X} - \bar{Y}) \pm z\sigma_{(\bar{X}-\bar{Y})}$
Difference of proportions	$z = \dfrac{(p_X - p_Y) - (P_X - P_Y)}{\sigma_{(p_X-p_Y)}}, \; \sigma_{(p_X-p_Y)} = \sqrt{\sigma_{p_X}^2 + \sigma_{p_Y}^2}$	$(p_X - p_Y) \pm z\sigma_{(p_X-p_Y)}$

9.8 EXERCISES

1. Assume the mean height of adult men is 70 inches, with a standard deviation of 3 inches, and that the mean height of adult women is 64 inches with a standard deviation of 2.5 inches. Describe the sampling distribution of proportions for samples of size 60 and 50, respectively. (How is it formed? What is its mean? its standard error? its shape?)

2. What assumptions about (a) the populations, (b) the samples, must be satisfied if you are to find a confidence interval for the difference of means?

3. In a study* of audience interest in mass media messages about lung cancer in Vermont, interest in a variety of smoking concepts was rated. Among them were the following:

	Smokers ($n = 44$)		Non-Smokers ($n = 105$)	
	Mean	S. D.	Mean	S. D.
(a) Children affected by smokers	78.1	30.8	84.8	29.1
(b) Fumes from end of cigarette	73.2	32.1	87.3	24.7
(c) Smoke in public places harms people with lung problems	61.0	34.4	84.1	30.2

For each of these three concepts, find a 90% confidence interval for the difference in means of smokers and non-smokers.

---

*Worden, J. K., Sweeney, R. R., and Waller, J. A. "Audience Interest in Mass Media Messages About Lung Disease in Vermont, *American Journal of Public Health, 68,* 1978, pp. 378–382.

4. An economist finds that a random sample of 100 steelworkers have a mean weekly wage of $300 with a standard deviation of $50, while 80 automobile workers chosen at random have a mean wage of $250 with a standard deviation of $45. Find a 95% confidence interval for the difference in wages of steelworkers and automobile workers.

5. A political scientist finds the mean age of 50 voters who favor Jerry P. Goodman, a liberal, for Congressman is 36 with a standard deviation of 4 years, while 80 voters who favor Alexander J. Whitehead, a conservative, have a mean age of 46 years with a standard deviation of 5 years. Both are random samples. Find the differences in ages of those who favor the liberal and the conservative, at a 98% confidence level.

6. Professor Think taught the same psychology course 2 years in a row, the first year at 8 A.M. and the second year at 1 P.M. A random sample of 50 morning students had a mean quiz grade of 82 with a standard deviation of 8, while the mean was 87 with a standard deviation of 6 in a random sample of 60 from the afternoon class. Find a 95% confidence interval for the difference in grades between the two classes.

7. At Johnson University, 200 students chosen randomly were signed up for a mean of 15 credits per semester with a standard deviation of .4. At Jameson University, 150 students chosen randomly were taking an average of 16 credits per semester with a standard deviation of .3. What is the difference in mean number of credits taken per semester at the two universities, at a 90% confidence level?

8. The proportions in two populations are .60 and .50, respectively. Describe the sampling distribution of differences of proportions for samples of size 400 from the first population and 900 from the second. (How is it formed? What is its mean? its standard error? its shape?)

9. In the *American Journal of Psychiatry* study quoted earlier (see Exercise 10, page 184), it was reported that Berkowitz et al. treated 40 smokers by hypnosis and Spiegel treated 615. 25% of Berkowitz's patients and 19.7% of Spiegel's patients quit smoking for at least 6 months. Find a 95% confidence interval for the difference in the proportion of smokers who stopped smoking after treatment by the two men.

10. A random sample of 1,000 Democrats shows that 400 favor capital punishment, whereas 250 of a random sample of 500 Republicans favor capital punishment. Find a 90% confidence interval for the difference in the proportion of Democrats and Republicans who favor capital punishment.

11. A random sample of 800 adults and 400 teenagers at 9 P.M. on November 16 shows that 200 adults and 50 teenagers are watching TV. Find the difference in the proportion of all adults and all teenagers who were watching TV at that time, at a 95% confidence level.

12. In a random sample of 800 city residents, 480 favor day care centers, but only 200 of a random sample of 500 suburban residents favor day care centers. Find a 98% confidence interval for the difference in the proportion of urban and suburban residents who favor day care centers.

13. Study of a random sample of 1,200 male chemistry majors shows that 480 went to work for large corporations immediately after graduation, but the same is true of only 60 of 300 female chemistry majors, also a random sample. Find the difference in the proportion of men and women chemistry majors who get such jobs, with 90% confidence.

ANSWERS

1. Take a sample of 60 adult men and find the mean height of men in the sample ($\bar{X}$), then a sample of 50 adult women and find their mean height ($\bar{Y}$). The difference, $\bar{X} - \bar{Y}$, is the first score in the sampling distribution. Repeat this for all pairs of samples of sizes 60 and 50, respectively, to get all the scores. The mean of the sampling distribution is $70 - 64 = 6$ inches, the standard error is $\sqrt{\dfrac{3^2}{60} + \dfrac{2.5^2}{50}} = .52$. The sampling distribution of differences of means is normal since both samples are large (over 30), whether or not the populations are normal.

2. (a) Both populations must be normal unless the sample sizes are large (at least 30). (b) The samples must be random and independent, and the sample sizes must both be at least 30 if the standard deviations of the populations are not known.

3. (a) $78.1 - 84.8 \pm 1.64 \sqrt{\dfrac{30.8^2}{44} + \dfrac{29.1^2}{105}} = -6.7 \pm 1.64(5.44) = -15.6 \text{ to } +2.2.$

   (b) $73.2 - 87.3 \pm 1.64 \sqrt{\dfrac{32.1^2}{44} + \dfrac{24.7^2}{105}} = -14.1 \pm 1.64(5.41) = -23.0 \text{ to } -5.2.$

   (c) $61.0 - 84.1 \pm 1.64 \sqrt{\dfrac{34.4^2}{44} + \dfrac{30.2^2}{105}} = -23.1 \pm 1.64(5.96) = -32.9 \text{ to } -13.3.$

4. $(300 - 250) \pm 1.96 \sqrt{\dfrac{50^2}{100} + \dfrac{45^2}{80}} = \$36 \text{ to } \$64.$

5. $36 - 46 \pm 2.33 \sqrt{\dfrac{4^2}{50} + \dfrac{5^2}{80}}$; voters for the liberal are 8.1 to 11.9 years younger than voters for the conservative, at a 98% confidence level.

6. $82 - 87 \pm 1.96 \sqrt{\dfrac{8^2}{50} + \dfrac{6^2}{60}}$; the morning class has a mean grade 2.3 to 7.7 points lower than the afternoon class, with 95% confidence. (This may be caused by different students or different quizzes or different ways the professor teaches the class, not necessarily by the difference in the time the classes meet.)

7. $15 - 16 \pm 1.64 \sqrt{\dfrac{.4^2}{200} + \dfrac{.3^2}{150}}$; students at Johnson University are taking .94 to 1.06 fewer credits per semester than students at Jameson University, at a 90% confidence level.

8. Take samples of size 400 from the first and 900 from the second, and find the proportion in each sample; the difference is the first score in the sampling distribution of differences of proportions. Repeat this process for all pairs of samples, one from each population, to get the other scores. The mean of the sampling distribution is $.60 - .50 = .10$, the standard error is $\sqrt{\dfrac{.60(.40)}{400} + \dfrac{.50(.50)}{900}} = .030$. It is normal since $400(.6)$, $400(.4)$, and $900(.5)$ are all over 15.

9. $.25 - .197 \pm 1.96\sqrt{\dfrac{.25(.75)}{40} + \dfrac{.197(.803)}{615}} = .053 \pm .138 = -.085$ or $.191$. The propor-

tion of Berkowitz's patients who stop smoking for at least 6 months is 8.5% less to 19.1% more than the proportion of Spiegel's patients, with 95% confidence. The size of the sample of Berkowitz's patients is too small for this computation to be interesting; note that $n_X p_X = 10$, less than the 15 that gives a good normal approximation in the sampling distribution.

10. $\dfrac{400}{1,000} - \dfrac{250}{500} \pm 1.64\sqrt{\dfrac{.4(.6)}{1,000} + \dfrac{.5(.5)}{500}}$; from 6% to 14% fewer Democrats favor capital punishment at a 90% confidence level.

11. $\dfrac{200}{800} - \dfrac{50}{400} \pm 1.96\sqrt{\dfrac{(1/4)(3/4)}{800} + \dfrac{(1/8)(7/8)}{400}}$; 8.1% to 16.9%.

12. $\dfrac{480}{800} - \dfrac{200}{500} \pm 2.33\sqrt{\dfrac{(480/800)(320/800)}{800} + \dfrac{(200/500)(300/500)}{500}}$; 13.5% to 26.5% more city than suburban residents, with 98% confidence.

13. $\dfrac{480}{1,200} - \dfrac{60}{300} \pm 1.64\sqrt{\dfrac{.40(.60)}{1,200} + \dfrac{.20(.80)}{300}}$; the difference in the proportion of men and women chemistry majors who get such jobs is 1.56 to .244, with 90% confidence.

9.9 SIZE OF A SAMPLE

In the problems you have solved until now, the size of the sample has been stated. In real life, however, a research worker realizes that in general,* a larger sample will increase his accuracy—if the whole population is tested there is no need for statistical inference—but that testing a larger sample will take more time and cost more money. How large a sample is absolutely necessary for a given degree of accuracy is a very practical and necessary question for a statistician. How many people should be polled if a political scientist is to estimate within 1% and with 98% confidence the proportion who will vote for candidate X? How many rats should a psychologist test if he needs to know their mean reaction time within .6 second with 95% confidence?

Note that in each of these questions there is a statement of the expected accuracy of the answer ("within 1% ," "within .6 second"). Before sample size can be determined, the desired accuracy must be determined. But even when this is done, the desired accuracy cannot be guaranteed: "with 98% confidence" or "with 95% confidence" means that, for every 100 samples that might be chosen, we expect that the confidence intervals computed will in 2 or 5 cases **not** include the population parameter; in these cases, the desired accuracy will not be attained. The only way of guaranteeing the stated accuracy is to measure the whole population.

We shall take up determination of sample size for estimating a population mean and proportion. Do not get lost in the woods. There is a common pattern for both cases, and each involves the following steps:

---

* Surprisingly, there are a few very special cases when increasing the sample size will not increase the accuracy of the statistical inference. We shall not consider these cases, however.

1. Determine the required accuracy—or, to put it another way, decide what error E in the results is permissible. Do you need to know the population mean within 1 or 3 score units, or within .1 or .05 standard deviation units?
2. Set up an equation that involves both sample size n and the permissible error E. This equation will include population parameters that must be estimated.
3. Solve the equation for n.

9.10 SAMPLE SIZE FOR ESTIMATING A POPULATION MEAN

If I measure the length of this line ———————————— to the nearest inch, I might record its length as 2 inches $\pm$ .5 inches. This means the true length of the line is between 1.5 inches and 2.5 inches, or that the true length is 2 inches with a possible error of .5 inches (in either direction—that is, 2 inches may be too large or too small, but is within .5 inches of the actual length of the line).

Similarly, we have been finding confidence limits in the form $\mu = 70 \pm .8$; we have a given percent confidence that the population mean is 70 with a possible error of .8 in either direction. In the general case, $\mu = \bar{X} - z\sigma_{\bar{X}} = \bar{X} - z\dfrac{\sigma}{\sqrt{n}}$ where z has two possible values—one is positive, the other negative. If we give z only a positive value, then this equation can be written as $\mu = \bar{X} \pm |z|\dfrac{\sigma}{\sqrt{n}}$.* We have a given percent confidence that the error in the value of μ, then, is $|z|\dfrac{\sigma}{\sqrt{n}}$.

A psychologist, economist, or other scientist setting up a study decides the maximum error E she is willing to accept, and then solves the equation $|z|\sigma_{\bar{X}} = E$ to find the necessary sample size. But since $\sigma_{\bar{X}} = \dfrac{\sigma}{\sqrt{n}}$, this becomes

$$|z|\frac{\sigma}{\sqrt{n}} = E$$

$$z^2\frac{\sigma^2}{n} = E^2$$

$$n = \frac{z^2\sigma^2}{E^2}$$

Since n must be an integer (you can't have 44.2 scores), take the next larger whole number.

Example 1   You are a manufacturer of light bulbs, and know that the standard deviation of the lifetimes of your light bulbs is 100 hours. How large a sample must you test to be

———————————
*See page 69 for a reminder about absolute values.

95% confident that the error in the estimated mean life of your light bulbs is less than 10 hours?

$\sigma = 100$ hours; at a 95% confidence level, $z = \pm 1.96$; $E = 10$ hours.

$$n = \frac{z^2\sigma^2}{E^2} = \frac{(1.96)^2 100^2}{10^2} = 384.2; \; n = 385$$

Does this seem to be a very large number of light bulbs to test? It is because the error (10 hours) you are willing to allow is small. If the mean life of your bulbs is approximately 2,000 hours you are asking that it be determined within $\pm .5\%$. If you are willing to accept an error twice as large (20 hours), then the size of the sampling will be only one-fourth as large: you need test only 97 bulbs.

Example 2

You are still manufacturing light bulbs, but don't know the standard deviation of the lifetimes of your bulbs. How large a sample must you test to be 95% confident that the error in the estimated mean life of all your light bulbs is less than $.1\sigma$?

$$|z| \frac{\sigma}{\sqrt{n}} = .1\sigma = E$$

$$n = \frac{z^2\sigma^2}{E^2} = \frac{(1.96)^2\sigma^2}{(.1\sigma)^2} = \frac{3.842\sigma^2}{.01\sigma^2} = 384.2; \; n = 385$$

If σ is unknown, the appropriate sample size may be determined if the error is expressed in standard deviation units. The difficulty is that you don't know the error in score units until after you have chosen your sample, tested it, and determined s. Then $\sigma_X = \frac{\sigma}{\sqrt{n}} \approx \frac{s}{\sqrt{n}}$, so now σ (and therefore E) may be estimated: $\sigma \approx s$, and now $E = .1\sigma$ may be determined.

Example 3

Continue to manufacture light bulbs, again with σ unknown. How large a sample must you test to be 95% confident that the error in the estimated mean life of all your light bulbs is less than 10 hours?

$$n = \frac{(1.96)^2\sigma^2}{10^2}$$

help!!

Here the strategem that helped in the second example cannot be used. There are two unknowns, σ and n, and no knowledge of either. What is to be done? If you can estimate σ on the basis of previous experience, then you can use this estimated σ to determine n. If not, the answer is to make a small pilot study—test a sample of, say, 90 bulbs and determine s for that sample. Then use s as an approximation of σ. If 30 bulbs are tested and have a standard deviation of 90 hours, then

$$n \approx \frac{(1.96)^2 90^2}{10^2} = 311.2; \; n = 312$$

Note that the mean $\bar{X}$ of the pilot sample is not used.

You have already tested 30 and need to test 282 more. What if you don't have the time or money to test this many? You must either give up the experiment or you must realize that, if your sample is smaller, you cannot achieve the accuracy you had hoped for.

9.11 SAMPLE SIZE FOR PROPORTIONS

Using the same reasoning as for means, the confidence limits for P are given by $p \pm z\sigma_p$. The error term is $|z|\sigma_p$, so you would like to solve the equation

$$|z|\sqrt{\frac{PQ}{n}} = E \text{ for } n.$$

$$n = \frac{z^2 PQ}{E^2}$$

Again you have a population parameter, P, which is not known until after you have completed the experiment. There are three things you can do:

1. If you can estimate the range of P on the basis of previous experience, use the maximum value of PQ for P in this range. This means use for P that value in the range which is closest to 1/2. (If you think P is between .2 and .3, let $P = .3, Q = .7$.)

2. As before, you can make a pilot study, determine the proportion p in this sample, and use pq as an approximation to PQ in the equation, thus determining how much the sample should be enlarged.

3. The maximum value of PQ is 1/4. Algebra or calculus is needed to prove this, but a graph should be quite persuasive: recall that $PQ = P(1 - P)$.

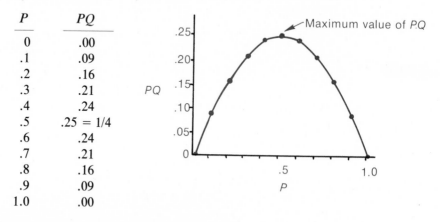

P	PQ
0	.00
.1	.09
.2	.16
.3	.21
.4	.24
.5	.25 = 1/4
.6	.24
.7	.21
.8	.16
.9	.09
1.0	.00

Since $PQ \leqslant 1/4$, then $n = \dfrac{z^2 PQ}{E^2} \leqslant \dfrac{z^2}{4E^2}$; use:

$$n = \frac{z^2}{4E^2}$$

as your estimate of n. Unless your sample proportion p turns out to be quite close to .5, the value of n you compute will be too large and you will do unnecessary work, as the next example illustrates. So use this method only if the first two cannot be used.

Example 1   A random sample of how many voters should be polled to be 95% confident that you know the percentage that will vote for candidate A for President, with an error less than 1%? Try to solve this before you read on.

At a 95% confidence level, $z \pm 1.96$; $E = .01$.

$$n = \frac{(1.96)^2}{4(.01)^2} = 9{,}604$$

If, on the other hand, a pilot study of 200 voters shows that only 10% will vote for A, then 3,458 is a large enough sample to be 95% confident that the error is less than 1%:

$$PQ \approx pq = .1(.9) = .09$$

$$n = \frac{(1.96)^2(.09)}{(.01)^2} = 3{,}458$$

9.12 EXERCISES

1.   In a test of driving skills at Morristown High School ten years ago, the mean score was 82.3 with a standard deviation of 9.0. How many students should be tested this year to find the mean score with 95% confidence that is known within 3 points?

2.   An anthropologist studying the height of adult natives of Madagascar wishes to know their mean height with 90% confidence that the error is less than 1 cm. How should he go about this?

3.   When the Deerfield History Motivation Test was used at Deerfield University the mean score was 75.2 with a standard deviation of 6.3. Russell Bingyou, a distinguished historian, has developed a new test of the same concepts. How many students should he test to know the mean within 2 points with 95% confidence?

4.   A patent medicine manufacturer wishes to know the proportion of people with eczema who will be relieved of symptoms for 10 hours or more after using his preparation. How many people with eczema should be tested to know this proportion within 5% with 90% confidence?

5.   A few years ago, 19% of those who needed glasses were fitted with contact lenses. You suspect the proportion has increased, and wish to determine the percent now with an error less than 4% and 95% confidence. How many people who wear glasses should be questioned?

6.   How large a sample should be tested to determine the proportion of those who stop smoking who gain at least 5 pounds in the first 6 months after stopping, with 90% confidence that the error is less than 4%?

ANSWERS

1. Assume σ is still 9.0; $\dfrac{(1.96)^2 9^2}{3^2} = 35$.

2. He can probably find similar studies of other groups of similar background and use an estimate of σ from one of these studies (or an average of several); if not, he must conduct a pilot study and use s from this as an estimate of σ. Once he estimates σ, his answer is

   $$\frac{(1.64)^2 \sigma^2}{1^2} = 2.69\ \sigma^2.$$

3. If the same range of grades (approximately) is expected on the new test, then assume $\sigma = 6$. $\dfrac{(1.96^2)6^2}{2^2} = 35$.

4. $\dfrac{(1.64)^2}{4(.05)^2} = 269$.

5. Estimate how much P has increased; use $P = .30$, say. Then $n = \dfrac{(1.96)^2(.30)(.70)}{(.04)^2} = 504$.

   If, after enquiring of 504 people, you find that P is closer to .5 you can determine n again and ask more.

6. $\dfrac{(1.64)^2}{4(.04)^2} = 420$.

9.13 VOCABULARY AND NOTATION

P	standard error of differences
Q	of means
p	$\sigma_{(\bar{X}-\bar{Y})}$
q	$\sigma_{(p_X-p_Y)}$
σ_p	standard error of differences
standard error of proportion	of proportions
	error term.

9.14 REVIEW EXERCISES

1. The proportion of voters who will vote for X for Senator is .60. Describe the sampling distribution of proportions for samples of size 400. How is it formed? What is its mean? its standard error? its shape?

2. A sampling distribution of proportions for samples of size 900 has a mean of .80. (a) What is its standard error? (b) Is it normally distributed? (c) What can you say about the population from which it comes?

3. If a random sample of size 1,600 is taken from a population in which $P = .50$, what is the probability that the proportion in the sample is between .49 and .52?

4. If 60% of the registered voters in New Jersey are Independents, what is the probability that in a random sample of 500 New Jersey voters between 290 and 320 are Independents?

5. The editors of *The Eagle Rises,* student newspaper of Oxo College, question a random sample of 200 subscribers about the editorials in the paper. 155 subscribers express dissatisfaction. At a 90% confidence level, what proportion of subscribers are dissatisfied with the editorials?

6. In a random sample of 100 city physicians, 77 have an income over $50,000. Find a 95% confidence interval for the proportion of all city physicians who have incomes over $50,000.

7. A drug treatment center tests a new withdrawal repressant on 100 heroin users, chosen at random. Of these, 36 patients display marked relief from withdrawal symptoms. Find a 95% confidence interval for the proportion of all addicts who would find relief.

8. A random sample of 150 General Motors workers shows that 105 own a General Motors vehicle. What proportion of all General Motors employees own a General Motors vehicle, at a 90% confidence level?

9. S. L. George and others* find that when therapy is stopped for 255 children who have had remissions from acute lymphocytic leukemia for $2\frac{1}{2}$, years, 55 have relapses. Find a 95% confidence interval for the proportion of all such children who have relapses.

10. Two populations have means of 83.0 and 80.5 and standard deviations of 4.0 and 5.1, respectively. Describe the sampling distribution of differences of means for samples of size 20 from the first and 12 from the second population. How is it formed? What is its mean? its standard error? its shape?

11. What is the probability that the difference in means of 2 random samples each of size 100 chosen at random from the populations described in Exercise 10 is between 2.0 and 3.1?

12. The mean age of 50 listeners to radio station WOLD is 62 with a standard deviation of 12, while the mean age of 60 persons listening to station WZAP is 19 with a standard deviation of 3. Both samples are random. Find a 95% confidence interval for the difference in ages of listeners to the two radio stations.

13. The mean length of 50 Beatle songs chosen randomly is 3 minutes with a standard deviation of 30 seconds. A random sample of 55 Rolling Stones' songs has a length of 4 minutes 6 seconds with a standard deviation of 45 seconds. At a

---

*George, S. L., Aur, R. J., Mauer, A. M., and Simone, J. V. "A Reappraisal of the Results of Stopping Therapy in Childhood Leukemia," *The New England Journal of Medicine, 300,* No. 6, 1979, pp. 269–273.

98% confidence level, what is the difference in song length between all songs sung by the Beatles and by the Rolling Stones?

14. One hundred basketball players have a mean height of 6.3 feet, with a standard deviation of .3, while 100 jockeys have a mean height of 4.9 feet with a standard deviation of .2. Find, with 98% confidence, the difference in heights of all basketball players and all jockeys.

15. Describe the sampling distribution of differences of proportions for samples of size 100 from a population in which the proportion who have a college education is .40, and of size 900 from a population in which this proportion is .30.

16. What assumptions must be made about (a) the populations and (b) the samples if a confidence interval for the difference of proportions is to be meaningful?

17. What is the probability that the difference of proportions of two samples of sizes 100 and 900 from the populations described in Exercise 15 is between (a) .05 and .08, (b) more than .12?

18. Harburg and others found that the mean systolic blood pressure of 171 Detroit whites whose fathers came from northern Europe was 126.5 with a standard deviation of 15.0, while 82 Detroit whites whose fathers came from the Mediterranean had a mean systolic blood pressure of 120.4, with a standard deviation of 12.7. Find a 90% confidence interval for the difference in systolic blood pressure of Detroit whites whose fathers came from these two regions.

19. In 1975, 1,210 state and local government employees were hired in Georgia, of whom 356 were black. Among these new jobs, 309 were administrative, professional, or technical, and 42 of these were filled by blacks. Find a 95% confidence interval for the difference in proportion of blacks in all jobs and in administrative, professional, and technical jobs.

20. Two countries plan to form one political unit, and appropriate resolutions are passed unanimously by the general assemblies in each country. In random polls of 1,000 people in each country, those in one country approve the proposal 9 to 1, but in the other country the choice is 3 to 1 for approval. Find the difference, with 95% confidence, in the proportion of those who favor the merger.

21. At a national fraternity convention, random samples were taken of 330 delegates from western states and 300 from eastern states. Among the westerners who were questioned, 231 wanted an increase in the number of chapters, but 180 easterners favored the opposite, a consolidation of chapters. Find a 95% confidence interval for the difference in the proportion of fraternity members who prefer consolidation of chapters.

22. An irate citizen passed around a petition to 212 homeowners and 44 store owners. Only 5.5% of the store owners signed, but 89.2% of the homeowners whom he approached did so. His petition demanded closing of their street to trucks. At a 95% confidence level, find the difference in the proportion of home-owners and merchants who favor his proposal.

23.   The makers of HELTHPUP, a diet supplement for dogs, wish to determine the mean gain in weight per month of 1-year-old Eskimo dogs whose diet includes HELTHPUP, with 95% confidence that the error is less than 2.5 ounces. It is known that the standard deviation of weight gain for such dogs is 6 ounces. How many dogs should be tested?

24.   The chairman of the French department at Norwich College wants to compare students in his department with those from other colleges who are studying French. A national examination has a mean of 500 and a standard deviation of 100. To how many of his students should he give the test if he wishes to know their mean grade within 35 points, with 95% confidence?

25.   At present only 60% of patients with leukemia survive 6 years. A doctor has developed a new drug, and wants to know with 95% confidence the 6-year survival rate with an error of less than 10%. On how many patients should he test the new drug?

ANSWERS

1.   Find the proportion p in each different sample of size 400; the set of p's from all these samples form the sampling distribution of proportions for samples of size 400. Its mean is .60; $\sigma_p = \sqrt{\dfrac{.60(.40)}{400}} = .0245$. It is normal, since .60(.400) and .40(400) are over 15.

3. $\sigma_p = \sqrt{\dfrac{.50(.50)}{1600}} = .0125, \quad z = \dfrac{p - .50}{.0125}.$ $\Pr(.49 < p < .52) = \Pr(-.80 < z < 1.60) = .288 + .445 = .73.$

5. $p = \dfrac{155}{200} = .775; .775 \pm 1.65\sqrt{\dfrac{.775(.225)}{200}};$ with 90% confidence, 73% to 82% of the subscribers are dissatisfied with the editorials. (The editors must be careful that the sample is random; voluntary responses are probably biased toward dissatisfaction.)

7. $p = \dfrac{36}{100} = .36, .36 \pm 1.96\sqrt{\dfrac{.36(.64)}{100}};$ a 95% confidence interval for the proportion of heroin addicts who will find relief is .27 to .45. Be careful not to generalize about all addicts if only heroin addicts have been tested.

9. $p = \dfrac{55}{255} = .216, .216 \pm 1.96\sqrt{\dfrac{.216(.784)}{255}} = .165$ to .267.

11. $\mu_X - \mu_Y = 83.0 - 80.5 = 2.5, \quad \sigma_{(\bar{X}-\bar{Y})} = \sqrt{\dfrac{4.0^2}{100} + \dfrac{5.1^2}{100}} = .648, \quad z = \dfrac{(\bar{X} - \bar{Y}) - 2.5}{.648};$
$\Pr(2.0 < \bar{X} - \bar{Y} < 3.1) = \Pr(-.77 < z < .93) = .279 + .324 = .60.$

13. $(3 - 4.1) \pm 2.33\sqrt{\dfrac{.5^2}{50} + \dfrac{.75^2}{55}};$ I have 98% confidence that Beatle songs take .81 to 1.39 minutes (49 to 83 seconds) less time than those sung by the Rolling Stones.

15. It consists of the difference of proportions for each pair of different samples from the two populations, the first of size 100 and the second of size 900. Its mean is $.40 - .30 = .10$, its standard error is $\sqrt{\dfrac{.40(.6)}{100} + \dfrac{.30(.70)}{900}} = .0513$, and it is normally distributed, since $.4(100)$ and $.3(100)$ are both over 15.

17. $\sigma_{(p_X - p_Y)} = .0513$ (see exercise 15). $z = \dfrac{(p_X - p_Y) - (.40 - .30)}{.0513}$;

    (a) $\Pr(.05 < p_X - p_Y < .08) = \Pr(-.97 < z < -.39) = .331 - .152 = .18$;

    (b) $\Pr(p_X - p_Y > .12) = \Pr(z > .39) = .500 - .152 = .35$.

19. Nonsense! These are data about all employees hired by state and local government departments in 1975 in Georgia, so there's no need for statistical inference.

21. $P_X = \dfrac{330 - 231}{330} = .30$. Notice the form of the question; P_X is the proportion of westerners who prefer consolidation (and oppose an increase). $P_Y = \dfrac{180}{300} = .60$;

    $\sigma_{(p_X - p_Y)} = \sqrt{\dfrac{.30(.70)}{330} + \dfrac{.60(.40)}{30}} = .0379$; $(.30 - .60) \pm 1.96(.0379) = -.226$ to $-.374$.

    With 95% confidence, 22.6 to 37.4% fewer westerners than easterners prefer consolidation.

23. $n = \dfrac{z^2 \sigma^2}{E^2} = \dfrac{1.96^2(6^2)}{2.5^2} = 23$ dogs (rounding up).

25. Assume $P = .60$. Then $n = \dfrac{z^2 PQ}{E^2} = \dfrac{1.96^2(.60)(.40)}{.10^2} = 93$. If no information about P for the new drug is assumed from present treatment, then use $PQ = .25$ and $n = \dfrac{1.96^2(.25)}{.10^2} = 97$.

10 HYPOTHESIS TESTING

ALL CH. 10

In this chapter you will learn how to state and test a hypothesis about the population mean or proportion, or about the difference between the means or proportions of two populations. You will learn how to formulate an alternative hypothesis which is accepted if your original one is rejected, how to decide which one to accept, and what is the probability of error. This is not difficult if you realize early that you are simply using your old friends, the sampling distributions, for a different purpose.

10.1 INTRODUCTION

In Chapters 8 and 9 you learned how to make an inference about the population mean or proportion from knowledge of the mean or proportion of a sample. For example, if a random sample of size 50 has mean 100 and standard deviation 5, we can then claim with 95% confidence that the interval 98.6 to 101.4 includes the population mean.

Now the emphasis will be turned around. We shall make a guess or assumption or **hypothesis** about the population mean or proportion (or about the difference between the means or proportions of two different populations), and then test that hypothesis by looking at a sample (or samples) drawn at random from the population(s). For example, we might form the hypothesis that a coin is fair (that $P = 1/2$). If you flip the coin 100 times and it comes up heads 50 times you will certainly accept the hypothesis that $P = 1/2$. If heads turn up 49 times you will probably say, "I expect some fluctuation between samples. If I flip the coin another 100 times, maybe it will come up heads 51 times. So I still accept the hypothesis that $P = 1/2$." But if heads turn up only once in 100 flips, you would no doubt be pretty suspicious of the coin and reject the hypothesis that $P = 1/2$. Note, however, that if $P = 1/2$ (if the coin is fair), the probability is not 0 that heads will turn up exactly once in 100 throws. It is $\dfrac{100!}{99! \ 1!} (1/2)^1 (1/2)^{99}$, which is an extremely small number but *not* 0. There is a very small probability that you are making an error if you reject the hypothesis that $P = 1/2$ when heads turn up exactly once. Intuitively, you probably accept the hypothesis that $P = 1/2$ if 49 or 50 or 51 heads turn up in 100 throws; you reject it if 1 or 0 or 99 heads turn up. What do you do if heads turn up 37 times? 59 times? 43 times? It is clear that a method of making the decision whether or not to accept the hypothesis $P = 1/2$ must be decided upon.

In general, hypothesis testing in statistics involves the following steps:

1. Choose the hypothesis that is to be questioned.
2. Choose an alternative hypothesis which is accepted if the original hypothesis is rejected.
3. Choose a rule for making a decision about when to reject the original hypothesis and when to fail to reject it.
4. Choose a random sample from the appropriate population and compute appropriate statistics; that is, mean, variance, and so on.
5. Make the decision.

You are already familiar with the fourth step. The fifth step will be simple if you thoroughly understand the first four.

10.2 CHOOSING THE HYPOTHESIS TO BE TESTED (H_0)

The hypothesis you wish to test is called the **null hypothesis,** since acceptance of it commonly implies "no effect" or "no difference." It is denoted by the symbol H_0. (Read "H naught"; "naught" is a word about as old-fashioned as a horse and carriage. Both have their uses on certain occasions, and here is the place for "naught," which means "zero.")

In the coin-flipping experiment in the preceding section, the hypothesis that the coin is fair may be expressed symbolically:

$$H_0: P = \frac{1}{2}$$

H_0 is **always** a statement about a parameter [mean, proportion, etc., of population(s)]. It is not about a sample, nor are sample statistics used in formulating the null hypothesis. H_0 might be $\mu = 100$ or $\mu_1 - \mu_2 = 4$ or $P = .6$ or $P_1 - P_2 = 0$, but **not** $\bar{X} = 4$, $p = .36$, $\bar{X}_1 - \bar{X}_2 = 0$, etc. "A random sample of size 50 has $\bar{X} = 100$, $s = 4.5$. At a .05 level of significance,* can μ be 101.5?" The numbers 100 and 4.5 will **never** occur in H_0. Here, H_0 is "$\mu = 101.5$." H_0 is usually easy to determine.

H_0 is an equality ($\mu = 14$) rather than an inequality ($\mu \geq 14$ or $\mu < 14$). For the present there are four possible choices of H_0:

$$\mu = \underline{\quad}$$

$$P = \underline{\quad}$$

$$\mu_X - \mu_Y = \underline{\quad}$$

$$P_X - P_Y = \underline{\quad}$$

where the blanks are filled in by constants. Why should H_0 be an equality? The reason is that H_0 usually presents the mean or proportion which is claimed or which is the case under present treatment. If you are handed a fair coin with which you are to gamble, you question the claim that it is fair; H_0 is "$P = 1/2$." Study the following examples, and more will be said about this in Section 10.3.

Example 1   You manufacture hammocks and buy ropes for them from the Nylon Coil Company. A sample of 36 ropes chosen at random from a Nylon Coil delivery has a

---

*The meaning of the phrase "level of significance" will be explained later (Section 10.5).

mean breaking strength of 127 pounds, with a standard deviation of 5 pounds. The desired mean breaking strength is 130 pounds. Should you accept the shipment, at a .05 level of significance?

What is H_0?

You are questioning the Nylon Coil Company's claim that the breaking strength is 130 pounds; H_0 is $\mu = 130$ pounds.

If you answered $\bar{X} = 127$ you forgot that H_0 is an assumption about populations; if $\mu = 127$, you forgot that sample figures do not determine the form of the null hypothesis; if $\mu \geqslant 130$, you forgot our convention that inequalities are not allowed in H_0.

Example 2

At present only 60% of patients with leukemia survive 6 years. A doctor develops a new drug. Of 40 patients, chosen at random, on whom the new drug is tested, 26 are alive after 6 years. Is the new drug better than the former treatment (.10 level of significance)?

You are questioning whether the proportion of patients who recover under the new treatment is still .60 (and hope that it will be improved; this hope will be reflected in our choice of H_1 in the next section). The correct answer is H_0: $P = .60$.

$p = 26/40$ refers to a sample; $P = 26/40$ depends on sample values; $P \leqslant .60$ is an inequality.

Example 3

In developing new strains of rice, a botanist has gradually improved the yield in Madras, India, by 30 bushels per acre. Now she thinks she has a better strain which will improve the yield by even more than 30 bushels per acre. Seeds of the new strain and of the kind presently sown in Madras are randomly selected, and 60 acres of roughly the same fertility, moisture, and exposure are divided into one-acre plots. Thirty of these are randomly chosen; when planted with the new strain they produce 180 bushels per plot with a standard deviation of 4 bushels. The remaining 30 acres, planted with the old strain, produce an average of 148 bushels per plot with a standard deviation of 5 bushels. Does the new strain increase production by more than 30 bushels per acre, at a .05 level of significance?

Let the new strain be X, the old Y. $\mu_X - \mu_Y = 32$ uses sample values, $\bar{X} - \bar{Y} = 32$ refers to samples, $\mu_X \geqslant \mu_Y + 30$ is an inequality, $\mu_X = \mu_Y$ is not the hypothesis that is to be tested. The correct answer is H_0: $\mu_X - \mu_Y = 30$.

Example 4

A machine fills milk bottles; the mean amount of milk in each bottle is supposed to be 32 ounces with a standard deviation of .06 ounces. In a routine check to see that the machine is operating properly, 36 filled bottles are chosen at random and found to contain a mean of 32.03 ounces. At a .05 level of significance, is the machine operating properly?

Look only at the information about the population and at the question being asked when deciding on H_0. H_0: $\mu = 32$.

Example 5

For some time past, the mean salary of professors in public colleges has been $1,000 more than in private colleges. A random sample of 200 public college professors shows a mean salary of $25,000 with a standard deviation of $1,000, while 100 private college professors selected at random have a mean salary of $23,700 with a standard deviation of $900. At a .01 level of significance, is the mean salary of public college professors now greater by more than $1,000 than that of private college professors?

H_0: $\mu_X - \mu_Y = 1,000$. (Assume X refers to all public college professors, Y to all private college professors.)

10.3 CHOOSING THE ALTERNATIVE HYPOTHESIS (H_1)

The notation H_1 (read "H one") is used for the hypothesis that will be accepted if H_0 is rejected. H_1 must also be formulated before a sample is tested, so it, like H_0, does not depend on sample values. If the fairness of a coin is questioned, H_0 is "$P = 1/2$". H_1 probably is "$P \neq 1/2$", but other alternatives are "$P < 1/2$", "$P > 1/2$", or "$P = .1$". An example of the last will be given very shortly (page 213). With that exception the hypotheses that we shall consider are severely limited, since in practice the alternative hypothesis cannot be specified exactly. Here are the possible choices:

if H_0 is	then H_1 is
$\mu = A$	$\mu \neq A$ or $\mu < A$ or $\mu > A$
$P = B$	$P \neq B$ or $P < B$ or $P > B$
$\mu_X - \mu_Y = C$	$\mu_X - \mu_Y \neq C$ or $\mu_X - \mu_Y < C$ or $\mu_X - \mu_Y > C$
$P_X - P_Y = D$	$P_X - P_Y \neq D$ or $P_X - P_Y < D$ or $P_X - P_Y > D$

A, B, C, and D are constants.

If you are alert and intelligent you are asking, "Why do I have to choose an alternative hypothesis before I test the one I've got? Why can't I accept or reject the null hypothesis, and think of an alternative only if it is rejected?" The most important reason is that H_1 expresses your hopes (or fears) that the claim is rejected, that a new treatment is better, that you have found a statistically significant change from the presently accepted state of affairs. It is when H_0 is rejected and H_1 is accepted that you have found evidence for an original paper in a journal or found something worthy of further investigation. Another reason is that the form of the alternative hypothesis will affect the decision to continue to accept or to reject the null hypothesis. This matter will be discussed in detail in Section 10.5. Look again at the examples in the preceding section (Section 10.2); now try to determine H_1 in each example.

Answers

Example 1. H_0: $\mu = 130$; H_1: $\mu < 130$. If the sample mean is over 130, you will be happy to accept the shipment. But if the sample mean is much lower than the expected population mean—that is, lower not just because of sampling fluctuations, but because it is drawn from a whole shipment whose mean is less than 130—then you will reject the shipment.

Example 2. H_0: $P = .60$; H_1: $P > .60$. The doctor is trying to reach a decision on whether to make further tests on the new drug. If the proportion of patients who live at least 6 years is not increased under the new treatment or is increased only by an amount due to sampling fluctuation, he will look for another drug. But if the proportion who are aided is significantly larger—if he is able to conclude that the population proportion is greater than .60—then he will continue his tests.

Example 3. $H_0: \mu_X - \mu_Y = 30; H_1: \mu_X - \mu_Y > 30$. Here μ_X is the mean yield per acre of the new strain. Unless it is more than 30 bushels per acre higher than the yield of the strain used now, the botanist will continue with the strains he already has and will not encourage the expense of producing large quantities of the new strain or the trouble of inducing farmers to use it. If, however, production is increased by significantly more than 30 bushels per acre, he will be willing to encourage the expense and trouble.

Example 4. $H_0: \mu = 32; H_1: \mu \neq 32$. Look only at the information about the population and at the question being asked when deciding on H_0 and H_1. You are deciding whether the amount of milk differs from 32 ounces, and will get the machine adjusted if it is too high or too low.

Example 5. $H_0: \mu_X - \mu_Y = 1,000; H_1: \mu_X - \mu_Y > 1,000$.

10.4 EXERCISES

State H_0 and H_1 for each of the following:

1. Is the average height of New York policemen 5 foot 11 inches or is it more?

2. Is the average height of New York policemen 5 foot 11 inches or is it something different?

3. Do cats have 9 lives, on the average, or do they have less than 9?

4. Do men and women get lung cancer in equal proportions, or do a higher proportion of men get lung cancer?

5. Is the mean height of men just 5 inches greater than that of women?

6. Is the mean height of men 5 inches greater than that of women, or is it greater by more than 5 inches?

7. A home owner needs to determine the pH of her soil; she will add lime if it is less than 6.5. She takes 36 samples from cores whose location is randomly chosen, and finds the average pH of the samples is 6.3 with a standard deviation of .2. Must she add lime to the soil?

8. A random sample of 42 male freshmen who live in a dormitory are weighed when they enter college in September and again on April 1. Have male freshmen in this dormitory gained weight between September and April?

9. The Speed Reading Institute claims that adults will read 400 words per minute faster after completing its course. A random sample of 24 adults are found to increase their reading speed by 385 words per minute, with a standard deviation of 50 words per minute. Is the Institute's claim substantiated?

10. Two countries plan to form one political unit. In random polls of 400 people in each country, 300 people in one and 256 in the other approve the proposal. Are the proportions who favor the proposal the same in the two countries?

11. 400 of 1,000 Democrats who are polled favor capital punishment, and 350 of 700 Republicans favor it. Do fewer Democrats favor capital punishment?

ANSWERS

1. H_0: $\mu = 5$ foot 11 inches; H_1: $\mu > 5$ foot 11 inches.

2. H_0: $\mu = 5$ foot 11 inches; H_1: $\mu \neq 5$ foot 11 inches.

3. H_0: $\mu = 9$; H_1: $\mu < 9$.

4. H_0: $P_M - P_W = 0$; H_1: $P_M - P_W > 0$.

5. H_0: $\mu_M - \mu_W = 5$; H_1: $\mu_M - \mu_W \neq 5$.

6. H_0: $\mu_M - \mu_W = 5$; H_1: $\mu_M - \mu_W > 5$.

7. H_0: $\mu = 6.5$; H_1: $\mu < 6.5$.

8. H_0: $\mu_X - \mu_Y = 0$? But then you will need the sampling distribution of differences of means, and the samples have to be chosen independently. But here you have two measurements on each member of one sample. H_0: the mean of the differences in weight = 0; H_1: the mean of the difference (April minus September weight) is greater than 0.

9. H_0: $\mu = 400$; H_1: $\mu < 400$.

10. H_0: $P_X - P_Y = 0$; H_1: $P_X - P_Y \neq 0$.

11. H_0: $P_D - P_R = 0$; H_1: $P_D - P_R < 0$.

10.5   CHOOSING A RULE FOR MAKING A DECISION

A method for making a decision must be agreed upon. Should H_0 be rejected or not? If H_0 is rejected, then H_1 is accepted. Every sample chosen at random will not have the same mean or proportion as the population from which it is taken. It seems reasonable still to accept the null hypothesis if a sample value is relatively close to the value predicted for it if the null hypothesis is true. But if the outcome of tests on a sample is highly improbable on the basis of sheer chance, we say the difference is significant and reject the null hypothesis. The method of making a decision should be decided on **before** the test on a sample is carried out.

Note that the words "the hypothesis is true" are always preceded by the word "if." In the seventeenth century, Newton formulated a law (or hypothesis) about gravitation. It was tested for over 200 years and always appeared to be true. Then Einstein developed his theory of relativity which suggested further tests, and it was found necessary to reject Newton's hypothesis under certain conditions. When a scientific test rejects a hypothesis, we conclude the hypothesis is false and look for an alternative. But if the test appears to confirm the hypothesis, then we fail to reject it—but that doesn't necessarily prove it to be true. Statistical tests are similar: we shall either "reject" or "fail to reject" the null hypothesis. (But sometimes, when "fail to reject" is awkward or has been repeated too many times, we shall "continue to accept" or "still accept.")

But how is a "significant" difference defined? A null hypothesis is either true or false, and it is either rejected or not rejected. No error is made if it is true and we fail to reject it, or if it is false and rejected. An error is made, however, if it is true but rejected, or if it is false and we fail to reject it.

> **Definitions: A Type I error** is made when H_0 is true but is rejected.
> **A Type II error** is made when H_0 is false but we fail to reject it.

Time out to learn two more Greek letters:
α (alpha) is the Greek lower case a.
β (beta) is the Greek lower case b.

> **Notation:** α is the probability of a Type I error; α is called the **level of significance.**
> β is the probability of a Type II error.

The following table summarizes these relationships.

		If H_0 is	
		True	False
When we	Fail to Reject	No error	Type II error (β)
	Reject	Type I error (α)	No error

Suppose you plan to test whether a coin is fair ($H_0: P = 1/2$; $H_1: P \neq 1/2$) by tossing it 4 times. You must decide on what basis H_0 should be rejected or still accepted. You will soon learn how to make a wise decision. At the moment, let us consider four different arbitrary choices and the effect of each on α and β if H_1 is (i) $P = .3$ or (ii) $P = .1$.

Example 1   Find α and β when a coin is tossed 4 times if H_0 is $P = .5$ and H_1 is either $P = .3$ or $P = .1$, with each of 4 decision choices: H_0 is never rejected, rejected if 0 or 4 heads come up, rejected when 0, 1, 3, or 4 heads come up, or always rejected.

	(a)	(b)	(c)	
			β	
Reject H_0	α	$H_1: P = .3$	$H_1: P = .1$	
Never	0	1.0	1.0	(Note 1)
If 0 or 4 heads	.125	.753	.344	(Note 2)
If 0, 1, 3 or 4 heads	.625	.265	.049	(Note 3)
Always	1.00	0	0	(Note 4)

Note 1: Since you never reject H_0, the probability is 0 that it is rejected when true. $\beta = 1$ since you continue to accept H_0 whether it is true or false, so you don't reject it when false.

Note 2: $\alpha = $ Pr(getting 0 or 4 heads in 4 tosses if $P = .5$) $= .062 + .062 = .125$. (See Table 2 in Appendix C.) If $P = .3$, $\beta = $ Pr(1, 2, or 3 heads when $P = .3$) $= .412 + .265 + .076 = .753$. If $P = .1$, Pr(1, 2, or 3 heads when $P = .1$) $= .292 + .049 + .004 = .345$.

Note 3: $\alpha = $ Pr(0, 1, 3, or 4 heads when $P = .5$) $= 1 - $ Pr(2 heads when $P = .5$) $= 1 - .375 = .625$. If $P = .3$, $\beta = $ Pr(2 heads when $P = .3$) $= .265$. When $P = .1$, $\beta = $ Pr(2 heads when $P = .1$) $= .049$.

Note 4: $\alpha = 1.00$ since H_0 is always rejected, even when true. $\beta = 0$ since you always reject H_0, so you never continue to accept it even when it is false.

This example should have renewed your acquaintance with the binomial distribution, but also—and far more interesting—should illustrate some important points.

Note the following:

1. Once H_0 and H_1 are chosen, then a decision choice has to be made: how is one to decide whether or not to reject the null hypothesis? This, you will discover, is a straightforward matter for the type of problems you will be asked to solve. In this example, arbitrary decision choices were made to show the effect of different choices on α and β.

2. α can be easily determined. β can be determined **only** if H_1 is specified exactly ($P = .3, P = .1$, etc.) and varies as this specific choice varies; one value of β cannot be determined when H_1 is simply the general statement "$P \neq 1/2$." (The exceptions are cases in which H_0 is always accepted or never accepted, no matter what the results of the experiment—then there is no need to carry out the experiment at all.)

3. As α increases, β decreases. Compare column (a) in turn with columns (b) and (c): As the probability of rejecting H_0 when it is true increases, the probability of not rejecting H_0 when it is false decreases. It is not true in general, however, that $\beta = 1 - \alpha$. α is determined by assuming the population has a binomial distribution with $P = .5$, while β is computed under different assumptions about the population distribution (that $P = .3, .1$, respectively). The relation between α and β is much more complicated; this will be discussed in Section 10.9.

In this example, various decision choices were made arbitrarily: it was decided to fail to reject H_0 if the number of heads is between two limits, and to reject H_0 if the number of heads is outside those limits. Then α was determined for each decision choice. In practice, however, the emphasis is put the other way around: α is chosen arbitrarily and then limits for continuing to accept H_0 are determined; if a sample statistic is outside those limits, H_0 is rejected (and H_1 is accepted).

The form of H_1 will determine the kind of limits to be set up. Let us first consider how to make a decision about rejecting or failing to reject H_0 when H_1 includes the symbol "$\neq$": when H_1 is $\mu \neq \underline{}$, $P \neq \underline{}$, $\mu_X - \mu_Y \neq \underline{}$, or $P_X - P_Y \neq \underline{}$, where the blank is filled in with a constant. Then a decision is made in the following fashion:

1. α is arbitrarily chosen, equal to a small number (usually .01, .05, or .10). On page 227 this choice will be discussed. α is called the **level of significance.**

2. z values are determined so that the area in each of the tails of the normal distribution is $\alpha/2$. The most common values of z are, then,

α	.10	.05	.01
z	± 1.64	± 1.96	± 2.58

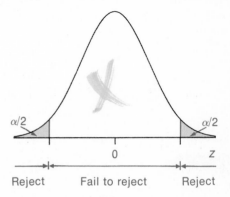

The form of H_1 we are considering now leads to a "two-tail test"; in Section 10.7 we shall find that when H_1 is of the form $\mu < \underline{}$, $P > \underline{}$, $\mu_X - \mu_Y > \underline{}$, etc., then a "one-tail test" will be used. Remember that H_1 is decided on before the experiment is carried out, and therefore a two-tail or one-tail test is also decided upon before the experiment is carried out.

3. The experiment is carried out and the z value of the appropriate sample statistic ($\bar{X}$, p, $\bar{Y} - \bar{Y}$, or $p_X - p_Y$) is determined. If this computed z value falls within the limits determined in Step 2 above, we fail to reject H_0; if the computed z value is outside those limits, H_0 is rejected (and H_1 is accepted). Since they separate the "fail to reject" and "reject" regions, the limits determined in Step 2 will be referred to as **the critical values of z.**

Study the following examples with care, to discover the logic behind this method of making a decision.

Example 1 Assume that the mean IQ of all high school seniors is 110 with a standard deviation of 10. A random sample of 49 seniors at Swampscott High School has a mean IQ of 112. Does the mean IQ of all seniors at Swampscott High School differ* from that of high school students in general, at a .05 level of significance?

---

*Questions such as "Is the mean IQ of Swampscott students **higher?**" will be discussed in Section 10.7.

H_0: $\mu = 110$. (The population consists of all seniors at Swampscott High School. The hypothesis to be tested states that their mean IQ is the same as the mean IQ of all high school seniors.)

H_1: $\mu \neq 110$

$\alpha = .05$

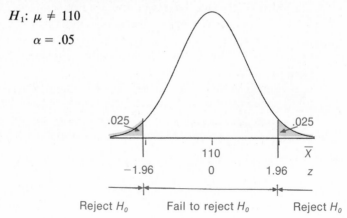

Sample: $n = 49$, $\bar{X} = 112$

Look at the sampling distribution of means, with mean of 110 and standard error $= \dfrac{\sigma}{\sqrt{n}} = \dfrac{10}{\sqrt{49}} = \dfrac{10}{7}$. The probability of rejecting H_0 when it is true is to be .05. The area of each shaded "tail" is .025, and the corresponding z scores at the boundaries are ± 1.96 (see Table 4, Appendix C). This means that H_0 is to be rejected if the z score of a sample mean is greater than 1.96 or smaller than -1.96: the probability of getting such a sample is so small (less than 5%) that it is unlikely to occur because of sampling fluctuation and likely to occur because H_0 is false. If, on the other hand, a sample mean has a z score which falls between -1.96 and $+1.96$, then we reason as follows: "Every sample will not have a mean exactly equal to the population mean. But if H_0 is true, there is a .95 probability that the mean of a sample will have a z score between -1.96 and $+1.96$. If this occurs for the particular random sample chosen, then H_0 cannot be rejected." The z score for the random sample of 49 Swampscott High School students is $\dfrac{112 - 100}{10/\sqrt{49}} = 1.40$.

This score falls inside the "fail to reject region" from -1.96 to $+1.96$, so H_0 is still accepted; the mean IQ of all seniors at Swampscott High School is the same as the mean IQ of all high school students.

Example 2

A political scientist knows that a certain district voted 54% Democratic in the last Congressional election. He also knows, however, that drastic changes in the types of employment in this district have occurred since the last election. He believes that these changes may have changed the percentage of voters who will vote for the Democratic candidate for Congress, but doesn't know which way it will be changed. He polls 1,000 voters in the district and finds that 480 of these plan to vote for the Democratic candidate. Is his belief supported, at a .01 level of significance?

Percentage

$$H_0: P = .54; \quad H_1: P \neq .54$$

Here $\alpha = .01$; that is, we find rejection regions such that the probability of rejecting H_0 if it is true is .01. This means that if the sampling distribution of proportions does have a mean of .54 (as is the case if H_0 is true), then we must find a region or regions such that the probability of a random sample's proportion falling there is .01. Since H_1 has the form "$P \neq .54$," we find two regions, each of area $.01/2 = .005$.

Referring to Table 4, Appendix C, we find that the z value corresponding to an area of $.5 - .005 = .495$ is 2.58. The null hypothesis will be rejected then (and H_1 accepted) if the z score of the sample proportion is greater than 2.58 or less than -2.58; it will not be rejected if this z score is between -2.58 and $+2.58$.

Of 1,000 voters polled, 480 will vote for the Democratic candidate; $p = 480/1,000 = .48$; $n = 1,000$.

$$\sigma_P = \sqrt{\frac{PQ}{n}} = \sqrt{\frac{.54(.46)^*}{1,000}} = .0158$$

$$z = \frac{p - P}{\sigma_p} = \frac{.48 - .54}{.0158} = -3.80$$

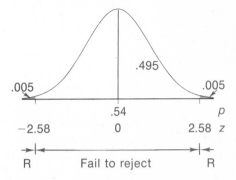

This z score for the proportion in the random sample falls in the rejection region (it is less than -2.58), so H_0 is rejected; it is concluded that there is a change in the proportion of voters who will vote Democratic, at a .01 level of significance.

Note, however, that if 100 different random samples were drawn, probably 1 of them would have a z value for the proportion voting Democratic that would be less than -2.58 or more than $+2.58$ **if the null hypothesis were true**; the probability is .01 that H_0 is rejected even if it is true. But, faced with uncertainty, the political scientist says "I am aware of this probability of error, but I think that, if my sample proportion falls in a tail, it is more likely to do so because H_0 is not true; the difference from the population proportion is so great that I think it arises not from sampling fluctuation but because the sample comes from a different population."

Example 3   An economist in 1970 compared the wages of electricians and of carpenters in New Jersey; she discovered that the mean monthly wage of electricians is $100 more than that of carpenters. She repeats the study in 1980, since she wonders

---

* Note that in hypothesis testing P is given in the null hypothesis, so there is no difficulty in determining σ_p; testing a hypothesis about a proportion differs in this way from finding a confidence interval for a proportion.

whether or not there is still the same difference in their pay. She finds that the mean monthly wage of 100 electricians, chosen at random, is $1,700, with a standard deviation of $200, while a random sample of 81 carpenters has a mean wage of $1,550, with a standard deviation of $100. At a .05 level of significance, should she conclude that the difference in wages of electricians and carpenters is unchanged?

H_0: $\mu_X - \mu_Y = 100$. (She is asking whether the mean monthly wage of all electricians in New Jersey is still $100 more than that of all carpenters in the state.)

H_1: $\mu_X - \mu_Y \neq 100$. (She doesn't know whether the difference in wages of electricians and carpenters has decreased or increased; she simply wants to find out whether it has changed.)

As in Example 1, the form of H_1 ("$\neq$") tells us to make a two-tail test: H_0 is rejected if the z score for difference of sample means is too high **or** too low. The stated level of significance ($\alpha = .05$) tells us to find the z score corresponding to an area of .475; $z = 1.96$. H_0 will not be rejected if the z score for the difference of means of samples taken from the two populations is between -1.96 and $+1.96$, and rejected if this z score is outside those limits.

$$n_X = 100, \ s_X = 200, \ \sigma_{\bar{X}} = \frac{200}{\sqrt{100}} = 20$$

$$n_Y = 81, \ s_Y = 100, \ \sigma_{\bar{Y}} = \frac{100}{\sqrt{80}} = 11.1$$

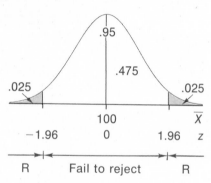

$$\sigma_{(\bar{X} - \bar{Y})} = \sqrt{20^2 + 11.1^2} = \sqrt{523.21} = 22.9$$

$$\bar{X} = 1700, \ \bar{Y} = 1550$$

$$z = \frac{(\bar{X} - \bar{Y}) - (\mu_X - \mu_Y)}{\sigma_{(\bar{X} - \bar{Y})}} = \frac{(1,700 - 1,550) - 100}{22.9} = \frac{50}{22.9} = 2.2$$

H_0 is rejected since $2.2 > 1.96$; we accept the alternative hypothesis that the mean difference in monthly wages of electricians and carpenters in New Jersey is no longer $100. On the basis of this test we do not conclude that the mean difference is now less than $100, nor that it is more; we only conclude that it is one or the other.

10.6 EXERCISES

1. Assume that you have only two choices: (i) α small and β large, (ii) α large and β small. Which would you prefer?

    (a) H_0: This man just died. H_1: He is still alive. (Assume you are a doctor about to make out a death certificate.

(b) H_0: All police are corrupt. H_1: That is not true. (Assume your wife or husband is a member of a police force.)

(c) H_0: Deaf children have the same anxiety as hearing children. H_1: Deaf children have less anxiety. (You hope to publish a paper if H_0 is rejected, but want to be very persuasive about your conclusions.)

(d) H_0: Deaf children have the same anxiety as hearing children. H_1: Deaf children have less anxiety. (You hope to publish a paper if H_0 is rejected, and will publish at any cost.)

2. A machine fills milk bottles; the mean amount of milk in each bottle is supposed to be 32 ounces with a standard deviation of .06 ounces. In a routine check to see that the machine is operating properly, 36 filled bottles are chosen at random and found to contain a mean of 32.03 ounces. At a .05 level of significance, is the machine operating properly? (See Example 4, page 209.)

3. It is known that 1-year-old dogs have a mean gain in weight of 1.0 pound per month with a standard deviation of .40 pound. A special diet supplement, HELTHPUP, is given to a random sample of 50 1-year-old dogs for a month; their mean gain in weight is 1.15 pounds, with a standard deviation of .30 pound. Does HELTHPUP affect weight gain in 1-year-old dogs? (.01 level of significance).

4. An economist working for the Bureau of Labor Statistics knows that, last month, 7.1% of those in the labor force in the United States were unemployed. This month he discovers that 350 are unemployed in a random sample of 5,000. At a .05 level of significance, has there been a change in the unemployment rate?

5. Criticize the underlined words if necessary:
   (a) After a test of H_0: $\mu_X - \mu_Y = 0$; H_1: $\mu_X - \mu_Y \neq 0$, we fail to reject H_0. This shows that there is no significant difference between the means of the X and Y populations.
   (b) If $\alpha = .05$, then β must be .95.
   (c) If H_0: $P = .40$ is rejected when $a = .05$, then it is also rejected when $\alpha = .10$.
   (d) If H_0: $P = .40$ is rejected when $\alpha = .05$, then it is also rejected when $\alpha = .01$.

6. The results of a reading test are standardized so the mean is 50 with a standard deviation of 10. In a random sample of 86 students from private schools in New Jersey the mean is 51.2. Is there a real difference in New Jersey private school students on this test, or is the difference due just to sample fluctuations? ($\alpha = .05$).

7. 40% of all adults over 30 wear glasses. In a random sample of 342 college graduates over 30 in Utah, 158 wear glasses. Is the proportion who wear glasses different among Utah college graduates over 30, or is this difference explained by sample variation, at a .01 level of significance?

8. The registrar of Peterson University is comparing the grade-point averages of married and unmarried students. He finds that 200 married students, cho-

sen at random, have a mean GPA of 2.85 with a standard deviation of .4, while a random sample of 100 unmarried students have a mean GPA of 2.73, with a standard deviation of .3. At a .10 level of significance, is there a difference between the grade-point averages of married and unmarried students?

9. A political scientist is investigating the difference in the proportion of registered Republicans among well-to-do and poor voters. He takes random samples of 1,000 voters whose families have incomes over $32,000 and of 1,000 whose families have incomes under $9,000. He finds that in these two groups 25.3% and 22.0%, respectively, are registered Republicans. At a .05 level of significance, do well-to-do and poor voters register Republican in different proportions?

10. The chairman of the French department at Norwich College finds that a random sample of 100 Norwich students who have studied French for 1 year have a mean score of 470 on a language achievement test; the national average on this test after 1 year of French in college is 500 with a standard deviation of 100. Do students at Norwich differ from the national average? (.01 level of significance).

11. The mean number of years of school completed by adults over 21 in the United States is 10.4 with a standard deviation of 2.0 years. The Board of Education in Madison, New Jersey, surveys a random sample of 200 adult residents and finds that their mean number of years of school completed is 11.3 with a standard deviation of 1.8 years. At a .05 level of significance, do Madison adults differ from the national average in years of schooling completed?

12. Is the proportion of students applying for financial aid the same among men and women students? A university Dean of Admissions randomly chooses from his files the folders of 60 male and 50 female applicants for admission. He finds that, in this sample, 36 men and 35 women had applied for aid. What is his conclusion, at a .01 level of significance?

ANSWERS

1. (a) (ii) (If he rejects H_0 when it is true, he can just wait a while); (b) (ii); (c) (i); (d) (ii) (Fortunately, most journals have referees, so it is unlikely to get published.)

2. H_0: $\mu = 32.0$; H_1: $\mu \neq 32.0$. $\alpha = .05$; H_0 is accepted if z is between -1.96 and $+1.96$.
$$z = \frac{32.03 - 32.0}{.06/\sqrt{36}} = 3.0.$$ H_0 is rejected; the machine needs adjustment.

3. H_0: $\mu = 1.0$; H_1: $\mu \neq 1.0$. $\alpha = .01$; H_0 is rejected if $z > 2.58$ or if $z < -2.58$.
$$z = \frac{1.15 - 1.0}{.4/\sqrt{50}} = 2.65.$$ Reject H_0.

HELTHPUP does affect weight gain. Note: use $\dfrac{\sigma}{\sqrt{n}}$, not $\dfrac{s}{\sqrt{n}}$ if σ is known.

4. H_0: $P = .071$; H_1: $P \neq .071$. $\alpha = .05$; H_0 is rejected if $z > 1.96$ or if $z < -1.96$.
$$z = \frac{.070 - .071}{\sqrt{(.071)(.929/5,000)}} = -.28.$$ Accept H_0; compare with Example 2, page 216.

5. (a) You are testing for any difference, and fail to find any.
(b) It is not true in general that $\alpha + \beta = 1$.

(c) This statement is correct.

(d) Not true, since the "fail to reject" region is wider when $\alpha = .01$ than when $\alpha = .05$, the computed z value may fall in between.

6. H_0: $\mu = 50$; H_1: $\mu \neq 50$. Critical values of z are ± 1.96. The computed value of $z = \dfrac{51.2 - 50}{10/\sqrt{86}} = 1.11$, so we fail to reject H_0; the difference, in the sample can be explained by chance variation.

7. H_0: $P = .40$; H_1: $P \neq .40$. Critical values of z are ± 2.58. $\sigma_p = \sqrt{\dfrac{.40(.60)}{342}} = .0265$, computed $z = \dfrac{(158/342) - .40}{.0265} = 2.34$, so we fail to reject H_0; again, the sample difference can be explained by chance variation.

8. H_0: $\mu_X - \mu_Y = 0$; H_1: $\mu_X - \mu_Y \neq 0$. $\alpha = .10$; H_0 is rejected if $z > 1.64$ or if $z < -1.64$. $z = \dfrac{(2.85 - 2.73) - 0}{\sqrt{.4^2/200 + .3^2/100}} = 2.9$; reject H_0. Compare with Example 3, page 217.

9. H_0: $P_X - P_Y = 0$; H_1: $P_X - P_Y \neq 0$. $\alpha = .05$; H_0 is rejected if $z > 1.96$ or if $z < -1.96$.

$$z = \dfrac{(.253 - .22) - 0}{\sqrt{\dfrac{(.253)(.747)}{1,000} + \dfrac{(.22)(.78)}{1,000}}} = 1.74.$$

There is no difference in the proportion of registered Republicans in the two groups.

10. H_0: $\mu = 500$; H_1: $\mu \neq 500$. $\alpha = .01$; reject H_0 if $z > 2.58$ or < -2.58.

$z = \dfrac{470 - 500}{100/\sqrt{100}} = -3.00$; reject H_0; the students at Norwich do differ from students in general on this test.

11. H_0: $\mu = 10.4$; H_1: $\mu \neq 10.4$. $\alpha = .05$; the critical values of z are ± 1.96.

$z = \dfrac{11.3 - 10.4}{2/\sqrt{200}} = 6.36$. Reject H_0.

12. H_0: $P_M = P_W$ (M = men, W = women); H_1: $P_M \neq P_W$. $\alpha = .01$, so fail to reject H_0 if $-2.58 < z < 2.58$.

$$P_M = \dfrac{36}{60} = .6, \; n_M = 60, \; P_W = \dfrac{35}{50} = .7, \; n_W = 50,$$

$$\sigma_{(P_M - P_W)} = \sqrt{\dfrac{.6(.4)}{60} + \dfrac{.7(.3)}{50}} = .091 \qquad .$$

$z = \dfrac{(.6 - .7) - 0}{.091} = -1.10$; H_0 is not rejected. There is no difference in the proportion of men and women applying for aid upon admission.

10.7 ONE-TAIL TESTS

In Section 10.5, you learned how to set up a null hypothesis H_0 and an alternative hypothesis H_1, and how to reach a decision on whether or not to reject the null hypothesis at a given level of significance when H_1 has one of the following forms: $\mu \neq \underline{\quad}$, $P \neq \underline{\quad}$, $\mu_1 - \mu_2 \neq \underline{\quad}$, $P_1 - P_2 \neq \underline{\quad}$. This decision was reached by following a specific convention: Let the area in each tail of the sampling distribution be $\alpha/2$; this determines two values of z which separate a "fail to reject" region from two rejection regions. Then the z value of a statistic is computed for a random sample, and H_0 is still accepted if this z value falls in the "fail to reject" region previously determined; otherwise, H_0 is rejected.

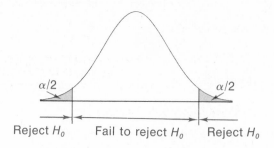

$\alpha/2$ $\alpha/2$

Reject H_0 Fail to reject H_0 Reject H_0

Now let's look at "one-tail tests" in which H_1 has the form $\mu < \_\_$, $P > \_\_$, $\mu_1 - \mu_2 > \_\_$, etc.

Suppose H_0 has the form $\mu = 100$. If H_1 has the form $\mu > 100$, this means that it is reasonable to believe the observed sample is from a population with mean of 100 rather than from a population whose mean is greater than 100 if the mean of the sample is itself less than 100. The rejection region is therefore concentrated in one tail of the sampling distribution of means, rather than separated into two parts or tails as when H_1 has the form $\mu \neq 100$. The area of this one tail is now α, rather than the $\alpha/2$ used previously, and the critical value of z changes accordingly.

Two-Tail Test	One-Tail Test
H_0: $\mu = 100$; H_1: $\mu \neq 100$; $\alpha = .05$	H_0: $\mu = 100$; H_1: $\mu > 100$; $\alpha = .05$

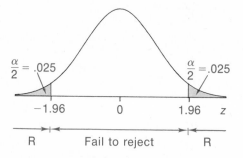 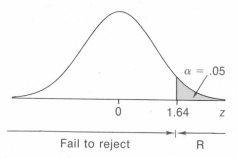

The most frequently used values of α and the corresponding critical values of z are:

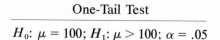

		z	
	Two-Tail	One-Tail, $<$	One-Tail, $>$
$\alpha = \begin{cases} .10 \\ .05 \\ .01 \end{cases}$	± 1.64	-1.28	1.28
	± 1.96	-1.64	1.64
	± 2.58	-2.33	2.33

Remember that α is still being emphasized and chosen arbitrarily, as with two-tail tests; try to keep an open mind on which value of α to use, and on whether this is even a good way to decide to accept or reject H_0, until you have read Section 10.9.

Example 1   It is known that 1-year-old dogs have a mean gain in weight of 1.0 pound per month, with a standard deviation of .40 pound. A special diet supplement, HELTHPUP, is given to a random sample of 50 1-year-old dogs for a month; their mean gain in weight per month is 1.15 pounds, with a standard deviation of .30 pound. Does weight gain in 1-year-old dogs increase if HELTHPUP is included in their diet? (.01 level of significance.)

H_0: $\mu = 1.0$; H_1: $\mu > 1.0$; $\alpha = .01$; one-tail test.

Reject H_0 if $z > 2.33$.

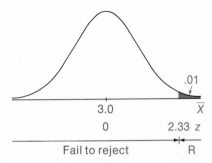

$z - \dfrac{1.15 - 1.0}{.4/\sqrt{50}} = 2.7$. Reject H_0. At a .01 level of significance, weight gain is increased if pumps have HELTHPUP.

Example 2   An economist working for the Bureau of Labor Statistics knows that last month 7.1% of those in the labor force in the United States were unemployed. This month he discovers that 350 are unemployed in a random sample of 5,000. At a .05 level of significance, has unemployment decreased this month?

H_0: $P = .071$; H_1: $P < .071$; $\alpha = .05$; one-tail test.

Reject H_0 if $z < -1.64$.

$z = \dfrac{.070 - .071}{\sqrt{(.071)(.929)/5,000}} = -.28$

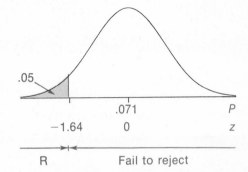

Do not reject H_0; the unemployment rate for all those in the United States labor force has not changed this month.

Example 3 The registrar of Peterson University is comparing the grade-point averages of married and unmarried students. He finds that 100 married students, chosen at random, have a mean GPA of 2.85 with a standard deviation of .4, while a random sample of 100 unmarried students has a mean GPA of 2.73 with a standard deviation of .3. At a .10 level of significance, do married students have a higher GPA?

$H_0: \mu_X - \mu_Y = 0$; $H_1: \mu_X - \mu_Y > 0$; $\alpha = .10$; one-tail test.

H_0 is rejected if $z > 1.28$.

$$z = \frac{(2.85 - 2.73) - 0}{\sqrt{.4^2/100 + .3^2/100}} = 2.4$$

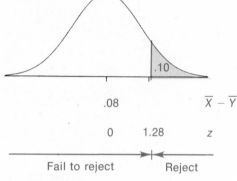

H_0 is rejected; married students do have a higher GPA.

Example 4 A political scientist is investigating the difference in the proportion of registered Republicans among well-to-do and poor voters. He takes random samples of 1,000 voters whose families have incomes over \$32,000 and 1,000 whose families have incomes under \$9,000. He finds that in the two groups 25.3% and 22.0%, respectively, are registered Republicans. At a .05 level of significance, is the proportion of all registered Republicans higher among well-to-do than among poor voters?

$H_0: P_X - P_Y = 0$; $H_1: P_X > P_Y$; $\alpha = .05$; one-tail test.

H_0 is rejected if $z > 1.64$.

$$z = \frac{.253 - .22}{\sqrt{\dfrac{(.253)(.747)}{1,000} + \dfrac{(.22)(.78)}{1,000}}} = 1.74$$

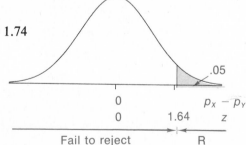

H_0 is rejected; more well-to-do-voters are registered Republicans.

10.8 EXERCISES

1. Twelve out of 100 randomly selected West Side retail stores were robbed in 1980, as were 24 out of 150 randomly selected East Side stores. At a .10 level of significance, do you conclude that robbers prefer the East Side?

2. A test of a random sample of 100 True Blue cigarettes by the Madder Laboratories shows an average of 17.36 mg. of tar, with a standard deviation of 2 mg. (a) Do these results support the hypothesis that the mean amount of tar in True Blue cigarettes does not differ from 17 mg.? (b) Do these results support the claim that the mean amount of tar is at most 17 mg.? (.05 level of significance.)

3. In 1970, 25% of the fire alarms in Edgerton were false alarms. A reporter for the Edgerton News finds that this year 42 of a random sample of 150 fire alarms have been false. Has the proportion of alarms which are false increased since 1970? ($\alpha = .01$.)

4. A flu epidemic caused many employees of the Firethorn Tire Company to be absent in January. In a random sample of 400 employees it was found that the mean number of days absent for 150 employees who had had flu shots was 2.4 with a standard deviation of .3, while the mean number of days absent for the rest of the sample (who had had no flu shots) was 4.5 days with a standard deviation of .6. Did the flu shots decrease the mean number of days absent by 2 days or more? ($\alpha = .05$.)

5. It is suspected that the mean salary of professors in public colleges is $1,000 more than in private colleges. A random sample of 100 public college professors shows a mean salary of $25,000 with a standard deviation of $1,000, while 100 private college professors selected at random have a mean salary of $23,800 with a standard deviation of $900. At a .01 level of significance, is the salary of public college professors more than $1,000 greater than that of private college professors?

6. Supermarkets throughout the United States sell an average of 460 quarts of milk per day. Over a sample of 36 days chosen at random, a supermarket sells an average of 470 quarts per day, with a standard deviation of 20 quarts. Does the store sell more milk than the national average, at a .05 level of significance?

7. A group of 40 sophomores and a group of 40 seniors were given a geography test; the means are 79.8 and 77.5, and the variances 102.5 and 108.7, respectively. Assume both groups were chosen at random. Did seniors do less well than sophomores on the test, at a .01 level of significance?

8. A pharmaceutical company claims that a drug which it manufactures relieves cold symptoms for a period of 10 hours in 90% of those who take it. In a random sample of 400 people with colds who take the drug, 350 find relief for 10 hours. At a .05 level of significance, is the manufacturer's claim correct?

9. You are suspicious that a certain die is weighted. You roll the die 1,200 times; a 3 comes up 230 times. At a .05 level of significance, are your suspicions justified?

10. A reading test is given to random samples of 200 boys and 100 girls in the third grade in Michigan public schools. The boys have a mean score of 60 with a standard deviation of 10; the girls have a mean of 65 with a standard deviation of 8. Do boys read less proficiently than girls in the third grade in Michigan schools, at a .01 level of significance?

ANSWERS

1. W = West Side, E = East Side. H_0: $P_W = P_E$; H_1: $P_W < P_E$. $\alpha = .10$, so $z = -1.28$. $p_W = \dfrac{12}{100} = .12$, $p_E = \dfrac{24}{150} = .16$.

$$\sigma_{(p_W - p_E)} = \sqrt{\frac{.12(.88)}{100} + \frac{.16(.84)}{150}} = .044$$

$z = \dfrac{(.12 - .16) - 0}{.044} = -.91$, which falls in the "fail to reject" region; robbers treat West and East Sides equally, at a .10 level of significance.

2. (a) H_0: $\mu = 17.0$; H_1: $\mu \neq 17.0$. $\alpha = .05$; critical z values are ± 1.96. $z = \dfrac{17.36 - 17.0}{2/\sqrt{100}} = 1.80$; H_0 is still accepted.

(b) H_0: $\mu = 17.0$; H_1: $\mu > 17.0$. $\alpha = .05$; critical value of z is now 1.64. Computed z value is the same, so H_0 is now rejected. Note that H_0 can be rejected with a one-tail test when we continue to accept it with a two-tail test.

3. H_0: $P = .25$; H_1: $P > .25$, $p = .28$, $\sigma_p = \sqrt{\dfrac{.25/(.75)}{150}} = .0354$, $z = \dfrac{.28 - .25}{.0354} = .85$, < 2.33. We fail to reject H_0; the proportion of false alarms has not increased.

4. It's probably easier to let the X population be those without flu shots: H_0: $\mu_X - \mu_Y = 2$; H_1: $\mu_X - \mu_Y > 2$. $\sigma_{(\bar{X} - \bar{Y})} = \sqrt{\dfrac{.3^2}{150} + \dfrac{.6^2}{250}} = .0452$, $z = \dfrac{(4.5 - 2.4) - 2}{.0452} = 2.21$. Critical $z = 1.64$, so H_0 is rejected; flu shots decrease the mean number of days absent by more than 2 days. (If X = those with flu shots, then H_0: $\mu_X - \mu_Y = -2$; H_1: $\mu_X - \mu_Y < -2$; computed $z = -2.21$, critical $z = -1.64$, and H_0 is, of course, still rejected.)

5. X = public colleges, Y = private colleges. H_0: $\mu_X - \mu_Y = 1,000$; $\mu_X - \mu_Y > 1,000$. $\alpha = .01$, $z = 2.33$

$$\sigma_{(\bar{X} - \bar{Y})} = \sqrt{\frac{1,000^2}{100} + \frac{900^2}{100}} = 134.5$$

$$z = \frac{(25,000 - 23,800) - 1,000}{134.5} = 1.49$$

H_0 is not rejected; professors at public colleges are paid $1,000 more than those at private colleges.

6. H_0: $\mu = 460$; H_1: $\mu > 460$; $\alpha = .05$, one-tail test. Reject H_0 if $z > 1.64$.

$$z = \frac{470 - 460}{20/\sqrt{36}} = 3.00$$

Reject H_0; the Madison store does sell more milk.

7. H_0: $\mu_X = \mu_Y$ (or $\mu_X - \mu_Y = 0$); H_1: $\mu_X > \mu_Y$; $\alpha = .01$, one-tail test. Reject H_0 if $z > 2.33$.

$$n_X = 40; \ \sigma_{\bar{X}}^2 = \frac{102.5}{40} = 2.56$$

$$n_Y = 40, \ \sigma_{\bar{Y}}^2 = \frac{108.8}{40} = 2.72$$

$$\sigma_{(\bar{X} - \bar{Y})} = \sqrt{2.56 + 2.72} = 2.30$$

$$z = \frac{(79.8 - 77.5) - 0}{2.30} = 1.00$$

Do not reject H_0; the difference between seniors' and sophomores' grades can be explained by sampling fluctuation at a .01 level of significance.

8. $H_0: P = .90; H_1: P < .90; \alpha = .05$, one-tail test. Reject H_0 if $z < -1.64$.

$$\sigma_p = \sqrt{\frac{(.90)(.10)}{400}} = .015; \ p = \frac{350}{400} = .875$$

$$z = \frac{.875 - .90}{.015} = -1.67$$

H_0 is rejected: The manufacturer's claim is not upheld.

9. $H_0: P = \frac{1}{6}; H_1: P \neq \frac{1}{6}. \ \alpha = .05$, two-tail test. Reject H_0 if $z > 1.96$ or if $z < -1.96$.

$$p = \frac{230}{1,200}; \ \sigma_p = \sqrt{\frac{(1/6)(5/6)}{1,200}} = .011$$

$$z = \frac{230/1200 - 1/6}{.011} = 2.3$$

Reject H_0; your suspicions are well-founded.

10. $H_0: \mu_X = \mu_Y; H_1: \mu_X < \mu_Y; \alpha = .01$, one-tail test. Reject H_0 if $z < -2.33$.

$$n_X = 200, \bar{X} = 60, s_X = 10, \sigma_{\bar{X}}^2 = \frac{100}{200} = .50$$

$$n_Y = 100, \bar{Y} = 65, s_Y = 8, \sigma_{\bar{Y}}^2 = \frac{64}{100} = .64$$

$$\sigma_{(\bar{X}-\bar{Y})} = \sqrt{.50 + .64} = 1.07$$

$$z = \frac{(60 - 65) - 0}{1.07} = -4.67. \ \text{Reject } H_0; \text{ boys read less well than girls in third}$$

grade in Michigan public schools.

10.9 TYPE I AND TYPE II ERRORS

In our convention for testing H_0 using the sampling distribution of a statistic, it is only a Type I error (whose probability is measured by the level of significance, α) that is important; β (the probability of a Type II error) is ignored. Usually this is done because it is H_1 that you hope can be supported—so you hope that H_0 will be rejected—but at the same time you wish to persuade others. You wish to say to them "the probability is so small that my sample statistics would occur if H_0 is true that it is unlikely my data are accounted for by sample fluctuations, but rather because my new theory is true. H_1 gives a better explanation of what has happened than H_0 does, so let us agree that H_0 should be rejected and replaced by H_1." But you are hardly persuasive unless the probability is indeed small; that means α must be small. This makes the choice of emphasis on α or β sound clear-cut, but unfortunately it's not as simple as that. You must look further into the relationship of α and β, and think clearly about which to minimize when testing a hypothesis in your own research.

A doctor who accepts the hypothesis that a new drug has the same effect on a disease as the drug presently used, when in fact the new drug is better (Type II error), may be missing a new cure for his patients and may cause great loss to a pharmaceutical company. If, on the other hand, he concludes that the new drug looks promising when it is really no improvement (Type I error), he may use much energy, time, and money on further tests which prove fruitless.

A manufacturer of television sets who rejects a shipment of TV tubes, because after testing a sample he concludes their mean life is too short, is making a Type I error if this is not the case; he may have production problems for lack of an essential part and he may have poor relations in the future with his supplier. But if he accepts the shipment when the mean life is indeed less than he expects (Type II error), he may have trouble with his guarantee to his customers.

These examples indicate that emphasis on a Type I or a Type II error depends on the hypothesis that is being tested and the reason for testing it.

Some examples will illustrate further the relationship between α and β in testing H_0 using the sampling distribution of a statistic. As you study these examples, concentrate on the following results:

1. As α increases (or decreases), β decreases (or increases). The relationship is not a simple one, however. It is not true that doubling α will cut β in half, for example.
2. For fixed α, β is easily determined if the alternative H_1 has the form $\mu = 80$, $P = .40$, etc., but the value of β depends upon the actual constant selected. For the kind of alternative hypotheses we have been testing ($\mu < 80$, $P \neq .40$, etc.) one value of β cannot be determined.
3. For fixed α, the value of β can be decreased by increasing the size of the sample.

Example 1

A manufacturer claims his light bulbs have a mean life of 1,000 hours with a standard deviation of 40 hours. A big department store tests a random sample of 64 bulbs from a large shipment.

(a) The shipment should be accepted if the mean life of the bulbs in the sample is over what value, if the level of significance is (i) .05, (ii) .01?

(b) What is β if H_1 is $\mu = 987$ when (i) $\alpha = .05$. (ii) $\alpha = .01$?

(c) What happens to β as α decreases from .05 to .01?

(a) H_0: $\mu = 1,000$; H_1: $\mu < 1,000$; one-tail test.

(i) $\alpha = .05$, so $z = -1.64$, $\sigma_{\bar{X}} = \dfrac{40}{\sqrt{64}} = 5$.

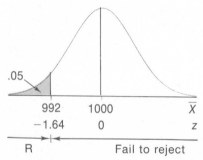

$$z = \frac{\bar{X} - \mu}{\sigma_{\bar{X}}}; \; -1.64 = \frac{\bar{X} - 1000}{5}; \; \bar{X} = 991.8 \text{ hours}$$

The shipment is accepted if $\bar{X} > 991.8$ hours.

(ii) $\alpha = .01$, so $z = -2.33$.

$$-2.33 = \frac{\bar{X} - 1{,}000}{5}; \; \bar{X} > 988.4 \text{ hours}.$$

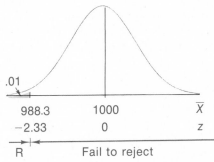

The shipment is accepted if $\bar{X} > 988.4$ hours.

Note that, up until now, we have determined a critical value or values of z which separate a "fail to reject" region from a rejection region; then the z score for a sample statistic was computed to find in which region it falls. Here the critical value is expressed not in z or standard deviation units, but rather in original score units (hours).

(b) Remember that β is the probability of continuing to accept H_0 when it should be rejected; but this is also the probability of rejecting H_1 when it should be accepted.

(i) If μ is 987 (if H_1 is true), then the sampling distribution of means is moved to the left by 13 hours, and now the correspondence between z and $\bar{X}$ is given by $z = \dfrac{\bar{X} - \mu}{\sigma_X} = \dfrac{\bar{X} - 987}{5}$ (since we assume that σ has not changed).

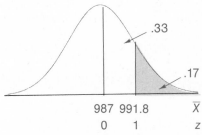

In (a) we found that H_0 was not rejected if $\bar{X} > 991.8$ hours when $\alpha = .05$.

$$\beta = \Pr(\bar{X} > 991.8 \text{ hours when } \mu = 987) = \Pr\left(z > \frac{991.8 - 987}{5} = .96\right) = .5 - .33 = .17.$$

(ii) When $\alpha = .01$, we found that the shipment is accepted if $\bar{X} > 988.3$ hours. $\beta = \Pr(\bar{X} > 988.3 \text{ if } \mu = 987) = \Pr\left(z > \dfrac{988.3 - 987}{5} = .26\right) = .5 - .102 = .40.$

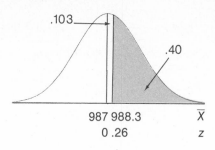

.103

.40

987 988.3 $\overline{X}$
0 .26 z

(c) α β β increases as α decreases.

.05   .17
.01   .40

To find β when H_1 is an equality ($\mu = 987$, $P = .34$, etc.):
   1. Find the critical value(s) of z which define the "fail to reject" region for H_0 for the given α.
   2. Using H_0 value in the z formula, translate this (these) critical value(s) into the original score units (hours, proportion(s), etc.), thus finding the region in which H_1 is rejected.
   3. Find the probability that a sample statistic ($\overline{X}$, p, etc.) from a random sample will fall in the region determined in Step 2 if the population parameter (μ, P, etc.) is that stated in H_1. This probability $= \beta$.

Example 2   Continue Example 1(b), finding β if H_1 is (a) $\mu = 977$, (b) $\mu = 982$, (c) $\mu = 987$, (d) $\mu = 992$, (e) $\mu = 997$. Remember that H_0 is $\mu = 1{,}000$, that $\sigma = 40$, $n = 64$, $\alpha = .05$, and that H_0 is not rejected if $\overline{X}$ is greater than 991.8 hours.

$H_1: \mu =$	Reject H_1 if $z >$	β
(a) 977	$\dfrac{991.8 - 977}{5}$	.002
(b) 982	$\dfrac{991.8 - 982}{5}$	.025
(c) 987	$\dfrac{991.8 - 987}{5}$	.17 ← [Example 1(b)]
(d) 992	$\dfrac{991.8 - 992}{5}$	.52
(e) 997	$\dfrac{991.8 - 997}{5}$	.85

These results can serve as the basis for graphing an **operating characteristic curve** (OC curve), which shows the probability of a Type II error under various forms of H_1 if H_0 is $\mu = 1{,}000$ and $\alpha = .05$.

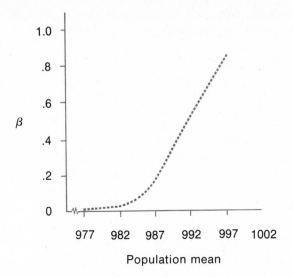

It is seen from this curve that the probability of a Type II error is close to 1 if the true population mean is between 997 and 1,000, but rapidly diminishes if the true population mean is less than 997; β is almost 0 if the true population mean is 977 or less but the null hypothesis states that this mean is 1,000.

The value of β, then, depends on the form of the particular alternative hypothesis that you will accept if H_0 is rejected. In practice, however, this is seldom known. As a form for H_1, "$\mu > 100$" is far more likely than "$\mu = 108$." For purely practical reasons, α is emphasized rather than β. But bear in mind that α is often emphasized rather than β for reasons of safety, too. The new drug being tested is rejected unless it almost certainly is an improvement, rather than being accepted because it might be an improvement. Remember constantly that choice of α for continuing to accept or rejecting the null hypothesis is a convention; always consider in a particular experiment whether it is really a small α or a small β that is needed.

10.10   WARNINGS

1. Under our rules of law, a defendant who is found not guilty by a jury cannot be tried for the same crime again. The same kind of thing is not true with hypothesis testing. If the manufacturer claims the mean life of his light bulb is 1,000 hours with $\sigma = 40$ hours, a test of a random sample of 64 light bulbs showing their mean life is 998 hours gives a result consistent with the hypothesis at a .05 level of significance. The result is, however, also consistent with the hypothesis that the mean life of all light bulbs is 998 or 996 hours. Such an outcome, then, does not prove that the null hypothesis is true; it can only be concluded that the null hypothesis may be true.

If, on the other hand, the random sample of 64 bulbs has a mean life of 960 hours, then it can be shown that such a sample is extremely unlikely to be chosen if the null hypothesis is true; we conclude that H_0 is probably false. Failing to reject H_0 implies that it **may be** true; rejection of H_0 implies that it is very likely to be false.

2. In testing hypotheses, the assumption is always made that the sample being tested is a random sample. If a sample is deliberately chosen with only high scores, for example—that is, the sample mean occurs in the extreme right tail of the distribution of sampling means—then no conclusions can be drawn about the probability of such a mean occurring at random, and no fair test of the null hypothesis has been made.

3. It is assumed that the sampling distribution of the statistic found is known. This distribution may, however, be known only if certain assumptions are satisfied. Up until now, we have assumed the sampling distribution is normal. In the next chapter we shall find another kind of distribution if σ is not known and n is small.

4. Always remember that H_0, H_1, and α are determined before the test is carried out, and that the form of H_1 determines whether a one-tail or two-tail test is carried out.

10.11 EXERCISES

1. State two ways in which you can decrease β when testing a null hypothesis.

2. In Chapter 12 you will study a type of hypothesis (H_0: two populations have the same distribution; H_1: their distributions are different) where the researcher frequently hopes that H_0 can be accepted. If this is the case, should α or β be small? Why?

3. H_0: $\mu = 80$ pounds; H_1: $\mu > 80$ pounds, $\sigma = 4$, $n = 100$, $\alpha = .05$.
   (a) What is the critical value of z that separates the "continue to accept" region from the "reject H_0" region?
   (b) H_1 is rejected (and H_0 is not) if $\overline{X}$ of a random sample is over what value (in pounds)?
   (c) What is β if H_1 is (i) $\mu = 80.4$, (ii) $\mu = 81.2$?

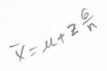

$\overline{X} = \mu + z \frac{\sigma}{n}$

4. (a) H_0: $P = .20$, H_1: $P > .20$, $n = 900$, $\alpha = .01$. Find, in terms of proportions, the region in which H_0 continues to be accepted (and H_1 is rejected.)
   (b) Find β if H_1 is (i) $P = .24$, (ii) $P = .23$, (iii) $P = .22$, (iv) $P = .21$, (v) $P = .201$.
   (c) Draw the operating characteristic curve.

5. What happens to β as sample size increases? The light bulb manufacturer of Example 1 is still claiming a mean life of 1,000 hours with a standard deviation of 40 hours. If $a = .05$, find β for the alternative hypothesis $\mu = 987$ if the size of the sample is (a) 64, (b) 100.

6. The manufacturer has given up the making of light bulbs and has become

president of a company that makes electrical resistors. A supplier who purchases parts from him is wary of resistors that are either too high or too low. The manufacturer claims that part P36483E has a mean resistance of 1,000 ohms with a standard deviation of 40 ohms. The supplier chooses a random sample of 64 from his order, and is testing the manufacturer's claim at a .05 level of significance.

(a) The manufacturer's claim should be accepted if the sample mean is between what values?

(b) What is the probability of a Type II error if the mean resistance of all the P36483E resistors is actually 980 ohms? 985? 990? 995? 999? 1,005? 1,010? 1,015? 1,020?

(c) Draw the operating characteristic curve.

ANSWERS

1. Increase either sample size or α.

2. β should be small (and therefore increase α) if the researcher is to be persuasive, since the probability of accepting H_0 when it is false should be small.

3. (a) $z = 1.64$ (one tail, right).

(b) $1.64 = \dfrac{\bar{X} - 80}{4/\sqrt{100}}$; $\bar{X} = 80.7$ pounds. Reject H_1 if $\bar{X}$ is less than 80.7 pounds.

(c) (i) $\Pr(\bar{X} < 80.7 \text{ if } \mu = 80.4) = Pr\left(z < \dfrac{80.7 - 80.4}{.4} = .75\right) = .5 + .27 = .77.$

(ii) $\Pr(\bar{X} < 80.7 \text{ if } \mu = 81.2) = Pr\left(z < \dfrac{80.7 - 81.2}{.4} = -1.25\right) = .5 - .39 = .11.$

4. (a) $\sigma_p = \sqrt{\dfrac{.20(.80)}{900}} = .0133$, $z = \dfrac{\bar{X} - .20}{.0133}$. $.01 = \Pr(z > 2.33) = \Pr(\bar{X} > .231)$. H_1 is rejected if p is less than .231.

(b)

$P =$	Reject H_1 if $z <$	β
.24	$\dfrac{.231 - .24}{.0133} = -.68$	.25
.23	$\dfrac{.231 - .23}{.0133} = .08$	.53
.22	$\dfrac{.231 - .22}{.0133} = .83$	.80
.21	$\dfrac{.231 - .21}{.0133} = 1.58$	.92
.201	$\dfrac{.231 - .201}{.0133} = 2.26$	.99

(c)

5. (a) $\beta = .17$ if $n = 64$ (see Example 1(b) of Section 10.9).

(b) $\sigma_{\bar{X}} = \dfrac{40}{\sqrt{100}} = 4$, $-1.64 = \dfrac{\bar{X} - 1{,}000}{4}$; $\bar{X} = 993.4$. H_0: $\mu = 1{,}000$ is not rejected if the mean of a sample of size 100 is more than 993.4 hours. $\beta = \Pr(\bar{X} > 993.4$, given that $\mu = 987) = \Pr\!\left(z > \dfrac{993.4 - 987}{4} = 1.60\right) = .5 - .445 = .06$.

For fixed α, β gets smaller as n gets larger.

6. (a) H_0: $\mu = 1{,}000$, H_1: $\mu \neq 1{,}000$; $\sigma = 40$, $n = 64$, $\alpha = .05$, so $z = \pm 1.96$.

$$\pm 1.96 = \dfrac{\bar{X} - 1{,}000}{\dfrac{40}{\sqrt{64}}}; \bar{X} = 990 \text{ or } 1{,}010.$$

(b) If $\mu = 980$ (or 985 or ... or 1,020), what is the probability that a random sample of size 64 will have a mean between 990 and 1,010?

H_1: $\mu =$	$\Pr(990 < \bar{X} < 1{,}010) =$	β
980	$\Pr(2 < z < 6)$	.02
985	$\Pr(1 < z < 5)$	.16
990	$\Pr(0 < z < 4)$	.50
995	$\Pr(-1 < z < 3)$	.82
999	$\Pr(-1.8 < z < 2.5)$	.96
1,005	$\Pr(990 < z < 1)$	.82
1,010	$\Pr(-4 < z < 0)$	.50
1,015	$\Pr(-5 < z < -1)$	.16
1,020	$\Pr(-6 < z < -2)$	.02

(c)

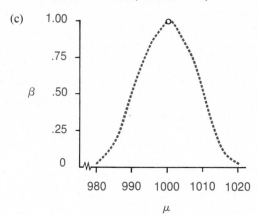

10.12 VOCABULARY AND NOTATION

null hypothesis	H_0	Type II error	β
alternative hypothesis	H_1	one-tail test	
level of significance		two-tail test	
Type I error	α	operating characteristic (OC) curve	

10.13   REVIEW EXERCISES

1. During the Salk polio vaccine trials (see Exercise 9, page 184), 200,745 children received vaccine; 33 of them caught paralytic and 24 nonparalytic polio. 201,229 received no vaccine; of these, 115 caught paralytic and 27 non-paralytic polio. Test the hypothesis that the vaccine reduces the risk of (a) paralytic polio, (b) non-paralytic polio, each at the 1% significance level. $\sigma_{(p_x - p_y)} = .0000605$ for (a), $= .0000355$ for (b).

2. In roulette, the probability of an even number coming up is 18/38. The roulette wheel has numbers 1 to 36, but also has 0 and 00 which don't count as either odd or even. Do you think the wheel should be checked if an even number comes up 402 times in 900 spins of the wheel? ($\alpha = .05$.)

3. Of 19 people who chose apple cider at the Ground Hog's Eve party mentioned earlier (Exercise 9, page 50), 14 preferred it hot; of 16 who chose other fruit juices 4 asked for these hot. (Actually two first asked for coffee, but let us ignore that complication.) Test the null hypothesis that the proportion of cider drinkers who like their drink hot is the same as for other fruit juice drinkers. ($\alpha = .05$.)

4. Jagannadjam studied zircon crystals found in several adjoining valleys in North Carolina. The ratio of width to length of zircon crystals in rock provides a clue to the origin of the rock: igneous (formed when the molten earth set) or sedimentary (washed by streams). A sample of 209 zircon crystals from one North Carolina valley had a mean ratio of 0.435 and standard deviation of 0.126, while a sample of 217 from another valley had a mean ratio of 0.494 and standard deviation of 0.134. Test the null hypothesis that the zircons in the two valleys have the same mean ratios (which might be a reasonable assumption if rocks from both locations have the same origin). ($\alpha = .05$.)

5. Iris Weisman (M.A. thesis, Dept. of Nutrition, University of Wisconsin, 1978) investigated changes in serum phosphate levels following treatment with the drug 5-fluorouracil of three groups of patients; $X =$ (phosphate level after) $-$ (phosphate level before treatment):

	n	$\bar{X}$	s
(a) Patients with Cancer of the Breast	39	-0.30	0.67 mg./100 ml.
(b) Patients with Cancer of the Colon	40	-0.38	0.52 mg./100 ml.
(c) Patients with Cancer of the Stomach	16	-0.79	0.48 mg./100 ml.

For each tumor site separately, test the null hypothesis that the treatment does not change average serum phosphate concentration, at the 1% significance level.

6. In Exercise 12, page 85, distributions of numbers of people sitting at Nancy's and Maria's tables at Marchant's Fine Restaurant were shown, and you found for Nancy's tables, number of mean people = 2.34, standard deviation = 0.82 people, and for Maria's tables mean = 2.43 people/table, standard deviation = 1.02. Test the null hypothesis, at the 5% significance level, that both waitresses get parties of the same size on the average.

7. Galanter, Rabkin, Rabkin and Deutsch reported in "The 'Moonies', a Psychological Study of Conversion and Membership in a Contemporary Religious Sect,"* that a standard questionnaire was given to 337 members of the Unification Church and 305 controls. The mean "well-being" scores for the two groups were 74.4 and 83.4, with standard deviations of 17.2 and 16.2, respectively. Test the null hypothesis of equal population means at the 1% significance level. State possible conclusions and interpretations.

8. Assume that a particular null hypothesis is true, and that it is tested by 20 different investigators with $\alpha = .05$. The 20 samples are random and independent. Comment on the expected results.

9. In 1977 there were 49,500 motor vehicle deaths in the United States; in 1978 there were 48,700. In Oregon there were 675 motor vehicle deaths in 1977 and 638 in 1978. At a .01 level of significance, does the rate of change in the death incidence in Oregon differ from that of the nation?

10. In 1970, a random survey of 1,000 American families with at least one child under 18 revealed that 100 families had four or more children. A random survey of 1,000 similar families in 1980 revealed that 98 had four or more children. At a .01 level of significance, did the proportion of United States families with four or more children fall during the decade 1970 to 1980?

11. A passenger railway has been carrying a mean of 12,000 passengers per day with a standard deviation of 2,000. The railway is reorganized and within the first year a random sample of 30 days shows a mean of 12,500 passengers a day, with a standard deviation of 1,000. Has the number of passengers carried per day increased, at a .02 level of significance?

12. Over a long period of time, the percentage of A's in Economics 121 at Bulwich University has been 20%. During this Fall term there were 24 A's in a class of 100 students. Has the proportion of A's increased? (.05 level of significance.)

13. A new cafeteria is planned for Claudius College. The manager thinks 1,800 students can be served dinner between 5:30 and 7:00 P.M. if the average time a student spends eating dinner is 30 minutes. He times a random sample of 64 students and finds their mean time for dinner is 32 minutes with a standard deviation of 6 minutes. Does the manager conclude that the cafeteria will serve at least 1,800 students in the given hours? (.02 level of significance.)

14. A piece of land is divided into 60 plots of equal size, fertility, sunlight, and so on. A new fertilizer, GROWCO, is tested by spreading it on 30 plots chosen at random, while 5-10-5 fertilizer is used on the other 30 plots as a control. Each plot is planted with the same number of tomato seedlings. The mean number of pounds of tomatoes from plots with GROWCO is 130 pounds, with a standard deviation of 10 pounds; the mean for the 5-10-5 plots is 125 pounds with a standard deviation of 8 pounds.

    (a) Is there a difference in the weight of tomatoes produced on the GROWCO and 5-10-5 plots, at a .05 level of significance?

---

*American Journal of Psychiatry, 136, 1979, pp. 165–170.

(b) Do the GROWCO plots produce more tomatoes, at a .05 level of significance?

15. Nationwide, the mean grade on a college entrance examination was 75 with a standard deviation of 8. A random sample of 144 students in New Mexico had a mean of 76 with a standard deviation of 8.

(a) At a .05 level of significance, did the students in New Mexico do better than the national average?

(b) What is the probability of accepting H_0: $\mu = 75$ when the mean of New Mexico students taking the test is actually (i) 78, (ii) 77, (iii) 76, (iv) 75, (v) 74?

(c) Draw the operating characteristic curve.

16. A scuba diver finds a trunk full of gold and silver coins. He samples 10 coins, with replacement, and has adopted the following decision rule: accept the hypothesis that half the coins are gold if 3 to 7 of the coins he samples are gold, and reject it otherwise.

(a) Find the probability of rejecting the hypothesis when it is true.

(b) Find the probability of accepting the hypothesis that half the coins are gold when the true proportion of gold coins is (i) .1, (ii) .2, (iii) .3, (iv) .4, (v) .6.

(c) Draw the operating characteristic curve by making a graph with P on the horizontal axis and β on the vertical axis.

ANSWERS

1. $X =$ vaccinated children. H_0: $P_X = P_Y$, H_1: $P_X < P_Y$, critical value of $z = -2.33$.

(a) $z = \dfrac{.000164 - .000571}{.0000605} = -6.73$, so H_0 is rejected (whew!).

(b) $z = \dfrac{.0001196 - .0001342}{.0000355} = -.41$, so we continue to accept H_0; this vaccine does not protect against non-paralytic polio.

3. Nonsense! If X are cider drinkers, Y other drinkers, then $n_X Q_X = 5$, $n_Y P_Y = 4$. Is the sampling distribution of differences of proportions a normal distribution?

5. In each case H_0: $\mu = 0$, H_1: $\mu \neq 0$, $\alpha = .01$, and the critical values of z are ± 2.58.

(a) $z = \dfrac{-0.30 - 0}{0.67/\sqrt{39}} = -2.80$. Hence, we reject H_0.

(b) $z = \dfrac{-0.38 - 0}{0.52/\sqrt{40}} = -4.62$. Again, we reject H_0.

(c) We are unable to test the hypothesis since the sample size of 16 is too small.

7. $X =$ Moonies, $Y =$ Controls. H_0: $\mu_X = \mu_Y$; H_1: $\mu_X \neq \mu_Y$, $\alpha = .01$; the critical values of z are $z = \pm 2.58$. $\bar{X} = 74.4$, $n_X = 337$, $n_Y = 305$, $\bar{Y} = 83.4$, $s_X = 17.2$, $s_Y = 16.2$.

$$\sigma_{(\bar{X} - \bar{Y})} = \sqrt{\frac{(17.2)^2}{337} + \frac{(16.2)^2}{305}} = 1.32$$

$$z = \frac{(74.4 - 83.4) - 0}{1.34} = -6.72. \text{ Reject } H_0.$$

The difference is significant, but it's not clear whether this means joining makes you feel less well, or feeling less well makes you join, or something else. Also, the method of selecting subjects (who submits to the questionnaire?) and especially how the control group was chosen should be checked very critically.

9. These are census rather than sample figures; statistical inference is not needed. There was a 5.5% decrease in motor vehicle deaths in Oregon, which differs from the 1.6% decrease in the United States.

11. H_0: $\mu = 12,000$; H_1: $\mu > 12,000$, $\alpha = .02$, so H_0 is accepted if $z < 2.05$.

$$z = \frac{12,500 - 12,000}{2,000/\sqrt{30}} = 1.37$$

H_0 is not rejected; there has been no increase in the number of passengers.

13. H_0: $\mu = 30$; H_1: $\mu > 30$, $\alpha = .02$; the critical value of z is 2.05.

$$z = \frac{32 - 30}{6/\sqrt{64}} = 2.67$$

H_0 is rejected. He concludes that 1,800 students cannot be served between 5:30 and 7 P.M.

15. (a) H_0: $\mu = 75$, H_1: $\mu > 75$, $\sigma_X = \dfrac{8}{\sqrt{144}} = \dfrac{2}{3}$; $\alpha = .05$, so $z = 1.64$.

$$z = \frac{76 - 75}{2/3} = 3/2 = 1.5 < 1.64$$

H_0 is not rejected; the students in New Mexico did not differ.

(b) If $\alpha = .05$, H_0: $\mu = 75$ is accepted if $z < 1.64$, which implies $\overline{X} < 76.1$.
   (i) $\beta = \Pr(\overline{X} < 76.1/\mu = 78) = \Pr(z < -2.85) = .002$
   (ii) $\beta = \Pr(\overline{X} < 76.1/\mu = 77) = \Pr(z < -1.35) = .09$
   (iii) $\beta = \Pr(\overline{X} < 76.1/\mu = 76) = \Pr(z < .15) = .56$
   (iv) $\beta = \Pr(\overline{X} < 76.1/\mu = 75) = \Pr(z < 1.65) = .95$
   (v) $\beta = \Pr(\overline{X} < 76.1/\mu = 75) = \Pr(z < 3.15) = 1.00$

(c)

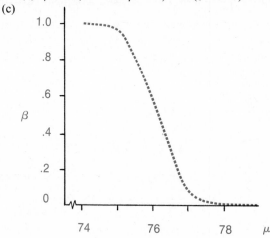

11

STUDENT'S *t* DISTRIBUTIONS

You know how to make inferences about the population mean if it can be assumed that the sampling distribution of means is a normal distribution. What can be done, however, if σ is unknown and n is small so that this assumption of normality cannot be made? The question will be answered in this chapter.

You will meet a family of t distributions (a different one for each value of n), but fear not, because you will soon discover that, as n gets larger, the corresponding t distribution gets very close to the normal distribution, and, even when n is small, a t distribution is used in much the same familiar way as a normal distribution.

You will learn when to use z scores and when to use z scores, and how to use t scores, both for a population mean and for the differences in means of two populations. Finally, you will discover how to treat paired scores in samples which are not independent.

11.1   THE NEED FOR *t* DISTRIBUTIONS

By now you should be quite comfortable with the sampling distribution of means, and you know that it is a normal distribution if (a) the population is normally distributed or (b) n is sufficiently large ($n \geqslant 30$). It is essential to recognize that if the sample means are normally distributed, then their standard or z scores $\dfrac{\overline{X} - \mu}{\sigma_{\overline{X}}}$

are normally distributed, and the table which gives areas under the standard normal curve (Table 4, Appendix C) may be consulted. But if a distribution is not normal, then the corresponding distribution of z scores is not normal either. For example, the probability distribution which consists of the scores 0, 3, 12, each

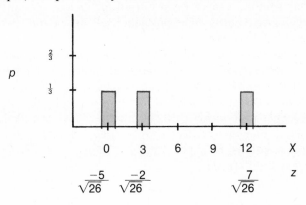

with probability 1/3, has mean 5, and standard deviation $\sqrt{26}$, and is not normally distributed. This distribution has *z* scores $-5/\sqrt{26}$, $-2/\sqrt{26}$, $7/\sqrt{26}$, which are also not normally distributed.

In discussing a confidence interval for a mean or for a null hypothesis about a mean, four possible cases arise. You are familiar with three of them.

Case 1. If *n* is large and σ is known: There is no problem. The sample means $\bar{X}$ are normally distributed and so are their standard scores $\dfrac{\bar{X} - \mu}{\sigma/\sqrt{n}}$. In practice "*n* is large" means $n \geqslant 30$. If the population distribution is very much skewed (incomes of all heads of household in the U.S., say) it would be wise if you, as a psychologist or economist setting up an experiment, were to increase the size of your sample—if you can afford the time and money.

Case 2. If *n* is small (< 30) and σ is known: Here one relies on the statement that the sampling distribution of means is normal if the population has a normal distribution (even if *n* is small). Thus, there is again no difficulty **if** the population is normally distributed. (If it is not, however, then you have no choice but to increase the size of your sample, regardless of time or expense. If you don't, your inferences are meaningless.)

Example 1

Past studies have shown that the heights of all American 18-year-olds are normally distributed, that σ is always 2.8 inches, but that mean height is gradually increasing. In 1980, a sample of 16 American 18-year-olds has a mean height of 70.0 inches. Find a 95% confidence interval for the height of all American 18-year-olds in 1980.

At a 95% confidence level, $z = \pm 1.96$.

$$\sigma_{\bar{X}} = \frac{\sigma}{\sqrt{n}} = \frac{2.8}{\sqrt{16}} = .70 \text{ inches.}$$

1.96 standard deviation units $= 1.96(.70) = 1.4$ inches. $\bar{X} \pm z\sigma_{\bar{X}} = 70.0 \pm 1.96(.70) = 70.0 \pm 1.4 = 68.6$ or 71.4 inches.

A 95% confidence interval for the mean height of all American 18-year-olds in 1980 is 68.6 to 71.4 inches.

This solution cannot be criticized for its use of standard scores and areas under the normal curve. [The problem itself can be questioned, however, on two grounds: First, how is the random sample selected? (Is it truly *random?*) Instead of "all Americans" aged 18, the population may really consist of "all seniors at Ishkosh High School" or "all apprentice plumbers in Essex County, Massachusetts." The second reason for questioning the problem is the size of the confidence interval when *n* is small. If the same example is carried out ($\bar{X} = 70$ inches, $\sigma = 2.8$ inches) but with $n = 10,000$, the new 95% confidence interval is 69.07 to 70.03 inches (compared to 68.6 to 71.4 inches when $n = 16$). If you wish to make a comparison with the height of American 18-year-olds in 1940, a small sample may hide any real difference.]

Case 3. If n is large and σ is unknown: In this case, $\sigma_{\bar{X}}$ cannot be computed by using $\dfrac{\sigma}{\sqrt{n}}$. Earlier (page 170) we discovered that s^2 is an unbiased estimator of σ^2 and that the scores $\dfrac{\bar{X} - \mu}{s/\sqrt{n}}$ are approximately normally distributed if $n \geq 30$. You will better understand the reason for this approximation shortly.

Case 4. If n is small and σ is unknown: Here s is known or can be computed. Take all possible samples of size n, compute the mean of each sample, and then for each compute a "t score."

$$t = \frac{\bar{X} - \mu}{s/\sqrt{n}}$$

If the population is normally distributed (and only in this case!), the distribution of all these t scores for different samples of size n is called a t distribution; it is not a normal distribution. (An Irishman named William Gosset worked out the t distribution. The brewery for which he worked would not allow him to publish his results, so he did so anonymously with the pseudonym "Student"; hence the term "Student's t distribution," or, simply, "t distribution.") Actually, it would be much better to use a name other than "t distribution." You can determine a t score for any sample, and all possible t scores for different samples certainly have a distribution, but not necessarily a t distribution. You will see the distinction more clearly if you think of z scores (such as $-5\sqrt{26}$, $-2/\sqrt{26}$, $7/\sqrt{26}$ as on page 240; they have a distribution which might be called a z distribution, but it is not a normal distribution. The phrase "t distribution" for t scores is used so commonly by statisticians, however, that its use will be continued here. But always remember that t scores must be taken from a normally distributed population if they are to form a t distribution.

Example 2   Two samples, each of size 100, are taken from a population with $\mu = 100$, $\sigma = 10$. The samples have the same mean: $\bar{X}_1 = \bar{X}_2 = 110$, but different standard deviations: $s_1 = 9$, $s_2 = 8$. Compute the z and t scores for each sample.

$$z = \frac{\bar{X} - \mu}{\sigma/\sqrt{n}}: z_1 = \frac{110 - 100}{10/\sqrt{100}} = 10,\ z_2 = \frac{110 - 100}{10/\sqrt{100}} = 10$$

$$t = \frac{\bar{X} - \mu}{s/\sqrt{n}}: t_1 = \frac{110 - 100}{9/\sqrt{100}} = 11.1,\ t_2 = \frac{110 - 100}{8/\sqrt{100}} = 12.5$$

Note that $z_1 = z_2$, but $t_1 \neq t_2$.

This example illustrates that if two samples from the same distribution have the same mean, they have the same z scores (even if they have different standard deviations). But two samples with the same mean and different standard deviations will have different t scores. The z scores of means of all possible samples are normally distributed if the population is normal or if n is large. The corresponding *t* scores are not identical with the *z* scores; the *t* scores are **not** normally distributed. We must look at *t* distributions to see what kind of curves they have, just as earlier we looked at normal distributions and normal curves.

11.2 CHARACTERISTICS OF *t* DISTRIBUTIONS

Student's *t* distributions have the following characteristics:

1. There is not just one *t* distribution but a different one for each value of *n*. There is one standard normal curve, and a one-page table can give areas under this curve for most of the values of *z* that we are interested in. There is a whole family of "standard" *t* curves, however. If a table were given for areas under the *t* curve for *each n* (comparable to the table for areas under the standard normal curve), it would take quite a volume. The *t* tables will be very much abbreviated. In Section 11.5 you will learn what they look like and how to use them.

2. Every *t* curve is symmetric about 0. Because of this, the **mean of every *t* distribution is 0.**

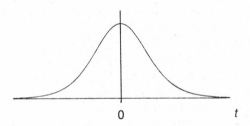

3. The highest point on the curve occurs when *t* = 0.

4. As *n* gets larger, the *t* curves get closer and closer to the normal curve. It's easy to see this by looking at some superimposed graphs:

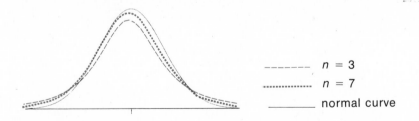

- - - - - - - *n* = 3

·················· *n* = 7

――――― normal curve

If *n* is 30 or larger, a *t* distribution and the standard normal distribution are sufficiently close so that areas under the latter may be used to approximate areas under the former. It is this closeness of *t* and *z* curves for large *n* which justifies the use of $s/\sqrt{n}$ to approximate $\sigma_{\bar{x}}$, when *n* is large and σ is unknown (Case 3, page 241).

5. Every *t* distribution is a probability distribution—that is, the area under the whole curve is 1, and the probability that a *t* score is between *a* and *b* equals the area under the curve between the lines *t* = *a* and *t* = *b*.

11.3 WHEN IS A *t* DISTRIBUTION USED?

A *t* distribution is used when:

> (a) the population is normally distributed **and**
> (b) σ is unknown (but *s* is known or can be computed) **and**
> (c) $n < 30$. ← ALL OTHERS

If the population is not normally distributed, then you will be making an error. How large is the error? It depends on how far from being a normal distribution the population is. If the distribution is almost normal, then your error is small. If it is very far from being normal, then the *t* distribution is pretty useless. In practice, the problem is that you often don't know whether or not the population follows a normal curve.

What do you do then? One answer is to use a different test—one that does not require the assumption that the population is normal. Such tests are discussed in Chapter 16. Another answer is to increase the size of the sample. The less confident you are that the population is normal, the larger your sample size should be. Be very careful not to use very small samples unless you are quite certain that the population is normally distributed. This is the main trouble with *t* tests; they are most questionable when they are most useful (that is, with small samples), but many a researcher has forgotten to question the population distribution. Beware!

If σ is known, then use standard or *z* scores and the normal distribution.

If $n > 30$, then the *t* and *z* curves are so close together than it doesn't much matter which you use if the population is normal.

11.4 DEGREES OF FREEDOM

How is a *t* distribution used? Before going on, we must sidetrack for a moment and consider degrees of freedom.

Imagine three children playing a very simple game: three cards are marked, respectively, 0, 10, and 20. The cards are shuffled, and each child in turn chooses one card. Jimmy is impulsive and grabs a card first; he gets 10 points. Mary chooses next, and gets 0. Tommy takes the last card; he must get 20 points. If, however, Jimmy drew 0 and Mary 20, then Tommy would get 10. The point is that once two children have chosen their cards, the third child's card is fixed. Two of the children choose freely: there are two **degrees of freedom.**

If four scores have a mean of 50, how many of the scores can be freely chosen? Try it and see: fill in the blanks with whatever scores you like, so long as their mean is 50:

___ ___ ___ ___

I'm going to choose 40 for the first score, then 30 for the second, and 70 for the third. I choose these numbers arbitrarily; because 40 happened to be first, I feel no inhibitions about choosing 30 or 40 or 1,000,000 or -32 for the second. The same is true for the third. But once three have been chosen, then the fourth is fixed by the requirement that the mean be 50. If my first three choices are 40, 30, 70, then my fourth choice can *only* be 60:

$$\underline{40} \ \underline{30} \ \underline{70} \ \underline{60} \qquad \bar{X} = 50$$

If I try anything else for the fourth score, then the mean no longer equals 50:

$$\underline{40} \ \underline{30} \ \underline{70} \ \underline{20} \qquad \bar{X} = 40$$

I have freedom in choosing three of the four scores, and then the fourth is determined. There are three **degrees of freedom.**

If n scores have a mean $\bar{X}$, $n - 1$ can be freely chosen, and then the last is determined; there are $n - 1$ **degrees of freedom.**

For finding a confidence interval or testing a hypothesis about the mean using a t distribution for one sample of size n, the number of degrees of freedom is $n - 1$. The letter D is used to denote the number of degrees of freedom:

$$D = n - 1$$

11.5   TABLE OF *t* DISTRIBUTIONS

For the standard normal distribution, one table (Table 4, Appendix C) supplies enough information to find the area under any normal curve. But t distributions differ for different numbers of degrees of freedom, and each requires a separate table if the same amount of information is to be given. As a result, information about t distributions is given in very concentrated form.

Table 5, Appendix C, shows values of t for selected areas under the t curve. Different values of D appear in the first column. The table is adapted for efficient use for either one- or two-tail tests. Note very carefully, however, that areas under t curves are given for one or both tails, whereas areas under the normal curve are given from 0 to z.

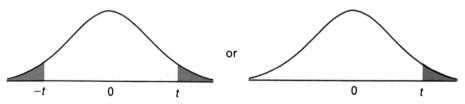

Values of t (See Table 5, Appendix C)

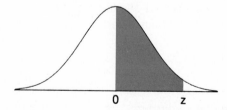

Areas Under the Normal Curve
(See Table 4, Appendix C)

Example 1   If $D = 8$, 5% of t scores are above what value?

Look in Table 5, Appendix C, along the row labeled "One tail" to the value .05; the intersection of the .05 column and the row with 8 in the D column gives the value of t: $t = 1.86$.

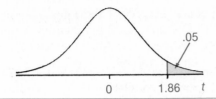

Example 2   Find t_0 if $n = 13$ and 95% of t scores are between $-t_0$ and $+t_0$.

$D = n - 1 = 13 - 1 = 12$. If 95% of t scores are between $-t_0$ and $+t_0$, then 5% are in the two tails. Look at Table 5, Appendix C along the row labeled "Two tail" to the value .05; the intersection of this .05 column and the row with 12 in the D column gives $t_0 = 2.18$.

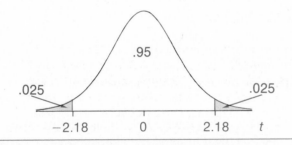

Example 3   If $D = 5$, what is the probability that a t score is above 2.02 or below -2.02?

Two tails are implied. Look along the "$D = 5$" row to find the entry 2.02. The probability is .10.

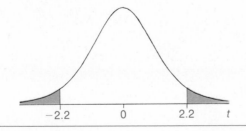

11.6   CONFIDENCE INTERVALS FOR, AND TESTS OF HYPOTHESES ABOUT, THE MEAN

Relax; confidence intervals and tests of hypotheses about the mean are carried out with the t distribution just as for the normal distribution, except that you must consider the number of degrees of freedom and use Table 5 instead of Table 4.

Finding a confidence interval with a *t* distribution will, of course, always involve a two-tail test. It is necessary to subtract the confidence coefficient from 1 to find the area in the two tails. The confidence interval is wide for small values of *n*, and therefore a *t* distribution is used more commonly for testing a hypothesis.

Example 1   A light bulb manufacturer claims his bulbs have a mean life of 1,000 hours. Is his claim justified, at a .05 level of significance, if a random sample of 25 bulbs has a mean life of 994 hours with a standard deviation of 30 hours? Assume the distribution of burning life for all bulbs is normal.

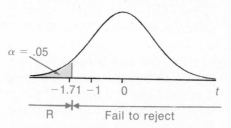

$H_0: \mu = 1,000$; $H_1: \mu < 1,000$; one-tail test; $\alpha = .05$.

$D = 24$; H_0 is rejected if $t < -1.71$.

$$t = \frac{\bar{X} - \mu}{s/\sqrt{n}} = \frac{994 - 1,000}{30/\sqrt{25}} = -1.00$$

His claim is justified; H_0 is still accepted.

Example 2   Find a 98% confidence interval for μ if a random sample of size 9 from a normally distributed population has a mean of 14 miles and a standard deviation of 2 miles.

$\bar{X} = 14$ miles, $s = 2$ miles, $n = 9$, $D = 8$.

With a .98 confidence coefficient when $D = 8$, $t = \pm 2.90$ (note that the area in the two tails is $1 - .98 = .02$).

Since $t = \dfrac{\bar{X} - \mu}{s\sqrt{n}}$, then $\mu = \bar{X} - t\dfrac{s}{\sqrt{n}}$; $\dfrac{s}{\sqrt{n}} = \dfrac{2}{\sqrt{9}} = .667$; 2.90 *t* units = 2.90(.667) = 1.93 miles.

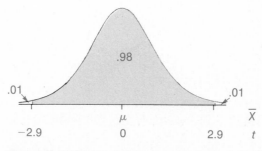

The probability is .98 that the mean of a random sample is within 2.90 *t* units or 1.93 miles of μ. This sample is random, with $\bar{X} = 14.0$. Therefore a 98% confidence interval for μ is $14.0 - 1.93$ to $14.0 + 1.93 = 12.1$ to 15.9 miles.

In a confidence interval for the population mean using a t distribution, the confidence limits for μ are

$$\bar{X} - t \frac{s}{\sqrt{n}} \text{ and } \bar{X} + t \frac{s}{\sqrt{n}},$$

where t is determined by the confidence coefficient and $D = n - 1$ degrees of freedom.

Example 3
The specifications for a certain drug call for 30% aspirin in each pill. Sixteen pills are chosen at random and analyzed; their mean content of aspirin is 30.4% with a standard deviation of .8%. Does the drug satisfy the specifications at a .01 level of significance? Assume the errors are normally distributed.

H_0: $\mu = .30$; H_1: $\mu \neq .30$. Two-tail test; $\alpha = .01$.

$\bar{X} = .304$, $s = .008$, $n = 16$, $D = 15$.

$D = 15$, $\alpha = .01$; H_0 is rejected if t is > 2.95 or < -2.95.

$$t = \frac{.304 - .30}{.008/\sqrt{16}} = 2.00.$$

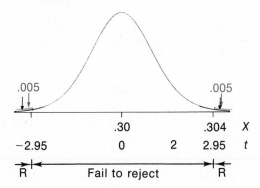

We fail to reject H_0; the specifications are satisfied at a .01 level of significance.

11.7 EXERCISES

1. Describe a t distribution for samples of size 10 from a population whose mean is 54.0: How is it formed? What requirement(s) exist for the population? What is its mean? How many degrees of freedom?

2. 5% of the scores in a t distribution with 6 degrees of freedom are over what value?

3. Find t_0 if $\Pr(t < t_0) = .10$ and $D = 9$.

4. Find t_0 if $\Pr(-t_0 < t < t_0) = .98$ and $D = 8$.

5. What is the probability that the mean of a sample of size 8 drawn at random from a normal population will have a t score between -1.90 and $+1.90$?

6. In a *t* distribution with 17 degrees of freedom, 1% of *t* scores are above what score?

7. The mean hourly pay of all kitchen workers in Ajax Miniburger Company's restaurants is $3.64. What is the probability that a random sample of 16 kitchen workers have a mean hourly pay greater than $4.09 if the standard deviation of the sample is $.70? Assume that the pay of all kitchen workers is normally distributed in order to answer the question; is this likely to be the case?

8. For a new fuse manufactured by the Portland Fuse Company, the mean time for the filament to melt is 5 seconds, according to specifications, with a standard deviation of 1 second. A random sample of 16 fuses has a mean time of 5.8 seconds with a standard deviation of .95 second. Does the batch of fuses from which the sample was taken meet the specifications at a .01 level of significance? Assume the times of all fuses in the batch are normally distributed.

9. The following numbers are taken in order from a random number table: 3, 9, 5, 2, 7, 1, 7, 6, 2, 4.

Test to see whether the table from which these numbers were taken is a random number table; $\alpha = .05$. (The mean of a table of random numbers is 4.5.)

10. An inspector from the Department of Weights and Measures weighs 10 1-pound samples of peanut butter; he finds their mean weight is 15.8 ounces with a standard deviation of .4 ounces

(a) Do the weights of packages of peanut butter sold by the shop from which these samples were taken differ from the announced weight? ($\alpha = .05$.)

(b) Are they significantly lighter than they should be? (.05 level of significance.)

11. A home owner needs to determine the pH of her soil; she will add lime if it is less than 6.5. She takes 16 samples from cores whose location is randomly chosen, and finds the average pH of the samples is 6.3 with a standard deviation of .3. Must she add lime to the soil? ($\alpha = .05$.)

12. Ten journeyman typesetters in a printing office have an average of 2.3 mistakes per column set. A new apprentice is taken on; a random sample of 10 of his columns has a mean of 2.6 errors with a standard deviation of .5. Does the apprentice have significantly more mistakes than the experienced printers?

13. While visiting New Jersey for a week, Peter took the train from Summit to Madison, twice. The first time the train was 20 minutes late, the second time 2 minutes late. Find a 99% confidence interval for how late the train from Summit to Madison is on the average. State any misgivings you may have about this particular problem.

14. T. Rotatori tested the extent to which overweight, mentally retarded adults lose weight, using a behavioral weight-reduction treatment he developed (see Exercises 1–8, page 84). Ten patients lost weight as follows during a 7-week treatment period: 10, 10, 10, −1 (that's a weight gain), 6, 5, 4, −1, 4, −3. ($\overline{X} = 4.4$, $s = 4.8$; check these.)

(a) Find a 95% confidence interval for the mean weight loss in overweight, mentally retarded adults with this treatment.

(b) Test H_0: $\mu = 0$ at the 5% level. Should the test be a one- or two-tail test? Why?

ANSWERS

1. The population must be normal (or approximately so). Each t score is a number found by subtracting 54.0 from the mean of a sample of size 10 and dividing by the standard deviation of the sample. The t distribution is the set of t scores for all different samples. Its mean is 0. $D = 15$.

2. 1.94.

3. −1.38.

4. 2.90.

5. $(D = 7)$ $1 - .10 = .90$.

6. 2.57.

7. $t = \dfrac{\overline{X} - \mu}{s/\sqrt{n}} = \dfrac{4.09 - 3.64}{.70/\sqrt{16}} = 2.6$; $D = 15$. $\Pr(t > 2.6) = .01$.

8. $\sigma(= 1)$ is known, so use z scores, not t scores! H_0: $\mu = 5$; H_1: $\mu \neq 5$; two-tail test; $\alpha = .01$. H_0 is rejected if the z score of the sample mean is > 2.58 or < -2.58.

$$\overline{X} = 5.8,\ n = 16,\ z = \dfrac{5.8 - 5}{1/\sqrt{16}} = 3.2.$$

H_0 is rejected; the specifications are not met.

9. Each of the numbers 0 through 9 should appear equally often in a random number table; this is called a uniform distribution. It is not a normal distribution or anything close to that, so don't try a t test.

10. (a) H_0: $\mu = 16.0$, H_1: $\mu \neq 16.0$. $\overline{X} = 15.8$, $s = .4$, $n = 10$, $t = \dfrac{15.8 - 16.0}{.4/\sqrt{10}} = -1.58$; $\alpha = .05$, $D = 9$, two-tail test: the critical values of t are ± 2.26. H_0 is still accepted.
(b) H_0: $\mu = 16.0$, H_1: $\mu < 16.0$. Same computed value of t as in (a). The critical value of t for a one-tail test with $D = 9$, $\alpha = .05$, is -1.83. H_0 is still accepted.

11. H_0: $\mu = 6.5$, H_1: $\mu < 6.5$; $t = \dfrac{6.3 - 6.5}{.3/\sqrt{16}} = -2.67$. $D = 15$, $\alpha = .05$, critical $t = -1.75$.
H_0 is rejected; she should add lime.

12. The mistakes of the new apprentice are unlikely to be normally distributed.

13. At the 99% level with $D = 1$, $t = \pm 63.7$, $\overline{X} = 11$, $s = 12.7$, $\mu = 11 \pm (63.7)\dfrac{12.7}{2} = -393.5$ or $+415.5$. A 99% confidence interval is -394 to 416 minutes (394 minutes early to 416 minutes late!) Clearly, one degree of freedom is too shaky. Peter would have to take that train more often to get a reasonable confidence interval.

14. (a) At a 95% confidence level with $D = 9$, $t = \pm 2.26$. $\mu = 4.4 \pm (2.26)\dfrac{4.8}{\sqrt{10}} = 1.0$ or

7.8. A 95% confidence interval is 1.0 to 7.8 pounds.

(b) $H_0: \mu = 0$, $H_1: \mu > 0$ (one-tail since we are only interested in their losing weight, not gaining it), $\alpha = .05$, $D = 9$, and the critical value for t is $t = 1.83$.

$$\text{Computed } t = \frac{4.4 - 0}{4.8/\sqrt{10}} = 2.90.$$

Hence, H_0 is rejected.

11.8   DIFFERENCES OF MEANS FOR INDEPENDENT SAMPLES

A t distribution can be used for testing hypotheses about differences of means for independent samples **if both populations are normal and have the same variance.**[*] If you suspect this is not the case, then sample sizes should be increased (at least 30 for each) so the sampling distribution of $\bar{X} - \bar{Y}$ approximates a normal distribution. It is also assumed that the samples are both random and are chosen independently.

Confidence intervals are determined and hypotheses are tested using t tests just as with z tests except that

(a) $\sigma_{(\bar{X}-\bar{Y})} = \sqrt{\dfrac{(n_X - 1)s_X^2 + (n_Y - 1)s_Y^2}{n_X + n_Y - 2}\left(\dfrac{1}{n_X} + \dfrac{1}{n_Y}\right)}$

This formula is arrived at by "weighting" the two sample variances to arrive at an estimate of the common population variance σ.[**]

(b) $D = (n_X - 1) + (n_Y - 1) = n_X + n_Y - 2$.

(c) Table 5 is used instead of Table 4.

(d) Confidence limits for $\mu_X - \mu_Y$ are $\bar{X} - t\sigma_{(\bar{X}-\bar{Y})}$ and $\bar{X} + t\sigma_{(\bar{X}-\bar{Y})}$.

Example 1   A random sample of 17 third graders who read poorly have a mean IQ of 98 with a standard deviation of 10; a random sample of 10 third graders who read well have a mean IQ of 101 with a standard deviation of 9. At a .05 level of significance, is

---

[*] There are special techniques that can be used if the variances differ, but we shall not consider them. If the samples are approximately the same size, the methods described here will work pretty well. If you are uncertain, then increase the size of both samples.

[**] You have already used this kind of weighting in finding the mean of repeated scores:

X	f	Xf	Variance	Degrees of Freedom	Product
2	9	2(9)	s_X^2	$n_X - 1$	$(n_X - 1)s_X^2$
2	4	5(4)	s_Y^2	$n_Y - 1$	$(n_Y - 1)s_Y^2$
		2(9) + 5(4)			$(n_X - 1)s_X^2 + (n_Y - 1)s_Y^2$

$$\mu = \frac{2(9) + 5(4)}{9 + 4} \qquad \sigma = \sqrt{\frac{(n_X - 1)s_X^2 + (n_Y - 1)s_Y^2}{(n_X - 1) + (n_Y - 1)}}$$

If both populations have the same variance, $\sigma_X = \sigma_Y = \sigma$, so

$$\sigma_{(\bar{X}-\bar{Y})} = \sqrt{\frac{\sigma_X^2}{n_X} + \frac{\sigma_Y^2}{n_Y}} = \sigma\sqrt{\frac{1}{n_X} + \frac{1}{n_Y}} = \sqrt{\frac{(n_X - 1)s_X^2 + (n_Y - 1)s_Y^2}{(n_X - 1) + (n_Y - 1)}}\sqrt{\frac{1}{n_X} + \frac{1}{n_Y}}$$

there a difference in mean IQ of poor and good readers? Assume the IQ's of both groups are normally distributed and have the same variance.

$H_0: \mu_X - \mu_Y = 0; H_1: \mu_X - \mu_Y \neq 0;$ two-tail test; $\alpha = .05.$

$\bar{X} = 98, s_X = 10, n_X = 17; \bar{Y} = 101, s_Y = 9, n_Y = 10$

$$\sigma_{(\bar{X}-\bar{Y})} = \sqrt{\frac{(16)(10^2) + (9)(9^2)}{17 + 10 - 2}\left(\frac{1}{17} + \frac{1}{10}\right)} = \sqrt{\left(\frac{2329}{25}\right)\left(\frac{27}{170}\right)} = 3.85$$

$$t = \frac{(\bar{X} - \bar{Y}) - (\mu_X - \mu_Y)}{\sigma_{(\bar{X}-\bar{Y})}} = \frac{(98 - 101) - 0}{3.85} = -.78$$

$D = 25, \alpha = .05,$ two-tail test, so $t = \pm 2.06.$

A t score of $-.78$ falls in the "fail to reject" region.

There is no difference in the mean IQ of poor and good third-grade readers at a .05 level of significance.

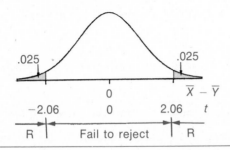

Example 2   Cotton threads are made by two different processes which are to be compared. A random sample of 20 threads manufactured by the first process has a mean breaking strength of 12 ounces with a standard deviation of 1.5 ounces, while a sample of 5 threads manufactured by the second process has a mean breaking strength of 10 ounces with a standard deviation of 2.0 ounces. Assume both populations are normal and have the same variance.

(a) Test the hypothesis that there is no difference in the mean breaking strength of cotton threads made by the two processes at a .05 level of significance.

(b) Find, with 95% confidence, the difference in mean breaking strengths of threads made by the two processes.

(a) $H_0: \mu_X - \mu_Y = 0; H_1: \mu_X - \mu_Y \neq 0.$

$\bar{X} = 12, s_X = 1.5, n_X = 20; \bar{Y} = 10, s_Y = 2.0, n_Y = 5$

$D = 20 + 5 - 2 = 23, \alpha = .05,$ so $t = \pm 2.07$

$$t = \frac{(\bar{X} - \bar{Y}) - (\mu_X - \mu_Y)}{\sqrt{\frac{(n_X - 1)s_X{}^2 + (n_Y - 1)s_Y{}^2}{n_X + n_Y - 2}\left(\frac{1}{n_X} + \frac{1}{n_Y}\right)}}$$

$$= \frac{(12 - 10) - 0}{\sqrt{\frac{19(1.5^2) + 4(2.0^2)}{23}\left(\frac{1}{20} + \frac{1}{5}\right)}} = \frac{2.0}{.80} = 2.5$$

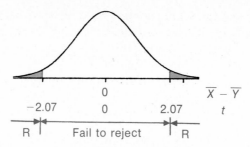

H_0 is rejected; there is a difference.

(b) 95% confidence, $D = 23$, so $t = \pm2.07$ (two-tail test with $\alpha = 1 - .95 = .05$). 2.07 t units $= 2.07(.80) = 1.66$ ounces.

There is a .95 probability that random samples from these populations have a difference of means that is within 1.66 ounces of $\mu_X - \mu_Y$; these samples are random and have a difference of means $= 12 - 10 = 2.0$ ounces. Confidence limits for μ are $2.0 - 1.66$ and $2.0 + 1.66 = .3$ to 3.7 ounces.

A 95% confidence interval for the difference in mean breaking strengths is .3 to 3.7 ounces.

11.9 EXERCISES

Assume populations are normal and have the same variance.

1. In the fall semester 5 students, chosen at random in a statistics course, have a mean grade of 75 with a standard deviation of 5. In the spring semester, a random sample of 10 students from a statistics course have a mean grade of 77 with a standard deviation of 5. At a .05 level of significance, do students get better grades in statistics in the spring semester?

2. Eight white lab coats were washed in Blue Suds and 6 in Whitelite detergent. After washing, the coats were rated for whiteness on a 1–10 scale (10 is whitest); the mean for the Blue Suds coats was 7.5 with a standard deviation of .3; the mean for Whitelite was 7.2 with a standard deviation of .4. Do the two detergents differ, at a .05 level of significance?

3. A high school teachers' group is investigating summer work patterns. It finds that the mean monthly income of 20 randomly selected teachers who teach in the summer is $834 with a standard deviation of $100, while a random sample of 10 teachers who sell real estate during the summer is $934 with a standard deviation of $50. At a .02 level of significance, is there any difference in the earnings of the two groups?

4. Two types of guns are to be tested at Picatinny Arsenal. Twelve of the first type and 10 of the second type, chosen at random, are tested. Those of the first type have a mean error of 3.0 cm. with a standard deviation of .40 cm., while those of the second type have a mean error of 2.1 cm. with a standard deviation of .30 cm. At a .05 level of significance, is the mean error of the second type 1 cm. less than that of the first type, or is the difference less than 1 cm.?

5. Two different machines are used for manufacturing automobile horns in a factory. The number produced in an hour on each machine is noted for hours chosen at random over a period of several months. Notes on the first machine, taken over 10 hours, show a mean production per hour of 100 horns with a standard deviation of 6.0. The second machine, noted also for 10 hours, has a mean production of 96 horns with a standard deviation of 5.0. Does the first machine produce at least 5 more horns per hour at a .05 level of significance?

6. A psychologist discovers that 10 orphans, viewing pictures (which have the same size) of friends and of unknown children, judge the friends to be 2 inches taller, on the average, with a standard deviation of .05 inch. Eight children living with both parents are given the same test and estimate friends to be 1.5 inches taller, with a standard deviation of .04 inch. Find a 95% confidence interval for the difference in heights estimated by orphans and by children living with both parents.

7. T. Rotatori tested the effect of a behavioral weight-reduction treatment on 10 obese, mentally retarded patients (see Exercises 1–8, p. 84). Their mean weight loss was 4.4 pounds, with a standard deviation of 4.8 pounds. Does the treatment help? Rotatori compared the results with weight losses for 8 similar adults who were only told to try to lose weight; their losses were −5 (a weight gain), 3, 6, 0, −4, −9, −2, −8 pounds. At a .05 level of significance, did the treated group lose more weight?

8. Perhaps fatter adults can lose weight more easily, so Rotatori also looked at the percentage of weight lost, with the following results (a negative percent is a weight gain).
Treated patients: 5.0, 5.9, 2.6, −0.6, 2.8, 2.5, −0.6, 2.5, −2.2, 2.6%.
Control group: −3.1, 1.7, 3.6, 0, −2.3, −5.8, −1.2, −5.1%.
Is the percentage weight loss for the treated group more than for the control group, at a .05 level of significance?

9. Do the behaviorally treated group (see Exercises 7 and 8) keep their weight down? Rotatori gave occasional further help for 7 more weeks, then checked weights after another 10 weeks (23 weeks after beginning of treatment). The mean percentage weight loss of the 10 treated adults was 2.01%, with a standard deviation of 2.42. After 23 weeks, the 8 controls had a mean percentage weight gain of 1.50%, with a standard deviation of 3.03.
   (a) Have the treated patients lost weight, at a (i) .01, (ii) .05 level of significance?
   (b) Have the treated group lost more weight (by percent) than the control group, again at a (i) .01 and (ii) .05 level of significance?

ANSWERS

1. $\bar{X} = 75$, $s_X = 5$, $n_X = 5$; $\bar{Y} = 77$, $s_Y = 5$, $n_Y = 10$. H_0: $\mu_X - \mu_Y = 0$; H_1: $\mu_X - \mu_Y < 0$. One-tail test; $\alpha = .05$, $D = 13$; H_0 is rejected if $t < -1.77$.

$$t = \frac{(75 - 77) - 0}{\sqrt{\frac{4(5^2) + 9(5^2)}{13} \left(\frac{1}{5} + \frac{1}{10}\right)}} = -.73; H_0 \text{ is still accepted; grades are no better in}$$

the spring.

2. $H_0: \mu_X - \mu_Y = 0; H_1: \mu_X - \mu_Y \neq 0, \bar{X} = 7.5, \bar{Y} = 7.2.$

$$t = \frac{(7.5 - 7.2) - 0}{\sqrt{\frac{7(.3^2) + 5(.4^2)}{8 + 6 - 2} \left(\frac{1}{8} + \frac{1}{6}\right)}} = 1.6.$$

$D = 12$, so the critical values of t are ± 2.18. H_0 cannot be rejected on the basis of these tests. The ratings of the two detergents do not differ.

3. $H_0: \mu_X - \mu_Y = 0; H_1: \mu_X - \mu_Y \neq 0, \bar{X} = 834, n_X = 20, s_X = 100; \bar{Y} = 934, n_Y = 10,$ $s_Y = 50.$

$$t = \frac{(834 - 934) - 0}{\sqrt{\frac{19(100^2) + 9(50^2)}{20 + 10 - 2} \left(\frac{1}{20} + \frac{1}{10}\right)}} = \frac{-100}{33.7} = -3.0.$$

$D = 28$; the critical values of t are ± 2.47, so H_0 is rejected; there is a difference in the mean monthly pay of those who teach and those who sell real estate.

4. $H_0: \mu_X - \mu_Y = 1.0; H_1: \mu_X - \mu_Y < 1.0.$

$$\sqrt{\frac{11(.40^2) + 9(.30^2)}{20} \left(\frac{1}{12} + \frac{1}{10}\right)} = .153.$$

$t = \dfrac{(3.0 - 2.1) - 1.0}{.153} = -.65. D = 12 + 10 - 2 = 20, \alpha = .05;$ the critical value of t is $-1.73.$ H_0 is not rejected; the mean error of the second type is 1 cm less.

5. $H_0: \mu_X - \mu_Y = 5; H_1: \mu_X - \mu_Y < 5, D = 10 + 10 - 2 = 18, \alpha = .05,$ critical value of $t = -1.73.$

$$\sqrt{\frac{9(6^2) + 9(5^2)}{18} \left(\frac{1}{10} + \frac{1}{10}\right)} = 2.47, t = \frac{(100 - 96) - 5}{2.47} = -.40.$$

H_0 is still accepted.

6. $\bar{X} = 2.0, s_X = .05, n_X = 10; \bar{Y} = 1.5, s_Y = .04, n_Y = 8; D = 10 + 8 - 2 = 16; 95\%$ confidence interval, so $t = \pm 2.12.$

$$\sqrt{\frac{9(.05^2) + 7(.04^2)}{16} \left(\frac{1}{10} + \frac{1}{8}\right)} = .022.$$

Confidence limits are $(2 - 1.5) \pm 2.12(.022) = .5 \pm .05 = .45$ or $.55$. 95% confidence interval is $.45$ to $.55$ inches. If the psychologist isn't sure that the two populations are normally distributed, he should have larger samples.

7. $\bar{X} = 4.4, s_X = 4.8, n_X = 10; \bar{Y} = -2.4, s_Y = 5.2, n_Y = 8. H_0: \mu_X - \mu_Y = 0; H_1:$ $\mu_X - \mu_Y > 0, \alpha = .05, D = 16,$ and the critical value of t is $t = 1.75.$

$$\text{Computed } t = \frac{(4.4 + 2.4) - 0}{\sqrt{\frac{9(4.8^2) + 7(5.2^2)}{16} \left(\frac{1}{10} + \frac{1}{8}\right)}} = 2.88.$$

Hence, H_0 is rejected.

8. $\bar{X} = 2.05$, $s_X = 2.52$, $n_X = 10$; $\bar{Y} = -1.53$, $s_Y = 3.23$, $n_Y = 8$.

$$\text{Computed } t = \frac{(2.05 + 1.53) - 0}{\sqrt{\frac{9(2.52^2) + 7(3.23^2)}{16} \left(\frac{1}{10} + \frac{1}{8}\right)}} = 2.65.$$

Hence, we reject H_0.

9. (a) H_0: $\mu = 0$; H_1: $\mu > 0$, $D = 9$. Computed $t = \dfrac{2.01 - 0}{2.42/\sqrt{10}} = 2.63$. For $\alpha = .05$, the critical value of $t = 1.83$ and H_0 is rejected. For $\alpha = .01$, the critical value of $t = 2.82$ and we fail to reject H_0.

(b) H_0: $\mu_X - \mu_Y = 0$; H_1: $\mu_X - \mu_Y > 0$, $D = 16$.

$$\text{Computed } t = \frac{(2.01 + 1.50) - 0}{\sqrt{\frac{9(2.42^2) + 7(3.03^2)}{16} \left(\frac{1}{10} + \frac{1}{8}\right)}} = 2.74.$$

For $\alpha = .01$, the critical value of $t = 2.58$ and we reject H_0. Therefore, H_0 is also rejected at a .05 level.

11.10   DIFFERENCES BETWEEN MEANS OF DEPENDENT SAMPLES

Sometimes you do not have independent samples: you want to take one sample and compare the people in it before and after a certain treatment, or you want to be sure certain factors (age, religion, or income, for example) do not affect the results by pairing people with the same age, religion, or income. The purpose in either case is to control variables other than the one being tested. A difference of means test cannot be used here, since you do not have independent samples. Instead, individuals in the two samples are paired, and a t test is made on the mean difference of the scores in each pair. (Note that the differences in scores of all pairs in the population(s) should be normally distributed.)

> **Notation:** d = difference in each pair
> $\bar{d}$ = the mean of these differences
> s_d = the standard deviation of these differences
> μ_d = the mean of the population differences
> $D = n - 1$ if there are n pairs of scores
>
> **Formula for hypothesis tests:**
>
> $$t = \frac{\bar{d} - \mu_d}{s_d/\sqrt{n}} \qquad s_d = \sqrt{\frac{(d - \bar{d})^2}{n - 1}}$$
>
> **Confidence limits for μ_d:**
>
> $$\bar{d} \pm t\frac{s_d}{\sqrt{n}}$$

Example 1 A random sample of 10 students takes a calculus quiz, and receives the grades shown below (X). Then a review session in algebra is held and a similar quiz given (Y). At a .05 level of significance, are the grades better on the second quiz?

Student	Quiz 1 (X)	Quiz 2 (Y)	$Y - X = d$	$d - \bar{d}$	$(d - \bar{d})^2$
1	80	84	+4	1	1
2	50	56	+6	3	9
3	78	81	+3	0	0
4	92	90	−2	−5	25
5	76	75	−1	−4	16
6	70	75	+5	2	4
7	62	72	+10	7	49
8	87	90	+3	0	0
9	95	93	−2	−5	25
10	68	72	+4	1	1
			+30		130

$$\text{Mean of differences} = \bar{d} = \frac{30}{10} = 3.0.$$

$$H_0: \mu_d = 0; \ H_1: \mu_d > 0$$

$$\text{Standard deviation of differences} = s_d = \sqrt{\frac{130}{9}} = 3.8.$$

A t test is applied to 10 differences in test scores: $\mu_d = 0$ implies that the mean difference in test scores for the whole population is 0; H_1 says a one-tail (right) test should be made.

$n = 10$, so there are 9 degrees of freedom; the number of differences (10) is used rather than the number of test scores (20). $\alpha = .05$, $D = 9$, one-tail test; the critical value of t is 1.83.

$$t = \frac{\bar{d} - \mu_d}{s_d/\sqrt{n}} = \frac{3 - 0}{3.8/\sqrt{10}} = 2.5.$$

H_0 is rejected; there is a significant difference in scores on the two tests. This does not show, however, that this difference is caused by the review of algebra. It might be due to learning from mistakes on the first quiz, for example.

Example 2 From 240 students registered for a statistics course, 8 pairs of students are selected who are well matched for number of years and grades of math courses in high school and college, college class (freshman, sophomore, and so on), and cumulative grade-point average up to the beginning of the statistics course. Then students in each pair are randomly assigned, one (X) to a section of 30 that meets 3 times a week, the other (Y) to 3 lectures each week for 210 students plus a recitation for 15 students once a week. The grades of the 8 pairs of students on the final examination common to both groups are shown below. At a .01 level of significance, do the mean grades on this examination differ?

Pair	X	Y	$X - Y = d$	$d - 3$	$(d - 3)^2$
1	90	82	+8	5	25
2	85	95	−10	−13	169
3	75	79	−4	−7	49
4	81	78	+3	0	0
5	95	88	+7	4	16
6	95	91	+4	1	1
7	60	50	+10	7	49
8	87	81	+6	3	9
			24		318

H_0: $\mu_d = 0$; H_1: $\mu_d \neq 0$

$\bar{d} = \dfrac{24}{8} = 3.0$, $s_d = \sqrt{318/7} = 6.74$.

$n = 8$, $D = 7$, $\alpha = .01$, two-tail test; the critical values of t are ± 3.5.

$t = \dfrac{3 - 0}{6.74/\sqrt{8}} = 1.26$.

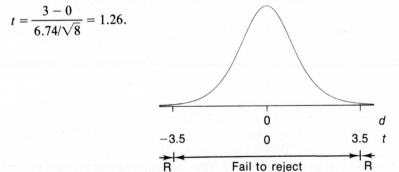

H_0 is not rejected; there is no difference in final grades.

11.11   EXERCISES

1. A test of anxiety state was given by C. Smoszyk to patients who knew they had cancer both before and after a talking-it-out treatment to relieve anxiety. (See Exercise 9, page 84.) The results were as follows:

Patient	A	B	C	D	E	F	G	H	I	J
Before	23	48	69	58	48	44	57	20	26	34
After	20	26	67	57	33	24	44	21	26	27
Change	−3	−22	−2	−1	−15	−20	−13	+1	0	−7

Does the treatment reduce the anxiety state of cancer patients, at a 5% significance level?

Use the following data in Exercises 2 to 6: Each of 15 volunteers with anxiety symptoms was given chlordiazepoxide for one week and a placebo (inactive pill) for one week, but didn't know in which order.* Anxiety was measured by the Hamilton Anxiety Rating Scale (HARS) and also by the Hopkins Symptoms Checklist (HSCL). The table below shows the difference between anxiety score at the end of the chlordiazepoxide week and at the end of the placebo week.

Subject	1	2	3	4	5	6	7	8	9	10	11	12	13	14	15
Sex	M	M	M	F	F	F	M	F	F	M	F	M	M	F	F
HARS	−5	−2	−7	−1	−5	−5	3	3	−12	−12	−20	−7	−4	−18	−11
HSCL	−6	−9	−10	−7	0	0	9	0	−1	−4	−9	−9	−6	−9	−1

2. Do people on the average have lower HARS anxiety scores after chlordiazepoxide treatment than after no treatment? ($\alpha = .05$.)

3. Do people on the average have lower HSCL anxiety scores after chlordiazepoxide treatment than after no treatment, at a .01 level of significance?

4. Do men and women differ in the reduction of anxiety scores associated with chlordiazepoxide treatment compared with placebos, as measured on the HARS scale? ($\alpha = .05$.)

5. Do women show less change than men in HSCL scores? ($\alpha = .01$.)

6. Do HARS and HSCL scores differ for all people (men and women together) in measuring the difference between treatment with chlordiazepoxide and with placebo? ($\alpha = .05$.)

7. Highly accurate measurements of tiny concentrations of metals in solution can be obtained by an atomic absorption spectrometer hooked up with a computer. More and more, mini-computers have been doing the job. Willmot and MacKenzie** studied how well an even smaller "micro" computer can do it. Measurements of 20 manganese solutions with a micro-computer and with a mini-computer were as follows (in parts per million):

Micro: n = 20, $\bar{X} = 750.6$, $s_X = 1.5$
Mini: n = 20, $\bar{Y} = 750.8$, $s_Y = 2.4$

Do the means differ, at a .01 level of significance?

---

*Lin, K. M., and Friedel, R. O.: Relationship of Plasma Levels of Chlordiazepoxide and Metabolites to Clinical Response. *American Journal of Psychiatry, 136* (1979), pages 18–23.
**Willmot, F. W., and MacKenzie, I.: Data Processing for Atomic Absorption Spectrometry with a Microcomputer. *Analytica Chimea Acta, 103* (1978), 401–408.

ANSWERS

1. H_0: $\mu_d = 0$, H_1: $\mu_d < 0$; $\alpha = .05$, $D = 9$ and for a one-tail test the critical value of t is
   -1.83. Computed $t = \dfrac{-8.2 - 0}{8.63/\sqrt{10}} = -3.0$, so H_0 is rejected.

2. H_0: $\mu_d = 0$, H_1: $\mu_d < 0$; $\bar{d} = -6.9$, $s_d = 6.8$, $t = \dfrac{-6.9 - 0}{6.8/\sqrt{15}} = -3.93$; $D = 14$, critical $t = -1.76$, so H_0 is rejected; scores are lower after treatment with chlordiazepoxide than after a placebo.

3. $\bar{d} = -4.1$, $s_d = 5.3$, $t = \dfrac{-4.1 - 0}{5.3/\sqrt{15}} = -3.00$; critical $t = -2.62$. Reject H_0.

4. (X = men) H_0: $\mu_{X_d} - \mu_{Y_d} = 0$, H_1: $\mu_{X_d} - \mu_{Y_d} \neq 0$, $\bar{d}_X = -4.9$, $s_{X_d} = 4.7$, $\bar{d}_Y = -8.6$,
   $s_{Y_d} = 8.1$; $t = \dfrac{-4.9 - (-8.6) - 0}{\sqrt{\dfrac{6(4.7)^2 + 7(8.1)^2}{13}\left(\dfrac{1}{7} + \dfrac{1}{8}\right)}} = 1.06$. Two-tail, $D = 13$, critical $t = \pm 2.16$;
   H_0 cannot be rejected.

5. $t = \dfrac{-5.0 - (-3.4) - 0}{\sqrt{\dfrac{6(6.5)^2 + 7(4.2)^2}{13}\left(\dfrac{1}{7} + \dfrac{1}{8}\right)}} = -.57$; 1-tail, critical $t = -2.65$, so H_0 cannot be rejected.

6. $t = \dfrac{-6.9 - (-4.1) - 0}{\sqrt{\dfrac{14(6.8)^2 + 14(5.3)^2}{28}\left(\dfrac{1}{15} + \dfrac{1}{15}\right)}} = -1.26$; $D = 28$; critical $t = \pm 2.05$; H_0 cannot be rejected.

7. Surely the same solutions were measured on both computers, so differences are needed, not the mean standard deviation for each computer separately.

11.12 VOCABULARY AND SYMBOLS

t distribution	t	independent samples	
(Student's t distribution)		paired differences	d
degrees of freedom	D	μ_d	
		s_d	

11.13 EXERCISES

1. What is the probability that the mean of a sample of size 17 drawn at random from a normal population will have a t score less than 2.58?

2. In a t distribution with 4 degrees of freedom, 10% of t scores are above what score?

3. You have a random sample of 12 students aged 15 with mean weight of 150 pounds with a standard deviation of 9 pounds, and wish to find a 95% confidence interval for the mean weight of all students aged 15.

(a) $D = ?$

(b) 95% of t scores with D degrees of freedom are between $-t_0$ and $+t_0$. Find t_0.

(c) t_0 in t units equals how many pounds?

(d) The probability is .95 that the mean of one random sample is within how many pounds of μ?

(e) This sample is random; $\bar{X} = ?$

(f) You have 95% confidence that $\bar{X}$ is within how many pounds of μ?

(g) Complete: A 95% confidence interval for $\mu = $ ___ to ___ pounds.

4. (a) Find a 90% confidence interval for the mean number of years chemists have been working for the same company if a random sample of 17 chemists have a mean of 5.6 years with a standard deviation of 2.0 years. Assume the years worked for the same company are normally distributed for all chemists.

(b) Do you think the assumption of normality in (a) is valid?

5. What assumptions must be made about the population(s) and sample(s) before a t test of (a) the mean, (b) the difference of means can be carried out?

6. Correct if necessary and explain why a correction is needed:

(a) For fixed sample size and level of significance, a null hypothesis is more likely to be rejected if a z test is used than if a t test is used.

(b) For samples of size 25 from a normal population for which σ is known, $\bar{X}$ has a normal distribution.

(c) For samples of size 25 from a normal population for which σ is unknown, $\bar{X}$ has a t distribution.

7. Nine workers, chosen at random from a work force of 250 workers in a factory, have a mean wage of $225 a week with a standard deviation of $12. Estimate the mean wages of all workers at a 95% confidence level. (Assume the distribution of all workers' wages is normal.)

8. A metallurgist made 8 measurements of the melting point of an alloy. His measurements were 850°, 848°, 852°, 848°, 851°, 850°, 851°, 850°. What is his 90% confidence interval for the melting point of the alloy? Assume his errors in measurement are normally distributed.

9. Five patients with hypertension are given a new drug which lowers their blood pressure by 4, 25, 13, 18, and 20 points, respectively. At a .05 level of significance, does the new drug lower blood pressure by at least 20 points? Assume the changes in blood pressure of all patients with hypertension are normally distributed.

10. The Speed Reading Institute claims that adults will read 400 words per minute faster after completing its course. A random sample of 12 adults are found to increase their reading speed by 350 words per minute, with a standard deviation of 50 words per minute. Is the Institute's claim substantiated? (.05 level of significance.)

11. Among a group of emotionally disturbed children, 14 were given "clown-workshop" therapy and 11 controls were not. A measure of imaginativeness was assigned to each child, with the following results:

	Clown-workshop			Controls		
	Before	During	After	Before	During	After
Mean	1.56	3.10	1.80	1.45	1.65	1.27
Standard Deviation	.63	.75	.84	.77	.38	.44
Number	14	14	14	11	11	11

Do t tests for differences between population means (a) before, (b) during, and (c) after treatment. ($\alpha = .01$.)

12. Test the difference in imaginativeness before and after "clown-workshop" therapy.

13. 60-yard, 300-yard, and 1,000-yard races were run during Borough Championships between Brooklyn and Queens. The times (in seconds) for these races were as follows:

Brooklyn: 6.5, 6.8, 6.8, 6.8, 6.8, 33.2, 33.4, 33.7, 34.2, 34.3, 140.7, 142.5, 146.5, 147.6, 150.6. ($n = 15$, $\Sigma X = 930.4$, $\Sigma X^2 = 111,957$)

Queens: 6.7, 6.8, 6.9, 6.9, 6.9, 33.5, 33.6, 33.9, 34.2, 34.5, 138.9, 139.2, 139.6, 141.1, 142.0. ($n = 15$, $\Sigma Y = 904.7$, $\Sigma Y^2 = 104,225$)

Do Brooklyn and Queens runners differ, or are they equally swift on the average? ($\alpha = .05$.) Comment on the result.

14. Those who attended the groundhog's eve party (see Exercise 9, page 50) were asked the usual transportation used to work or school. The heights of 10 who walked were 66, 69, 66, 69, 64, 63, 71, 69, 66, 72 inches. ($\Sigma X = 675$, $\Sigma X^2 = 45,641$) The heights of 8 who drove by car or motorbike were 70, 73, 71, 67, 67, 68, 62, 65 inches. ($\Sigma Y = 543$, $\Sigma Y^2 = 36,941$) Do those who walk and those who ride differ in height? ($\alpha = .05$.)

15. Professor Pike has given the same multiple-choice final exam to his Introduction to Psychology course for the past 12 years; the mean number of correct answers has been 153. This year, a mean number of correct answers by a random sample of 13 students is 162 with a standard deviation of 12. Has this year's class done better? ($\alpha = .05$.)

16. A sleep-efficiency index was determined* for each of 11 Israelis with combat neuroses resulting from the 1973 Yom Kippur War. The mean index was 84.8 with a standard deviation of 9.5. (a) Find a 90% confidence interval for the mean sleep-efficiency index of all Israeli veterans with combat neuroses; (b) test the null hypothesis that it is 100. ($\alpha = .05$.)

---

*Lavie, P.: "Long-Term Effects of Traumatic War-Related Events on Sleep," *American Journal of Psychiatry, 136* (1979), pp. 175–178.

17. Random samples are chosen at Dover High School from the first and second year typing classes. Twelve students in the first year typing class have a mean speed of 18 words per minute, with a standard deviation of 6 words per minute. For 9 students chosen from the second year typing class, the mean is 46 words per minute, with a standard deviation of 8. At a .01 level of significance, do second year students type 25 words per minute faster? Assume both populations are normal.

18. The structural constants of a crystalline substance are obtained by doing numerous X-ray diffractions from different angles. Two crystals of rutile were used in an experiment (see Exercise 14, page 86), and each was measured 4 times. In the first crystal, 4 measurements of a structure factor yielded $\bar{X} = 24.9$ and $s_X = 1.2$, while 4 similar measurements in a second crystal yielded $\bar{Y} = 26.1$ and $s_Y = 1.4$. Test the null hypothesis that both samples come from the same population (does $\mu_X = \mu_Y$?). ($\alpha = .05$.)

19. The mean of all 8 measurements in Exercise 18, you will agree, is 25.5. (a) Test the null hypothesis that $\mu = 25.7$ which, from other studies, is thought to be the true value. ($\alpha = .10$.) (b) Find a 90% confidence interval for μ. (c) Does this confidence interval include 25.7?

Note: the standard error of the sum of 2 variables equals the standard error of the difference: $\sigma_{(X+Y)} = \sigma_{(X-Y)}$

20. An economist is studying the differences in pay of workers in two mills operated by the same textile company. He takes a random sample of size 16 from each plant, pairs them for seniority with good matching, and then discovers the following hourly rates:

Pair:	1	2	3	4	5	6	7	8
X:	$2.50	3.40	3.20	2.60	3.00	2.75	4.00	3.75
Y:	$2.80	3.70	3.00	3.00	3.10	2.85	4.10	3.85
Pair:	9	10	11	12	13	14	15	16
X:	$5.20	4.50	3.00	4.05	4.20	4.15	4.30	4.60
Y:	$5.30	4.60	3.20	4.15	4.30	4.35	4.00	4.70

Test for differences in pay at a .01 level of significance. Assume the differences in pay are normally distributed.

21. A random sample of 12 male freshmen are weighed when they enter college in September and again on November 1, with the following results:

Student	1	2	3	4	5	6	7	8	9	10	11	12
Sept.	150	210	160	150	180	170	200	240	190	193	200	185
Nov. 1	155	210	153	150	185	165	205	230	193	198	205	189

Have male freshmen students in general gained weight in 2 months, at a .05 level of significance?

22. The water absorbency of 8 randomly chosen diapers is measured when brand new and again after washing 3 times, with the following results:

Diaper:	1	2	3	4	5	6	7	8
Absorbency when new (oz.):	3.4	3.3	3.5	3.2	3.5	3.4	3.6	3.6
Absorbency after washing (oz.):	3.7	3.5	3.5	3.4	3.9	3.6	4.0	4.0

After washing, does the absorbency of diapers increase significantly? ($\alpha = .01$.)

23. In Exercise 5, page 235, Weisman's data on changes in serum phosphate levels of patients with cancer of the stomach after treatment with 5-fluorouracil: $n = 16$, $\bar{X} = -0.79$, $s = 0.48$ mg/100 ml. You were unable to use a z-test of the null hypothesis that the treatment does not change average serum phosphate concentration because the sample was too small. Carry it out now with a t test. ($\alpha = .05$.)

ANSWERS

1. $D = 16$; $\Pr(t > 2.58) = .01$; $\Pr(t < 2.58) = .99$.

3. (a) $D = 12 - 1 = 11$; (b) 2.20; (c) $2.20 \left(\dfrac{9}{\sqrt{11}} \right) = 5.97$ pounds; (d) 5.97 pounds; (e) 150 pounds; (f) 5.97 pounds; (g) 150 ± 5.97 or 144.0 to 156.0 pounds.

5. (a) The population must be normal and the sample must be random. It is also assumed that σ is unknown (if it is known, a z test would be used).
(b) The populations must both be normal, and the samples must be random and independent. Again it is assumed that σ_X and σ_Y are unknown.

7. $\bar{X} = 225$, $s = 12$, $n = 9$, $D = 8$, $t = +2.31$. $\mu = 225 \pm 2.31 \left(\dfrac{12}{\sqrt{9}} \right) = 215.8$ or 234.2. A 95% confidence interval for the mean wages of all workers in the factory is \$215.80 to \$234.20.

9. $\bar{X} = 16$, $s = 8.0$, $D = 4$, H_0: $\mu = 20$, H_1: $\mu < 20$; critical value of t is -2.132. $t = \dfrac{16 - 20}{8.0/\sqrt{5}} = -1.1$; H_0 cannot be rejected. (But if you wish to be persuasive that the new drug is effective, do you need a small α or a small β?)

11. H_0: $\mu_X - \mu_Y = 0$, H_1: $\mu_X - \mu_Y \neq 0$, $D = 23$, critical $t = \pm 2.81$.

(a) $t = \dfrac{(1.56 - 1.45) - 0}{\sqrt{\dfrac{13(.63)^2 + 10(.77)^2}{23} \left(\dfrac{1}{14} + \dfrac{1}{11} \right)}} = .39$; fail to reject H_0.

(b) $t = \dfrac{(3.10 - 1.65) - 0}{\sqrt{\dfrac{13(.75)^2 + 10(.38)^2}{23} \left(\dfrac{1}{14} + \dfrac{1}{11} \right)}} = 5.83$. Reject H_0.

(c) $t = \dfrac{(1.80 - 1.27) - 0}{\sqrt{\dfrac{13(.84)^2 + 10(.44)^2}{23} \left(\dfrac{1}{14} + \dfrac{1}{11} \right)}} = 1.89$. Fail to reject H_0.

13. (X = Brooklyn) $\bar{X} = 62.03$, $s_X = 62.25$, $n_X = n_Y = 15$, $\bar{Y} = 60.31$, $s_Y = 59.56$. H_0: $\mu_X = \mu_Y$, H_1: $\mu_X \neq \mu_Y$. $D = 28$, critical $t = \pm 2.05$. Computed $t =$
$$\frac{(62.03 - 60.31) - 0}{\sqrt{\frac{14(62.25)^2 + 14(59.56)^2}{28}\left(\frac{1}{15} + \frac{1}{15}\right)}} = 0.08.$$ Hence, H_0 cannot be rejected. The variances are much too big to see anything, and no wonder, when you're mixing 60-yard and 1,000-yard running times together in one sample. No good! (With a "blocking" technique the test could be done very nicely, but that's beyond the scope of this book.)

15. H_0: $\mu = 153$, H_1: $\mu > 153$, $n = 13$, $\bar{X} = 162$, $s = 12$. $D = 12$, critical $t = 1.78$; computed $t = \dfrac{162 - 153}{12/\sqrt{13}} = 2.70$. The probability is less than .05 that this year's sample can be explained by sampling fluctuations, so we accept the hypothesis that this year's class has done better.

17. $t = \dfrac{(46 - 18) - 25}{\sqrt{\dfrac{8(8)^2 + 11(6)^2}{19}\left(\dfrac{1}{9} + \dfrac{1}{12}\right)}} = .98$. Critical $t = 2.54$. Fail to reject H_0.

19. $t = \dfrac{25.5 - 25.7}{\sqrt{\dfrac{3(1.2)^2 + 3(1.4)^2}{6}\left(\dfrac{1}{4} + \dfrac{1}{4}\right)}} = \dfrac{-.2}{.92} = -.22$; critical $t = \pm 1.94$. Continue to accept H_0.

(b) $\mu = 25.5 \pm 1.94(.92) = 23.7$ to 27.3.
(c) Yes.

21. H_0: $\mu_d = 0$; H_1: $\mu_d > 0$. d: 5, 0, −7, 0, 5, −5, 5, −10, 3, 5, 5, 4. $\bar{d} = 10/12 = .833$, $s_d = \sqrt{\dfrac{324 - 10^2/12}{11}} = 5.36$. $D = 11$; critical value of t is 1.80.

$$t = \frac{.833 - 0}{5.36/\sqrt{12}} = .54$$

H_0 is still accepted; there is no difference.

23. $t = \dfrac{-0.79 - 0}{0.48/\sqrt{15}} = -6.37$; $D = 15$, critical $t = \pm 2.13$. H_0 is rejected.

12
χ^2 (CHI SQUARE)

In the last four chapters, you learned to find a confidence interval or test a hypothesis about the population mean(s) or proportion(s). Inferences about the mean of course involve metric data (it is nonsense to talk about the mean of categorical data); inferences about a proportion can be made about categorical data, but then we are restricted to two categories. In this chapter you will learn to make statistical inferences about categorical data in which the number of categories may be larger than two.

A new family of probability distributions, χ^2, will be introduced; as in t distributions, there is a different member of the family for each different number of degrees of freedom. Two types of problems will be discussed: "goodness of fit" problems, in which you will determine how well an observed frequency distribution agrees with a theoretical frequency distribution, and secondly, problems in which you decide whether two variables are independent or related ("Does income at age 40 depend on amount of education or not?").

12.1 INTRODUCTION

First, become familiar with the Greek letter χ (chi, pronounced as the first two letters of "kite"). It is not equivalent to any English letter, but is similar to the German ch as in "ich." We shall see it **only** in the form χ^2 and shall refer to "chi-square distributions" or "chi-square tests."

You have met various families of probability distributions: the binomial, normal, and t distributions. Now you will meet another family of probability distributions; as with t distributions, there will be a different χ^2 distribution for each different number of degrees of freedom. But before becoming familiar with a χ^2 distribution, let us look at a problem in which there is a need for it.

Example 1
Studies made in 1950 showed that, among all men aged 30, 20% were college graduates, 50% were high school but not college graduates, and 30% were not high school graduates. Is the present population distribution the same? A study of 1,000 men, chosen at random from those now 30 years old, is carried out, and it is found that 250 are college graduates, 520 are high school but not college graduates, and 230 have not finished high school. Test at the .01 level of significance.

H_0: There is no change in the distribution between 1950 and now. (Note that H_0 is an assumption not about the mean or proportion of a population, but about the whole distribution of a population.)

H_1: The present population (namely, all men now 30 years old) has a distribution different from that in 1950.

265

Let O—the letter O, not the numeral zero—be used as abbreviation for observed frequency (at the present time) and E for expected frequency. If there is no change in the distribution between 1950 and now, the given information can be presented as follows:

	O	E
College graduates	250	200
High school graduates	520	500
Not high school graduates	230	300
	1,000	1,000

In 1950, 20% were college graduates. In a sample of 1,000, then, 20% or 200 men would be expected to be college graduates, 50% or 500 men to be high school graduates, and 30% or 300 to be "not high school graduates." Note that we start off by scaling the E column so that its sum is the same as that of the O column. It is quite inappropriate to express the O column in numbers of men (total 1,000) and the E column in percentages or proportions (total 100 or 1, respectively). Both must add up to the total number of scores in the sample.

A decision choice must now be made: What should be used as the criterion for accepting or rejecting H_0? Would $\Sigma(O - E)$ describe the difference between the two distributions? But $\Sigma(O - E) = \Sigma O - \Sigma E = n - n = 0$. Or, as shown in the $O - E$ column below, $50 + 20 - 70 = 0$.

	O	E	$O - E$	$(O - E)^2$	$\dfrac{(O - E)^2}{E}$
College graduates	250	200	50	2,500	12.5
High school graduates	520	500	20	400	.8
Not high school graduates	230	300	−70	4,900	16.3
	1,000	1,000	0	7,800	29.6

Do you remember that when we looked at measures of variability we considered $\dfrac{\Sigma(X - \bar{X})}{n}$, but it turned out that $\Sigma(X - \bar{X})$ always equals 0? Then we tried $\Sigma(X - \bar{X})^2$ and finally $\dfrac{\Sigma(X - \bar{X})^2}{n}$, the variance. So here we might see whether $\Sigma(O - E)^2$ reflects the difference between the two distributions. $\Sigma(O - E)^2 = 0$ only if there is a perfect fit between the observed and expected frequencies, and may be quite large if they are not in agreement.

Instead of simply using $\Sigma(O - E)^2$, however, we use $\Sigma\left[\dfrac{(O - E)^2}{E}\right]$. The reason for this is that a large entry in the $(O - E)^2$ column is more disturbing if it comes from a category with a small expected frequency than if the expected frequency is large, so the former case is weighted more heavily. An entry of 2,500 in the $(O - E)^2$ column becomes 12.5 if the expected frequency E is 200, but only 5.0 if the expected frequency is 500, and 2.5 if the expected frequency in that category is 1,000.

Now we define $\chi^2{}_P$:

$$\chi^2{}_P = \Sigma \left[\frac{(O - E)^2}{E} \right]$$

The subscript $_P$ stands for Karl Pearson, the statistician who developed this formula. $\chi^2{}_P$ is a statistic (it is based on sample data). In the next section you will study new kinds of distributions, called χ^2 distributions. The distribution of $\chi^2{}_P$ for all samples of size n is approximately a χ^2 distribution. In many books, the subscript $_P$ for the statistic is dropped. It will be retained in this text, however, to remind you of the difference between the statistic and the distribution.

In the given example, we fill in a column for $\dfrac{(O - E)^2}{E}$ and add, finding that

$\chi^2{}_P = 29.6$. But if we look at a χ^2 table (Table 6, Appendix C), we find, as in the case of t distributions, a column labeled D at the left. D again stands for **degrees of freedom.** To determine D, look at the column marked E. There are three entries here, but if any two are given, then the third is determined, since the sum of this column must be 1,000. Therefore $D = 3 - 1 = 2$.

	Probabilities (or areas under χ^2 curve above given values of χ^2)			
α	.10	.05	.025	$\downarrow$ .01
D	Values of χ^2			
1	2.71	3.84	5.02	6.63
$\rightarrow$ 2	4.61	5.99	7.38	(9.21)
3	6.75	7.82	9.35	11.3

$\alpha = .01$, $D = 2$, so we look on the $D = 2$ line under .01 and find $\chi^2 = 9.21$. This gives us the decision choice for the null hypothesis: we don't have enough evidence to reject H_0 if the computed value of $\chi^2{}_P$ is less than or equal to 9.21, and H_0 is rejected if the computed value is greater than 9.21. In our example, $\chi^2{}_P = 29.6$, so H_0 is rejected and H_1 is accepted. There is a difference in the distribution of 30-year-old men by education now as compared with 1950.

We shall return later (Sections 12.3 and 12.5) to a general discussion of χ^2 tests and to further examples. But now that you have a faint notion of the direction in which we are heading, let us digress and see how a χ^2 table is made.

12.2 χ^2 DISTRIBUTIONS

From a normal distribution of X scores, take a sample of size D. Suppose the scores in the sample are $X_1, X_2, \cdots, X_D$. Then determine

$$\frac{(X_1 - \mu)^2}{\sigma^2} + \frac{(X_2 - \mu)^2}{\sigma^2} + \cdots + \frac{(X_D - \mu)^2}{\sigma^2} = z_1{}^2 + z_2{}^2 + \cdots + z_D{}^2$$

This sum is the first score in the chi-square (χ^2) distribution for D degrees of freedom. Repeat this process for all other different samples of size D to get the whole χ^2 distribution for D degrees of freedom

There is a different χ^2 distribution for each different value of D, but all of them share the following characteristics:

1. Every χ^2 distribution extends indefinitely to the right from 0.
2. Every χ^2 distribution has only one (right) tail. We shall study only one-tail tests using χ^2.
3. As D increases, the χ^2 curves get more bell-shaped and approach the normal curve in appearance (but remember that a χ^2 curve starts at 0, not at $-\infty$).

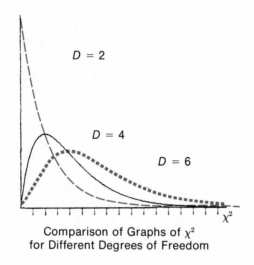

Comparison of Graphs of χ^2
for Different Degrees of Freedom

What is the relation between χ^2 as defined here and $\chi^2_P = \Sigma\left[\dfrac{(O - E)^2}{E}\right]$, used in Example 1, page 265? The answer is that χ^2_P is a statistic (sample data is used in computing it) whose distribution is approximately χ^2. The critical value of χ^2 for accepting or rejecting a null hypothesis is determined by consulting Table 6 for the appropriate value of α and degrees of freedom, and then a decision is made according to which region χ^2_P falls in. "The distribution of χ^2_P is **approximately χ^2**" implies some limitations: we shall ordinarily be using χ^2 tests for categorical data (discrete data), since more powerful methods can be used for metric data, but a χ^2 distribution is based on the normal curve. The approximation is close enough if **at least three-fourths of the entries in the E column are 5 or larger, and all are greater than 1.** If this is not the case, increase the size of the sample.

Using a χ^2 table (Table 6, Appendix C) will be easy, since it is similar to using a t table except that only one (right) tail is used since χ^2 is always greater than 0.

Example 1 (a) If $D = 6$, what is the probability that a randomly chosen χ^2 score is greater than 14.4?

(b) What is D if $\Pr(\chi^2 > 27.6) = .05$?

(a) Look at Table 6, Appendix C. Go down the D column to the number 6, look across the row until you find 14.4. It is in the .025 column. $\Pr(\chi^2 > 14.4) = .025$.

(b) Look down the .05 column until you find 27.6. $D = 17$.

12.3   TESTING "GOODNESS OF FIT" WITH χ^2

The comparison of distributions of 30-year-old men by education in 1950 and now, considered at the beginning of this chapter, is an example of a **"goodness of fit"** problem. In this use of χ^2 distributions, the observed frequencies in the distribution of a sample are used to test the hypothesis that the population from which the sample is taken does not differ in its distribution from that of some known population.

H_0 has the form "the population (A) from which the sample is taken has the same distribution as population B"; it is assumed that the distribution of B is known. H_1 states "populations A and B do not have the same distribution." A sample of size n is taken, and frequencies in various categories or of various scores are observed (O column). Then the frequencies expected for these categories or scores if the sample had the same proportion in each category or score as population B is computed (E column). Then $\Sigma \left[\dfrac{(O - E)^2}{E} \right]$ is determined.

For all χ^2 tests, the number of degrees of freedom D equals the number of entries in the E column minus the number of sample statistics used in determining the E column. In the examples of this section, only one sample statistic is used—namely, the total number of scores in the sample. Therefore $D = $ (number of E entries) $- 1$.

"Goodness of fit" tests:

$$\chi^2_P = \Sigma \left[\frac{(O - E)^2}{E} \right]$$

$D = $ (number of E entries) $- 1$

The critical value of χ^2 for accepting or rejecting H_0 is determined from Table 6, Appendix C, for the given α and D. This critical value is compared with the computed statistic $\Sigma \left[\dfrac{(O - E)^2}{E} \right]$, and a decision is reached.

Example 1   Last year, 30% of freshmen at Pitchen University lived on campus in single rooms, 50% on campus in double rooms, and 20% were commuters. In a random sample of 120 freshmen this year, 24 have single rooms, 65 have double rooms, and 31 are commuters. Has there been a change in campus living arrangements ($\alpha = .05$)?

H_0: The population distribution this year is 30%, 50%, and 20% for single rooms on campus, double rooms on campus, and commuters, respectively.

H_1: This is not the case.

	$O =$ This Year	Last Year	E	$O - E$	$(O - E)^2$	$\dfrac{(O - E)^2}{E}$
Single rooms	24	30%	36	-12	144	4.00
Double rooms	65	50%	60	5	25	.42
Commuters	$\underline{31}$	20%	$\underline{24}$	7	49	$\underline{2.04}$
	120		120	0		$6.46 = \chi^2_P$

$\chi^2_P = 6.46$

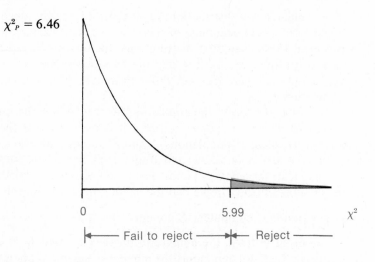

D = 3 − 1 = 2, α = .05; the critical value of χ² is 5.99, so H_0 is rejected; campus living arrangements have changed this year.

Example 2

The first digit in each of the first two columns of the random number table (Table 3, Appendix C) have been tallied below. If the numbers are random, each is likely to be chosen, so the probability is .1 for each number. Test to see whether these 100 numbers are random (α = .10).

Digit	O	Pr (digit)	E	$O - E$	$(O - E)^2$	$\dfrac{(O - E)^2}{E}$
0	7	.1	10	-3	9	.9
1	13	.1	10	3	9	.9
2	9	.1	10	-1	1	.1
3	8	.1	10	-2	4	.4
4	11	.1	10	1	1	.1
5	11	.1	10	1	1	.1
6	8	.1	10	-2	4	.4
7	10	.1	10	0	0	.0
8	15	.1	10	5	25	2.5
9	$\underline{8}$	.1	$\underline{10}$	-2	4	$\underline{.4}$
	100		100	0		$5.8 = \chi^2_P$

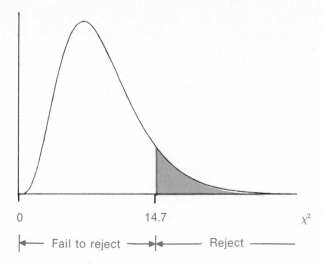

H_0: the numbers are random; H_1: they are not.

$\chi^2_P = 5.8$; $D = 10 - 1 = 9$, so the critical value of χ^2 is 14.7; H_0 cannot be rejected.

In many experiments, you are pleased if H_0 is rejected since you have found a significant difference from the usual result. In testing goodness of fit with χ^2_P, however, you are often testing a hypothesis which you hope is true. If you are a careful investigator, you want the probability β of continuing to accept H_0 when it is false to be small, and yet the convention of choosing α in reaching a decision is still used. In these cases, increase α to .20 or .25, since β is decreased as α is increased for fixed n.

12.4 EXERCISES

1. If $D = 4$, what proportion of χ^2 scores are > 7.78?

2. What is D if 1% of scores in a χ^2 distribution are over 32.0?

3. In a goodness of fit test with 8 entries in the O column, the computed value of χ^2_P is 14.4. If $\alpha = .05$, should the null hypothesis be rejected or still accepted?

4. (a) Describe a χ^2 distribution for 1 degree of freedom: how is it formed? What kind of numbers can you get?
   (b) Compare a χ^2 distribution for $D = 1$ with a standard normal distribution.
   (c) In a standard normal distribution, what proportion of z scores are below -1.96 or above $+1.96$?
   (d) What proportion of scores in a χ^2 distribution with $D = 1$ do you expect to be larger than $1.96^2 = 3.84$? (Figure it out, then compare with Table 6 in Appendix C.)

5. In a survey of 400 infants, chosen at random, it is found that 185 are girls. Are boy and girl births equally likely, according to this survey? ($\alpha = .05$)

(a) Use a χ^2 goodness-of-fit test.

(b) Use a z test to see whether the proportion of girls differs from .5.

6. A die is tossed 600 times with the following results:

Face	1	2	3	4	5	6
Frequency	95	110	90	85	112	108

Test the hypothesis that the die is fair at a .05 level of significance.

7. In a survey of 160 families with 4 children, it is discovered that in 6 families there are no girls, while 46 families have 1 girl, 56 have 2, 34 have 3, and 18 families have 4 girls. Are boy and girl births equally probable, according to this survey? (.05 level of significance.)

(a) Use a χ^2 test.

(b) Use a z test of H_0: $P_G = .5$. Note that the 160 families have 640 children; how many are girls?

8. In a cross between tall, potato-leaf tomatoes and dwarf, cut-leaf tomatoes, the following plants are observed:

Tall, potato-leaf	280
Tall, cut-leaf	100
Dwarf, potato-leaf	80
Dwarf, cut-leaf	20

Are these results consistent with the hypothesis that the ratios are $9:3:3:1$? (This means that we expect 9/16 to be tall, potato-leaf, 3/16 to be tall, cut-leaf, etc.) ($\alpha = .10$).

9. Upon observation of 1,000 dented fenders, it is found that 300 are the left front, 240 the right front, 210 the left rear, and 250 the right rear. At a .01 level of significance, are all fenders equally likely to be dented?

10. A manufacturer of television sets receives component transistors from 4 suppliers. Each month 1,000 transistors from each supplier are chosen at random and carefully tested. In January the numbers of defective transistors are 46, 53, 42, 59, from suppliers A, B, C, D, respectively. Is the number of defective transistors the same from each supplier? ($\alpha = .05$.)

11. On a true–false exam, a playful professor includes five questions he thinks can be answered only by guessing. The results on these five questions are as follows:

Number Correct	Number of Students
0	1
1	10
2	20
3	25
4	15
5	4

Is his belief correct, at a .10 level of significance?

12. The professor described in Exercise 11 tries the same experiment on a larger class; this time the numbers of students answering 0, 1, 2, 3, 4, 5 questions correctly are 2, 20, 40, 50, 30, 8, respectively. Is he correct, at a .01 level of significance?

ANSWERS

1. .10.

2. $D = 16$.

3. $D = 7$, critical value of $\chi^2 = 14.1$. Reject H_0.

4. (a) For each X score (each sample of size 1, whose mean equals the score itself) in a normal population, compute z^2. The sum of all these different z^2's forms the χ^2 distribution with $D = 1$. None of the scores is negative, since z^2 is never negative.
   (b) Since the population is normal, the sampling distribution of means for samples of size 1 is normal. Convert the scores in the sampling distribution to standard (z) scores and you get the standard normal distribution. Square these standard scores to get χ^2 with 1 degree of freedom.
   (c) .05.
   (d) $\Pr(z^2 > 1.96^2) = \Pr(\chi^2 > 1.96^2 = 3.84) = .05$.

5. (a)

	O	E	$O - E$	$(O - E)^2$	$\dfrac{(O - E)^2}{E}$
Girls	185	200	-15	225	1.125
Boys	215	200	15	225	1.125
					$\overline{2.25} = \chi^2{}_P$

   $D = 1$, critical value of χ^2 is 3.84. H_0 (that the proportion of girls and boys is the same) is not rejected.

   (b) H_0: P is 1/2; H_1: $P \neq 1/2$. The critical values of z are ± 1.96. $p = \dfrac{185}{200} = .4625$, $n = 400$, $\sigma_p = \sqrt{\dfrac{.50(.50)}{400}} = .025$; $z = \dfrac{.4625 - .50}{.025} = -1.5$. H_0 is again accepted.

   (More decimal places were carried than usual to demonstrate that the computed and critical values for the χ^2 test are the squares of the corresponding values for the z test: $\chi^2{}_P = 2.25 = (-1.5)^2 = z^2$, and $3.84 = 1.96^2$. A χ^2 test with $D = 1$ has given you nothing new.)

6. $\chi^2{}_P = .25 + 1.00 + 1.00 + 2.25 + 1.44 + .64 = 6.6$; $D = 5$, $\alpha = .05$, critical $\chi^2 = 11.1$. H_0 is not rejected; the die is fair.

7. (a) E column: 10, 40, 60, 40, 10; $\chi^2{}_P = 1.6 + .9 + .3 + .9 + 6.4 = 10.1$; $D = 4$, $\alpha = .05$, $\chi^2 = 9.49$. H_0 is rejected; the distribution of girls in families surveyed is not consistent with the distribution that would be expected if $P = 1/2$.

   (b) $\sigma_p = \sqrt{\dfrac{.5(.5)}{640}} = .0198$, $p = \dfrac{332 \text{ girls}}{640 \text{ children}} = .519$, $z = \dfrac{.519 - .5}{.0198} = 2.83$. The critical values of z are ± 1.96, so we reject H_0.

8. E column: 270, 90, 90, 30; $\chi^2{}_P = .37 + 1.11 + 1.11 + 3.33 = 5.93$. $D = 3$; $\chi^2 = 6.25$, > 5.93. Yes, the results are consistent with the hypothesis.

9. $\chi^2_P = 10.0 + .4 + 6.4 + 0.0 = 16.8$; $D = 3$, $\chi^2 = 11.3$. H_0 is rejected; fenders are not equally likely to be dented.

10. $\chi^2_P = .32 + .18 + 1.28 + 1.62 = 3.40$; $D = 3$, $\chi^2 = 7.81$; H_0 is not rejected.

11. Use Table 2 in the appendix with $n = 5$, $p = .5$ and multiply by 75 for the E column, which becomes 2.3, 11.7, 23.4, 23.4, 11.7, 2.3. The sample size should be increased if a χ^2 test is to be used, since only two-thirds of the entries are over 5. (If the class hears of the surmise, they will use stronger words than "playful" to describe the professor.) E column: 2.33, 11.70, 23.40, 23.40, 11.70, 2.33. $\chi^2_P = 3.74$; $D = 4$, $\alpha = .01$, $\chi^2 = 13.3$. His surmise is not rejected.

12. E column: 4.7, 23.4, 46.8, 46.8, 23.4, 4.7. $\chi^2_P = 7.45$. $D = 4$. $\alpha = .01$, $\chi^2 = 13.3$. His surmise is supported. Compare the answers to Exercises 11 and 12, and note what happens to χ^2_P when each frequency in the E column is doubled: χ^2_P is doubled.

12.5 RELATED AND INDEPENDENT VARIABLES; CONTINGENCY TABLES

In the last section, you learned how to use χ^2_P to test "goodness of fit"—that is, testing a distribution obtained from sample data to see whether the distribution of the population from which the sample was taken fits some theoretical distribution.

χ^2_P can be used not only to test "goodness of fit" but also to test whether two variables are related or are independent. Does a higher proportion of men than of women vote Republican, or is there no relation between sex and political party? Is there a relation between years of schooling after high school and income at age 40? Do different strains of rats have differing abilities to "solve" a maze? A χ^2 test will help in each of these problems.

Example 1   The following table shows the relation between the number of accidents in 1 year and the age of the driver in a random sample of 500 drivers between 18 and 50. Test, at a .01 level of significance, the hypothesis that the number of accidents is independent of the driver's age.

There are 75 drivers between 18 and 25 who have no accidents, 115 between 26 and 40 with no accidents, and so on.

Such a table is called a **contingency table.** Each "box" containing a fre-

Age of Driver

Number of Accidents	18–25	26–40	Over 40	
0	75	115	110	300
1	50	65	35	150
2	25	20	5	50
	150	200	150	500

quency is called a **cell**. This is a 3 × 3 [read "3 by 3"] table, since it has 3 rows and 3 columns. Note that the row and column totals are not counted in giving the size of a contingency table. The following is a 2 × 4 table.

Rows are counted first, then columns: $R \times C$.

The null hypothesis for this type of problem always states that the classifications are independent (here, that there is no relation between age of driver and number of accidents). H_1: the variables are dependent (or related). $\alpha = .01$.

What are the expected frequencies? Concentrate first on the upper left-hand cell. A total of 150 drivers are aged 18–25, and $\dfrac{300}{500} = \dfrac{3}{5}$ of all drivers have had no accidents. If there is no relation between driver age and number of accidents, we expect that $\dfrac{3}{5}(150) = 90$ drivers aged 18–25 would have no accidents. Note, however, that the same result is obtained by multiplying the column total (150) by the row total (300) and dividing by the grand total (500). Before you read on, find the expected number of drivers aged 26 to 40 who have had no accidents.

Answer   200 of 500 drivers $\left(\dfrac{2}{5}\right)$ are aged 26–40, and 300 have no accidents. So we expect $\dfrac{2}{5}(300) = 120$ will be 26–40 and have no accidents if there is no relation between age and number of accidents; or column total (200) × row total (300) divided by grand total (500) $= \dfrac{200 \times 300}{500} = 120$.

The results can be summarized in a table:

Frequencies expected if H_0 is true

Age of Driver

		18–25	26–40	41–50	Totals
Number of Accidents	0	$\dfrac{150\,(300)}{500}$ $=90$	$\dfrac{200\,(300)}{500}$ $= 120$	$\dfrac{150\,(300)}{500}$ $=90$	300
	1	$\dfrac{150\,(150)}{500}$ $=45$	$\dfrac{200\,(150)}{500}$ $=60$	$\dfrac{150\,(150)}{500}$ $=45$	150
	2	$\dfrac{150\,(50)}{500}$ $=15$	$\dfrac{200\,(50)}{500}$ $=20$	$\dfrac{150\,(50)}{500}$ $=15$	50
		150	200	150	500

Note that the marginal totals in this table are the same as the marginal totals in the table of observed frequencies on page 274. Once these marginal totals are copied from the earlier table, then the expected frequency is entered in each cell **without** reference to the observed frequency in that cell: **Multiply the row total by the column total, and divide by the grand total.**

Now the χ^2 test can be carried out with the same formula as before:

$$\chi^2_P = \Sigma \left[\frac{(O - E)^2}{E} \right]$$

Number of Accidents	Age of Driver	O	E	$O - E$	$(O - E)^2$	$\dfrac{(O - E)^2}{E}$
0	18–25	75	90	−15	225	2.5
0	26–40	115	120	−5	25	.2
0	41–50	110	90	20	400	4.4
1	18–25	50	45	5	25	.6
1	26–40	65	60	5	25	.4
1	41–50	35	45	10	100	2.2
2	18–25	25	15	10	100	6.7
2	26–40	20	20	0	0	0
2	41–50	5	15	−10	100	6.7
					0	$23.7 = \chi^2_P$

What about degrees of freedom, D? How many cells can be filled in freely before all are determined by the requirement that the row and column totals come out right? If any two cells in the first row are filled in, then the frequency in the third cell is determined by the requirement that the total for that row is 300. (Checks ✓ are used to indicate the two freely chosen cells, and an X to indicate the one which is then determined.)

✓	✓	X	300
✓	X	✓	150
X	X	X	50
150	200	150	

Similarly, if any two cells in the second row are filled in, then the third is determined by the requirement that the total for that row is 150. Finally, the frequency in each of the cells in the third row is now determined by the requirement that the column totals must be, respectively, 150, 200, and 150. You are free to fill in only four cells in the whole pattern (and no three of these can be in the same row or column), so $D = 4$.

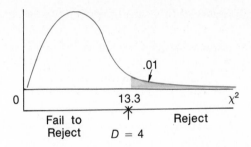

From the χ^2 table (Table 6, Appendix C), the critical value of χ^2 for $D = 4$, $\alpha = .01$, is 13.3. Since the computed value of χ^2_P, 23.7, is greater than 13.3, H_0 is rejected: There is a relationship between number of accidents and age of the driver.

Try a more complicated problem. See how many cells you can fill in freely in the 3×4 contingency table below.

(1)	(2)	(3)	(4)	25
(5)	(6)	(7)	(8)	50
(9)	(10)	(11)	(12)	25
10	20	30	40	100

If cells (1), (2), and (3) are filled in, (4) is determined, since the sum of these four numbers must be 25. If (5), (6), and (7) are filled in, then (8) is fixed. But now you can make no more choices, since the sums of the columns are given. By subtracting from the marginal totals, the frequencies in the last row and column are determined. In this example, then, the number of degrees of freedom is $(3 - 1)(4 - 1)$. **In general, the number of degrees of freedom in a contingency table with R rows and C columns is**

$$D = (R - 1)(C - 1)$$

Example 2   Is there a relationship between wearing of glasses and amount of education? It is found that, among 50 college graduates, 15 wear glasses and 35 do not. Seventy of 200 high school graduates wear glasses and 130 do not, while among 100 adults who attended only grade school, 28 wear glasses. Test at a .05 level of significance.

H_0: There is no relationship between wearing of glasses and amount of education.

H_1: There is a relationship between wearing of glasses and amount of education. (Note that, even if H_0 is rejected, no conclusion is drawn about the nature of the relationship between the two variables—not even whether people with more education are more or less apt to wear glasses, and certainly no cause-and-effect relation is shown.)

The observed frequencies give us the following 2×3 contingency table:

Amount of Education

	College	High School	Grade School	
Wear Glasses	15	70	28	113
No Glasses	35	130	72	237
	50	200	100	350

To find the expected frequencies if there is no relationship between wearing glasses and amount of education, we use the same marginal totals; the expected frequency in a cell is the product of the marginal totals in the row and column to which the cell belongs divided by the grand total. The contingency table for expected frequencies is

Amount of Education

	College	High School	Grade School	
Wear Glasses	$\dfrac{50\,(113)}{350} = 16.1$	$\dfrac{200\,(113)}{350} = 64.6$	$\dfrac{100\,(113)}{350} = 32.3$	113
No Glasses	$\dfrac{50\,(237)}{350} = 33.9$	$\dfrac{200\,(237)}{350} = 135.4$	$\dfrac{100\,(237)}{350} = 67.7$	237
	50	200	100	350

Now $\chi^2{}_P$ can be computed:

Glasses	Education	O	E	$O-E$	$(O-E)^2$	$\dfrac{(O-E)^2}{E}$
Yes	College	15	16.1	-1.1	1.21	.08
Yes	High school	70	64.6	5.4	29.2	.45
Yes	Grade school	28	32.3	-4.3	18.5	.57
No	College	35	33.9	1.1	1.21	.04
No	High school	130	135.4	-5.4	29.2	.22
No	Grade school	72	67.7	4.3	18.5	.27
				0		$1.63 = \chi^2{}_P$

In a 2×3 table, $D = (2 - 1)(3 - 1) = (1)(2) = 2$. From Table 6, $D = 2$ and $\alpha = .05$ implies $\chi^2 = 5.99$. Since 1.63 is less than 5.99, H_0 is still accepted. There is no relationship between wearing of glasses and amount of education.

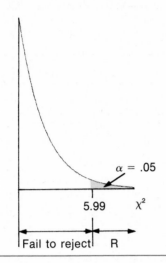

$\alpha = .05$

5.99 χ^2

Fail to reject   R

Example 3   A school psychologist finds that some students in his city are having problems because of their use of alcohol, glue, or marijuana. He carefully defines the term "drug user" and then collects information on a random sample. One hundred high school students who use drugs are classified by grade and type of drug as follows:

		Grade			
		Sophomore	Junior	Senior	
Type of Drug	Alcohol	16	12	12	40
	Glue	8	15	7	30
	Marijuana	6	8	16	30
		30	35	35	100

At a .01 level of significance, test whether there is a relation between grade in high school and the type of drug used.

H_0: There is no relation between grade in school and type of drug used.

H_1: There is such a relation.

Expected frequencies if there is no relation:

Grade

	Sophomore	Junior	Senior
Alcohol	$\dfrac{30\,(40)}{100} = 12$	$\dfrac{35\,(40)}{100} = 14$	$\dfrac{35\,(40)}{100} = 14$
Glue	$\dfrac{30\,(30)}{100} = 9$	$\dfrac{35\,(30)}{100} = 10.5$	$\dfrac{35\,(30)}{100} = 10.5$
Marijuana	$\dfrac{30\,(30)}{100} = 9$	$\dfrac{35\,(30)}{100} = 10.5$	$\dfrac{35\,(30)}{100} = 10.5$

Type of Drug

Type of Drug	Grade	O	E	$O-E$	$(O-E)^2$	$\dfrac{(O-E)^2}{E}$
Alcohol	Sophomore	16	12	4	16	1.33
Alcohol	Junior	12	14	−2	4	.29
Alcohol	Senior	12	14	−2	4	.29
Glue	Sophomore	8	9	−1	1	.11
Glue	Junior	15	10.5	4.5	20.25	1.93
Glue	Senior	7	10.5	−3.5	12.25	1.17
Marijuana	Sophomore	6	9	−3	9	1.00
Marijuana	Junior	8	10.5	−2.5	6.25	.60
Marijuana	Senior	16	10.5	5.5	30.25	2.88
				0		$9.60 = \chi^2{}_P$

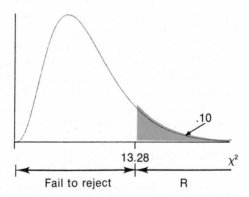

The contingency table is 3×3, so $D = (3 - 1)(3 - 1) = 4$. If $\alpha = .10$ and $D = 4$, then $\chi^2 = 13.28$. H_0 cannot be rejected; there is no relationship between grade in school and type of drug used.

12.6 EXERCISES

Compute the expected frequency for cells which are starred, then find the remaining E entries by using the row or column totals.

1. For divorced women, is age at the time of marriage related to the length of the marriage, at a .01 level of significance?

		Age at Time of Marriage		
		<19	19–22	Over 22
Duration of	<2	10*	24*	36
Marriage	2–4	6*	25*	39
(Years)	>4	4	26	50

2. Based on the following data (in thousands), is the type of accidental death independent of the year? ($\alpha = .05$.)

	Type of Accidental Death			
Year	Motor Vehicle	Burns	Drowning	Poison
1960	38*	8*	7*	2
1976	47	8	6	4

3. Test, on the basis of the following data and at a .05 level of significance, whether there is a relation between political affiliation and attitude towards a proposed law:

	Favor	Opposed	Undecided
Democrats	100	150	150
Republicans	150	100	150

4. Three brands of hair spray are tested. Among the blondes who use them, 20 prefer Brand A, 30 prefer Brand B, and 10 prefer Brand C. Among brunettes the number preferring the three brands are 30, 30, and 20, respectively, while among redheads who test the three brands, A is preferred by 10, B by 5, and C by 5. At a .025 level of significance, find out whether brand preference is related to hair color.

5. Lipsticks are manufactured by three different processes (A, B, and C.) Defective lipsticks are classified as those having defects in color, having defects in consistency, or being chipped. At a .05 level of significance, do the following data show that cause of defect is independent of the process used?

	Defect in Color	Defect in Consistency	Chipped
A	40	25	20
Process B	10	30	20
C	10	25	20

6. A professor claims that the grades in his course are normally distributed and that he gives A to those 1σ above the mean, B between .5 and 1σ above the mean, C if 1σ below the mean to $.5\sigma$ above, D if 1σ to 2σ below the mean, and F only if more than 2σ below the mean. Among 100 students in his class, there are

10 A's, 15 B's, 63 C's, 7 D's, and 5 F's. At a .05 level of significance, is his claim justified?

7. Some people (tasters) find phenylthiourea to be very bitter when tasted, while some (non-tasters) find it tasteless or almost so; it is hypothesized that the ratio of tasters to non-tasters is 2:1. In a test of 300 randomly chosen subjects, 87 are found to be non-tasters and 213 are tasters. Is the hypothesis supported by this evidence? (.10 level of significance.)
  (a) Use a χ^2 test;
  (b) Use a z test of proportions.

8. A maker of sleeping bags claims that 90% of those using his bags will be comfortable sleeping in them when the temperature is 10°F. Among 100 campers using them in winter, 15 are uncomfortably cold at 10°F. Is the maker's claim correct? ($\alpha = .10$.)

9. A geneticist is investigating the inheritability of a certain disease. He collects the following data on a random sample of 180 children:

		Child had Disease?	
		Yes	No
Mother had disease?	Yes	14	68
	No	4	94

At a .01 level of significance, are children more likely to have the disease if their mothers have had it?

10. A random sample of 50 people who hear the Crazy Eights perform in Wichita are queried about their reaction to the music, with the following results:

		Reaction	
		Like	Dislike
	Under 20	20	8
Age	20–25	10	12
	Over 25	5	5

At a .05 level of significance, is there a relation between the age of the listener and attitude towards the music?

11. A sociology major asks a random sample of 60 students to rate 40 statements from 1 ("I strongly agree") to 5 ("I strongly disagree"). On the basis of the answers, he classifies each of the students as showing or not showing racial prejudice. He also classifies students in 30 other two-way categories (upper or lower classman, father completed college or not, student belongs to a church or not, male or female, etc.) according to the answers to 30 other questions. Then he carries out 30 χ^2 tests at a .05 level of significance to see whether or not racial prejudice is related to any of these other categories.

(a) Assume that "H_0: Attitude toward race is independent of the sex of the student" is true; what is the probability that it will be rejected in the χ^2 test? What is the probability that it is not rejected?

(b) Assume that all 30 null hypotheses that he tests are true. What is the probability that all of them will fail to be rejected using χ^2 tests, if they are independent tests?

(c) Assume that all 30 null hypotheses are true, and that the tests are independent. What is the probability that at least one is rejected?

ANSWERS

1. E column: 6.4, 23.9, 39.8, 6.4, 23.9, 39.8, 7.3, 27.3, 45.5. $\chi^2_p = 2.03 + .00 + .36 + .03 + .05 + .02 + 1.49 + .06 + .45 = 4.49$; $D = 2(2) = 4$, $\chi^2 = 13.3$. We cannot reject the null hypothesis that age at time of marriage is independent of the length of the marriage.

2. Did you blindly find an E column of 39.0, 7.3, 6.0, 2.8, 46.0, 8.7, 7.0, 3.3, find $\chi^2_p = .86$, $D = 3$, $\chi^2 = 7.8$, and fail to reject H_0? That gives you good practice in using the formula, but consider: (1) the data are in thousands, and $\Sigma \left[\dfrac{(1,000\ O - 1,000\ E)^2}{1,000\ E} \right] = 1000\Sigma \dfrac{(O-E)^2}{E}$, so your computed figure is 860, not .86: (2) the data is given only to the nearest thousand; and (3) —most important of all—this is a census, not a sample. Just look at the data and draw conclusions such as "the proportion of accidental deaths due to motor vehicles is higher in 1976 than in 1960."

3. E column: 3.3, 32.6, 8.7, 5.4, 2.7, 27.4, 7.3, 4.6. $\chi^2_p = 2.34$. $D = (2 - 1)(4 - 1) = 3$, $\chi^2 = 7.8$. H_0 is accepted.

4. E column: 22.5, 24.4, 13.1, 30, 32.5, 17.5, 7.5, 8.1, 4.4; $\chi^2_p = 4.9$; $D = (3 - 1)(3 - 1) = 4$, $\chi^2 = 11.1$. H_0 cannot be rejected; preference in hair spray is independent of hair color.

5. E column (color, then consistency, then chips): 25.5, 18.0, 16.5, 34.0, 24.0, 22.0, 25.5, 18.0, 16.5, $\chi^2_p = 20.8$; $D = 4$, $\chi^2 = 9.5$. H_0 is rejected; cause of defect is related to the process used.

6. E column; 15.9, 15.0, 53.3, 13.6, 2.3; $\chi^2_p = 2.19 + 0.00 + 1.77 + 3.20 + 3.17 = 10.33$; $D = 4$, $\chi^2 = 9.49$. His claim is rejected.

7. (a) E column: 100, 200. $\chi^2_p = 2.53$. $D = 1$, critical value of $\chi^2 = 2.71$; cannot reject H_0.

(b) H_0: $P = 1/3$, H_1: $P \neq 1/3$, $p = \dfrac{87}{300} = .29$, $\sigma_p = \sqrt{\dfrac{1/3(2/3)}{300}} = .0272$, $z = \dfrac{1/3 - .29}{.0272} = 1.59$. Critical $z = \pm1.64$, so again we cannot reject H_0. Note that $1.58^2 = 2.5$, $1.6^2 = 2.69$ (within round-off); χ^2 with $D = 1$ is z^2.

8. $\chi^2_p = \dfrac{(-5)^2}{90} + \dfrac{(5)^2}{10} = 2.78$; $D = 1$, $\chi^2 = 2.71$. The maker's claim is rejected.

9. E column: 8.2, 73.8, 9.8 88.2; $\chi^2_p = 8.37$. $D = 1$, the critical value of $\chi^2 = 6.63$, so H_0 is rejected. You can conclude that there is a relation between a mother having the disease and her child having it, but from this test you cannot conclude that children are

more likely to have the disease if their mothers have had it. For that you have to go back and look at the proportions (compare 14/82 and 4/98). That's the trouble with χ^2 tests: they tell you nothing about what kind of relation there is when H_0 is rejected.

10. E column: 16.3, 11.7, 12.8, 9.2, 5.8, 4.2; $\chi^2_P = 3.7$. $D = 2$, the critical value of $\chi^2 = 5.99$, so we fail to reject H_0.

11. (a) Probability of rejection $= .05$ (this is measured by α, the level of significance). Probability of acceptance $= 1 - .05 = .95$.
    (b) $(.95)^{30} = .21$.
    (c) $1 - (.95)^{30} = 1 - .21 = .79$.
    More advanced methods are available to treat this data correctly.

12.7 VOCABULARY AND SYMBOLS

chi-square distribution	χ^2	goodness of fit
χ^2_P		related variables
observed frequency	O	contingency table
expected frequency	E	cell

12.8 REVIEW EXERCISES

1. In a χ^2 distribution with $D = 10$, what is the probability that a score is greater than 18.3?

2. In a χ^2 distribution with 8 degrees of freedom, (a) 1% of scores are over what value? (b) 90% are under what value?

3. How many degrees of freedom does a χ^2 distribution have if the probability is .05 that a score in the distribution is greater than 26.3?

4. A city is divided into four areas, of roughly equal size. The number of snow plows assigned to each area on March 13 is as follows: North 117, East 122, South 108, West 97. At a .10 level of significance, does each area get the same number of snow plows after a snow storm?

5. The number of students who register for five sections of an English Composition course are 30, 15, 27, 18, and 35. Are all sections equally popular? ($\alpha = .05$.)

6. Should beer advertisements be directed at a particular age group, or doesn't it matter? An advertising agency with a beer account asks 50 people in each of the age groups 18–22, 23–29, 30–39, and 40 and over whether they like beer. The numbers who say "yes" in each group are 40, 29, 18, and 33, respectively. Do all age groups like beer equally? ($\alpha = .05$.)

7. In a random sample of 100 blacks in Chicago, 50 have incomes less than $10,000, 40 earn between $10,000 and $15,000, and 10 earn over $15,000 per year. Among 100 whites, 30 earn under $10,000, 40 between $10,000 and $15,000, and 30 over $15,000 per year. At a .01 level of significance, are race and income related?

8. A chemical society sends a questionnaire to its members with the following results:

Degree	Number Returned	Number Not Returned	Total
B.S.	50	40	90
B.Chem.E.	100	90	190
M.S.	50	40	90
M.Chem.E.	100	90	190
Ph.D.	200	240	440

Is the proportion of questionnaires returned independent of the degree? ($\alpha = .10$.)

Use the following data in Exercises 9 to 13; it shows counts of state and local government employees newly hired in 1975 in Georgia and in Connecticut. A breakdown is given by the kind of employment and by minority status (SSA = Spanish-speaking American) and, for Georgia, also by sex. The percentage of the working-age population in each category is also included.

Type of Employment	Georgia					Connecticut			
	Total	Bl	Wh	M	F	Total	Bl	SSA	Wh
Administrative, technical or professional	309	42	267	130	179	1,309	86	21	1,202
Paraprofessional, skilled, or protection	107	30	77	64	43	1,771	242	74	1,455
Clerical	257	74	183	58	199	1,246	198	46	1,002
Service and maintenance	537	210	327	349	188	1,096	151	59	886
% of population	100	22.2	77.8	48.3	51.7	100	5.46	.08	94.46
Totals	1,210	356	854	601	609	5,422	677	200	4,545

9. Test the hypothesis that there is no relationship between race and type of employment in Georgia. ($\alpha = .01$.)

10. Test the hypothesis that there is no relationship between ethnic group and type of employment in Connecticut. ($\alpha = .01$.)

11. Test the hypothesis that there is no relationship (except random variation) between the sex of the new employee and the type of employment in Georgia. ($\alpha = .01$.)

12. Does the probability that the next new employee in Georgia is a woman equal the fraction of women in the population? At a .10 level of significance, use (a) a χ^2 test, (b) a z test. Check that your computed and critical values of χ^2 are the squares of the corresponding z values, within round-off.

13. At a .01 level of significance, is the distribution of jobs by type of employment the same in Georgia and in Connecticut? (Be careful how you set up your O and E columns; remember that they must have the same sum.)

14. In a study of "Skin Color, Ethnicity, and Blood Pressure" (see Exercise 18, page 204), high-stress areas of Detroit were defined in terms of crime, marital breakup, and poverty.

Distribution of White Ethnic Groups by Sex

Ethnicity of Father	Males	Females
Northern Europe	57	67
Central Europe	23	38
French	12	14
Mediterranean	28	15
	120	134

Is there a relation between sex of the person living in a high stress area and the ethnic background of that person's father? ($\alpha = .05$.)

15. (a) What is the probability one head and one tail will come up when two coins are tossed?

(b) If two fair coins are tossed 400 times, how many times do you expect to get one head and one tail?

(c) Two coins are tossed 400 times. Two heads come up 112 times, one head 204 times, and no heads 84 times. At a .05 level of significance, are these different from what you would expect due to sampling fluctuations, or do you think the coins are not fair?

16. At an (a) .10, (b) .05 level of significance, is regularity of exercise independent of difference from correct weight?

	Regular Exercise	No Regular Exercise
Overweight	12	68
Correct weight	47	83
Underweight	11	31

17. What assumptions about (a) the population, (b) the sample must be made in order to carry out a χ^2 test on the independence of two variables?

ANSWERS

1. .05

3. 16

5. $\chi^2_P = 1.00 + 4.00 + .16 + 1.96 + 4.00 = 11.12; D = 4, \chi^2 = 9.49.$ H_0 is rejected; registration is not the same.

7. $\chi^2_P = 2.5 + 0 + 5.0 + 2.5 + 0 + 5.0 = 15.0; D = 2, \chi^2 = 9.21.$ Reject H_0; there is a relation between income and race.

9. $\chi^2_P = 26.3 + 11.0 + .1 + .0 + .0 + .0 + 17.1 + 7.1 = 61.6; D = 3, \chi^2 = 11.3.$ H_0 is rejected; there is a relationship between race and type of employment in Georgia.

11. $\chi^2_P = 3.6 + 3.5 + 2.2 + 2.2 + 38.0 + 37.5 + 25.4 + 25.0 = 137.0; D = 3, \chi^2 = 11.3.$ H_0

is rejected; there is a relationship between race and sex and type of employment in Georgia. (Stop as soon as you come to that 38.0.)

13. Take the total number in Connecticut for the observed frequency, then multiply the number in each type of employment in Georgia by $\frac{5422}{1210}$ (why?) to get the E column. The contribution to χ^2_p made by "paraprofessional, skilled, or protection" is already 3,478, so clearly H_0 is rejected.

15. (a) $\frac{1}{2}$; (b) 200; (c) $\chi^2_p = 1.44 + .08 + 2.56 = 4.08$; $D = 2$, $\chi^2 = 5.99$. We cannot reject the null hypothesis that the coins are fair.

17. (a) None; (b) Sample must be random, at least three-fourths of the observed frequencies must be 5 or over, and all must be greater than 1.

13

ANALYSIS OF VARIANCE

In Chapters 10 and 11, we tested hypotheses about the differences between the means of two populations. An economist might use the methods discussed there to compare wages in a certain industry in New York and New Jersey; a psychologist might compare the performances of an experimental group and a control group. It is often desirable, however, to compare many means rather than just two: the economist wishes to compare wages in more than two states, or the psychologist wishes to compare performance of groups treated in a variety of ways with a control group. The procedure that is used to compare more than two means is called *analysis of variance*. An introduction to F distributions is a prerequisite for understanding this procedure. After becoming familiar with F distributions you will first use them to solve a simpler problem, that of comparing the variances of two populations, and then you will go on to the comparison of many means. Finally, if the hypothesis that all means are the same is rejected, you will learn how to find which pairs of means are significantly different.

13.1 F DISTRIBUTIONS

Imagine that you have two populations, that **both** are normally distributed, and that they have the same variance. They might have different means. From the first population you take a sample of size 9, and from the second a sample of size 21. Suppose $s_X^2 = 5.0$ and $s_Y^2 = 2.0$. Then the ratio $F = s_X^2/s_Y^2$ is computed: for these two samples it is $5.0/2.0 = 2.5$. Then these samples are replaced, and new samples are taken from each population—again of size 9 and size 21, respectively. This time $s_X^2 = 4.0$, $s_Y^2 = 8.0$, and $F = 4/8 = .5$. This process is continued for all possible different pairs of samples of size 9 from the first population and of size 21 from the second population selected independently. The resulting F scores form a new distribution. Since neither s_X^2 nor s_Y^2 is negative, an F score is never negative. Since the two populations have the same variance, you would expect many of your samples to have the same variances, resulting in F scores near 1. Some ratios are large, however (if the variance of the first sample is large and that of the second sample is small). Five percent of the F scores will be above 2.45.

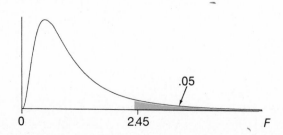

If the sizes of your samples are changed—say, 28 and 7 instead of 9 and 21—you will get a different F distribution. In general there will be a double family of F distributions; each member of the family will be determined by particular choices for the sizes of the samples from the two populations. Remember that an F distribution can be used **only** if both populations are normally distributed. This is true even if samples are large; there is no Central Limit Theorem which implies that the assumption that the populations are normally distributed can be relaxed if sample sizes are large enough.

All F distributions have certain things in common:

1. An F score is the quotient of two independently estimated population variances:

    $$F = \frac{s_X^2}{s_Y^2}$$

    where s_X^2 is the estimate of the variance of one population based on a sample of size n_X, and s_Y^2 is the same for another population based on a sample of size n_Y.

2. Each F distribution is a probability distribution—that is, the area under the entire curve is 1, and the probability that an F score falls between a and b is given by the area under the curve between a and b.

3. In every F distribution the range of F values is from 0 to $+\infty$.

4. F tables are given not in terms of n_X and n_Y but rather in terms of degrees of freedom:

    $$D_X = n_X - 1, \qquad D_Y = n_Y - 1$$

    The shape of any F distribution depends on the values of D_X and D_Y. All are skewed to the right, but become more symmetrical as the number of degrees of freedom gets larger.

13.2   HOW TO USE AN F TABLE

Since there are so many different F distributions, the values given for each are severely limited. Table 7, Appendix C, gives values of F when the right-hand tail of the distribution has an area of .05, .025, or .01. As a result, two-tail tests of hypotheses can only be found using these tables if $\alpha = .10, .05,$ or $.02$.

Example 1 $n_X = 15, n_Y = 7$. Five percent of F scores are above what value?

$D_X = 14, \quad D_Y = 6$

$D_X = 14, D_Y = 6$. In Table 7(a), Appendix C, look across the first row until you find the number 14; look down the first column to the number 6. In the intersection of the 14th column and the sixth row is found the number 3.96. $\Pr(F > 3.96) = .05$, or 5% of F scores are above 3.96 if samples of size 15 and 7, respectively, are chosen at random from two normal populations with the same variance. Here is a useful notation: $F_{.05}(14, 6) = 3.96$.

Notation: $F_{.05}(D_X, D_Y)$ is the value for F in the .05 table for D_X and D_Y degrees of freedom.

.05 may be replaced by .025 or .01, but note that the subscript is α for a one-tail test and $\dfrac{\alpha}{2}$ for a two-tail test.

Example 2 If $n_X = 4$, $n_Y = 11$, 97.5% of F scores are below what value?

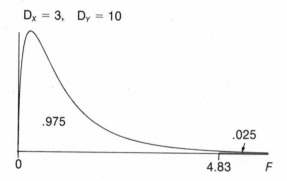

$D_X = 3$, $D_Y = 10$

.975

.025

0 4.83 F

$\alpha = 1 - .975 = .025$, so see Table 7(b), Appendix C. $D_X = 3, D_Y = 10$; 97.5% of F scores are below 4.83.

Two-tail tests using F distributions or one-tail tests using the left-hand tail are quite possible. Table 7 gives only values of F for percentages in a right-hand tail, so an extra computation would be required. The left-hand value of F is the reciprocal of the right-hand value for the same α but with D_X and D_Y interchanged:

$$F_{(left)}(D_X, D_Y) = \frac{1}{F_{(right)}(D_Y, D_X)}$$

But note that all the F values in Tables 7(a), 7(b), and 7(c) are larger than 1, so the reciprocals are less than 1. The use of left-hand tail values can **always** be avoided in hypothesis tests by labeling as the X population that one from which the sample with the larger variance is taken, so s_X^2/s_Y^2 is greater than 1. See Examples 3 and 4 on pages 292 and 293.

13.3   USE OF *F* DISTRIBUTIONS IN TESTING HYPOTHESES ABOUT VARIANCE

Example 1
Two machines are used to fill boxes with crackers. In a random sample of 25 boxes filled by the first machine, the mean weight of crackers is 1.03 pounds with a variance of .008; a random sample of 31 boxes filled by the second machine has a mean weight of 1.06 pounds with a variance of .006. At a .05 level of significance, does the first machine show more variability than the second? Assume weights are normally distributed for both machines.

$$H_0: \sigma_X{}^2 = \sigma_Y{}^2; \ H_1: \sigma_X{}^2 > \sigma_Y{}^2; \ \alpha = .05; \text{ one-tail test.}$$

$$n_X = 25, D_X = 24, s_X{}^2 = .008$$

$$n_Y = 31, D_Y = 30, s_Y{}^2 = .006$$

The mean weights are given, but are not relevant to the problem.

$$F = \frac{.008}{.006} = 1.33$$

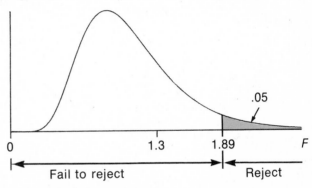

The critical value of F is $F_{.05}(24, 30) = 1.89$.
Since 1.3 is less than 1.94, we fail to reject H_0: there is only sample variation and no other difference in the variability of the two machines.

Example 2
A psychologist believes that scores of children who are under stress when they take an arithmetic test will be more variable because bright children do better but dull children will panic and do worse than when not under stress. He finds that 20 children under stress have a standard deviation of 15.0, while a control group of 11 children not under stress have a standard deviation of 8.6. At a .025 level of significance, do children under stress show more variability in their grades on this test than children not under stress?

$$H_0: \sigma_X{}^2 = \sigma_Y{}^2; \ H_1: \sigma_X{}^2 > \sigma_Y{}^2.$$

$$n_X = 19, D_X = 18, s_X{}^2 = 15^2 = 225$$

$$n_Y = 11, D_Y = 10, s_Y{}^2 = 8.6^2 = 74$$

$$F = \frac{225}{74} = 3.04$$

For $\alpha = .025$, the critical value of $F(18, 10) > F(20, 10) = 3.42$. Therefore H_0 is not rejected: Children under stress do not show more variability in their scores.

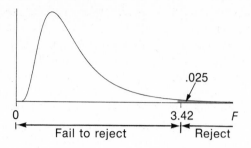

Example 3

Are the salaries of associate professors less variable than those of full professors at Great Western University? A random sample of 17 associate professors has a standard deviation of $1,000 in salaries, while 25 full professors chosen at random have a standard deviation of $2,500. Test at a .05 level of significance.

The necessity of finding $F(D_Y, D_X)$ and then taking its reciprocal can be avoided if the population whose sample has the larger variance is given the subscript X even if it is mentioned last in the statement of the problem. In this example,

H_0: $\sigma_X^2 = \sigma_Y^2$; H_1; $\sigma_X^2 > \sigma_Y^2$; right-tail test.

$s_X^2 = 2,500^2$, $s_Y^2 = 1,000^2$, $F = \dfrac{s_X^2}{s_Y^2} = 6.25$

$D_X = 24$, $D_Y = 16$, $F_{.05}(24, 16) = 2.24$, so H_0 is rejected.

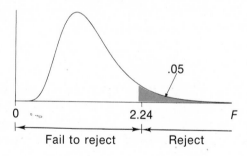

Note: The computations in Example 3 were carefully done and the samples were random. Nevertheless, the result may not be reliable. Why? (Think before you read on.) Are the salaries of associate and full professors really normally distributed? It depends on the practices at Great Western University. Suppose there are definite ranges of salary at each rank. If Great Western has expanded recently and promoted a large number of new professors, the salary curve for professors may be skewed to the right. If 30 assistant professors are appointed associate each year and tend to stay at this rank for about 5 years, with an increment each year, before becoming full professors, the distribution for associate professors may be uniform (the "curve" is a horizontal line) rather than normal. **Use of the F distribution is based on the assumption of normality,** even in large samples. Tests for variances differ in this way from z tests for means.

Example 4

A psychologist again wonders whether children under stress in a test show a different variance in their test scores than children not under stress. Bright children might do better but dull children panic and do worse, causing greater variability. On the other hand, perhaps stress has no effect on children who know the material well but causes those with less knowledge or ability to do better, so there is less variability in the results. He finds that a random sample of 15 children in a control group have a mean score of 77 with a standard deviation of 3.1, whereas 13 children under stress have a mean score of 82 with a standard deviation of 3.3. At a .05 level of significance, do the two groups have different variances?

H_0: $\sigma_X^2 = \sigma_Y^2$; H_1: $\sigma_X^2 \neq \sigma_Y^2$; two-tail test.

$s_X^2 = 3.3^2 = 10.89$, $n_X = 13$

$s_Y^2 = 3.1^2 = 9.61$, $n_Y = 15$

$$F = \frac{s_X^2}{s_Y^2} = 1.13$$

Note that X scores are children under stress (since this sample has the larger variance) so the computed F will be > 1.

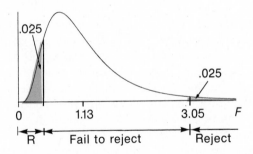

$D_X = 12$, $D_Y = 14$; $F_{.025}(12, 14) = 3.05$ when $\alpha = .05$. It is not necessary to determine the lower critical value for acceptance of H_0; it is certainly less than 1, whereas the computed F, 1.13, is greater than 1. H_0 is not rejected: the variances of the two groups of children do not differ except for sample fluctuations.

13.4   EXERCISES

1. Find the probability that $F(8, 10)$ is less than 5.06.

2. Find the probability that $F(7, 12)$ is more than 2.91.

3. If $D_X = 12$, $D_Y = 26$, what proportion of F scores are over 2.96?

4. The computed value of F in testing H_0: $\sigma_X^2 = \sigma_Y^2$, H_1: $\sigma_X^2 \neq \sigma_Y^2$ has been found to be 2.2. If $\alpha = .05$, $n_X = 25$, $n_Y = 22$, is H_0 rejected or not?

5. Test H_0: $\sigma_X^2 = \sigma_Y^2$, H_1: $\sigma_X^2 \neq \sigma_Y^2$, $\bar{X} = 12.5$, $s_X = 7.0$, $n_X = 31$, $\bar{Y} = 16.3$, $s_Y = 4.0$, $n_Y = 26$, $\alpha = .05$.

6. Test H_0: σ_Y^2, H_1: $\sigma_X^2 < \sigma_Y^2$, $s_X = 3.0$, $n_X = 9$, $s_Y = 4.0$, $n_Y = 7$, at a .05 level of significance. (What should you do first?)

7. Ten ropes made of nylon are chosen at random; their mean breaking strength is 800 pounds with a standard deviation of 50 pounds, whereas 12 ropes made of tufflon have a mean breaking strength of 900 pounds with a standard deviation of 28 pounds. At a .05 level of significance, test the hypothesis (a) that there is no difference in the variability of the two types of ropes in breaking strength, (b) that nylon ropes are more variable in their breaking strength than are ropes made of tufflon.

8. On a final examination, 30 statistics students taught by TV tapes have a mean score of 78 with a variance of 25, while 21 students taught "live" have a mean of 80 and a variance of 30 on the same examination. At a .01 level of significance, do students taught by TV tapes show less variability?

9. Twenty-five bottles of diet Cwik Cola chosen at random are tested for calorie content; the mean is 12.0 calories with a standard deviation of .60. Then 41 bottles of Pappy's Cola are found to have a mean calorie content of 20.0 calories with a standard deviation of .80. At a .02 level of significance, does Cwik Cola differ from Pappy's Cola in variability of calorie content?

10. Random samples of size 8 and 6, respectively, are chosen from among first graders in the Tacoma Elementary School and are given reading readiness tests. The first sample, of children who have had a year of kindergarten, has a mean of 104 and a standard deviation of 7 on the readiness test. For the second sample, of children who did not go to kindergarten, the mean on the reading readiness test is 95 with a standard deviation of 4. Assume that scores on the reading readiness tests of all children are normally distributed (both with and without kindergarten).

    (a) Do the test scores of children who have been to kindergarten have the same variance as those who have not? (.02 level of significance).

    (b) Is there a difference in the reading readiness scores of children who have not been to kindergarten? ($\alpha = .05$).

ANSWERS

1. $F_{.01}(8, 10) = 5.06$. Therefore the probability that $F(8, 10)$ is less than 5.06 is .99.

2. .05

3. 1%

4. Fail to reject H_0, since the critical value of $F_{.025}(24, 21) = 2.37$. (Two-tail test, so remember to use the .025 table for the upper tail.)

5. Computed $F = \dfrac{49}{16} = 3.1$, critical $F_{.025}(30, 25) = 2.18$. Reject H_0.

6. Interchange X and Y; H_1 becomes $\sigma_X^2 > \sigma_Y^2$. Computed $F = \dfrac{16}{9} = 1.8$; critical

$F_{.05}(6, 8) = 3.58$. Fail to reject H_0; the difference can be explained by sampling fluctuation.

7. (a) H_0: $\sigma_X^2 = \sigma_Y^2$; H_1: $\sigma_X^2 \neq \sigma_Y^2$ (two-tail test).

$n_X = 10$, $s_X^2 = 50^2$, $n_Y = 12$, $s_Y^2 = 28^2$.

$$F = \frac{s_X^2}{s_Y^2} = 3.2, D_X = 9, D_Y = 11, F_{.025}(9, 11) = 3.59.$$

H_0 is not rejected; there is no difference in variability.
(b) H_0: $\sigma_X^2 = \sigma_Y^2$; H_1: $\sigma_X^2 > \sigma_Y^2$ (one-tail test, right-hand tail). Computed F score is the same as in (a). $F_{.05}(9, 11) = 2.90$. H_0 is rejected.

8. Let students taught "live" be the X population.

H_0: $\sigma_X^2 = \sigma_Y^2$; H_1: $\sigma_X^2 > \sigma_Y^2$ (one-tail test, right-hand tail)

$n_X = 21$, $s_X^2 = 30$, $n_Y = 30$, $s_Y^2 = 25$; computed $F = 1.20$

$F_{.01}(20, 29) = 2.57$. H_0 is not rejected: There is no difference.

9. H_0: $\sigma_X^2 = \sigma_Y^2$; H_1: $\sigma_X^2 \neq \sigma_Y^2$; two-tail test. $n_X = 41$, $s_X = .80$, $n_Y = 25$, $s_Y = 60$;

$F = \dfrac{.80^2}{.60^2} = 1.78$. $F_{.01}(40, 24) = 2.49$. H_0 is accepted; Cwik and Pappy's have the same variability.

10. (a) $s_X = 7$, $n_X = 8$, $s_Y = 4$, $n_Y = 5$, $F = \dfrac{7^2}{4^2} = 3.06$; if $\alpha = .02$, two-tail test, $F_{.01}(7, 5) =$

10.46. The variances are the same.
(b) $\sigma_{(\overline{X}-\overline{Y})} = \sqrt{\dfrac{7(7^2) + 4(4^2)}{8 + 5 - 2}\left(\dfrac{1}{8} + \dfrac{1}{5}\right)} = 3.47$; $t = \dfrac{(104 - 95) - 0}{3.47} = 2.59$; $\alpha = .05$,

$D = 11$, so the critical values of t are ± 2.20. H_0 is rejected; there is a difference in scores of children who have gone to kindergarten and those who have not.

13.5 INTRODUCTION TO ANALYSIS OF VARIANCE

You have now learned how to use F distributions to answer the question, "Do two populations have the same variance?" If the answer is yes, then the F ratio for samples taken from the two populations is expected to be close to 1. If it is not close to 1, this difference may be due not just to sampling fluctuation but to different variances in the two populations.

Now F distributions will be used for quite a different kind of hypothesis testing: Do three or more populations have the same mean? Suppose tomato seeds can be treated with any one of three different chemicals or left untreated, and the yield of tomatoes under these four methods is to be compared. Four random samples of seeds are selected; the first three are each treated with a different chemical, and the fourth is left untreated. Then a field is divided into, say, 40 plots of the same size, and the 40 plots are randomly assigned to four groups; all the plots in a group are then planted with the same number of seeds from one of the samples.* The mean yield per plot of each of the four different samples is determined, with the following results:

---

*Both the seeds and the plots in which they are grown are selected at random. If this is not done, the effects of microclimate and different kinds of soil may have more influence on yield than the chemical treatment of the seed.

Chemical 1	Chemical 2	Chemical 3	Untreated
$\bar{X}_1 = 96$	$\bar{X}_2 = 99$	$\bar{X}_3 = 92$	$\bar{X}_4 = 101$

Are the yields so different that it can be concluded that different chemical treatments affect yield? Should we continue to accept or reject H_0: $\mu_1 = \mu_2 = \mu_3 = \mu_4$?

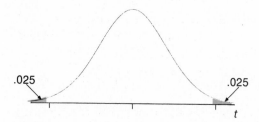

For any pair of means we could, of course, carry out a t test. But since there are four different means to be compared, six t tests would have to be carried out. Remember that if $\alpha = .05$, the probability is $1 - .05 = .95$ that given samples will have means whose difference falls in the "fail to reject" region if H_0: $\mu_X = \mu_Y$ is true. If k independent t tests of differences of means are made, the probability is $.95^k$ that the null hypothesis is still accepted in all k of them, so the probability is $1 - .95^k$ that H_0 is rejected at least once even when it is true. If six independent tests are carried out, this probability is .26, even though $\alpha = .05$ for each individual test. If you are testing four samples a pair at a time (six tests), you do not have independent t tests, so the probability computed above is only an approximation. This argument should persuade you, however, that t tests are not appropriate when more than two sample means are compared to see if the samples all come from the same population. A new technique is needed; this technique is called **analysis of variance,** often abbreviated **ANOVA.**

Suppose three samples, each of size 3, are taken from the same population and that there is no variability within the population. The sample scores then might be:

Sample 1	Sample 2	Sample 3
5	5	5
5	5	5
5	5	5

If, however, the samples come from populations which are affected by different treatments, but again there is no variability within a given population, the sample scores then might be:

Sample 1	Sample 2	Sample 3
4	6	5
4	6	5
4	6	5

But there is, inevitably, variability in each population, so the sample scores would more likely be:

Sample 1	Sample 2	Sample 3
3	5	3
4	6	7
5	7	5
$\bar{X}_1 = 4$	$\bar{X}_2 = 6$	$\bar{X}_3 = 5$

Note that differences exist not only between scores in a sample but between scores in different samples. If we are given three samples such as those in the box above and wish to decide whether or not they come from the same population, we must find out how to distinguish fluctuations in sampling from effects due to different treatments.

Suppose the scores in the box above represent the grades on one problem for students taught statistics by three different instructors. At a .05 level of significance, do the means of the populations taught by the three instructors differ?

H_0: $\mu_1 = \mu_2 = \mu_3$

H_1: Differences exist between at least some of the means.

If the populations from which the three samples were chosen are identical—if they have the same mean and standard deviation, and if all three are normally distributed—then the three groups can be considered to be three samples from one large population. The sample means (4, 6, and 5, respectively) are not the same, but we hardly expect them to be identical. Is the variation between the sample means what is expected if the samples come from the same population, or are the differences between them so great that it is likely to occur not from sampling fluctuations but rather because the three populations are not in fact identical?

We shall make two different estimates of the population variance, σ^2, under the assumption that the three populations are identical and therefore the three samples come from one large population. The first estimate will depend on the variations of the sample means about the grand mean that is obtained for all scores taken together; this estimate, then, will depend on the differences between samples. The second estimate will depend on the spread of scores in a sample about the mean of the sample; this estimate, then, will depend upon the differences within each sample, but will not compare one sample with another. The first estimate will depend on sampling fluctuation and on real differences between samples; if the samples come from populations with different means (if H_0 is not true), then this estimate will be significantly larger than the second estimate, which depends on sampling fluctuation alone. But if H_0 is true and the three samples can be thought of as three slices of the same population, then the two estimates differ only because of sampling fluctuations, and their ratio (their F score) should be approximately 1. A right-tail F test will be used.

The first estimate is based on the variability among the sample means: $X_1 =$

4, $\bar{X}_2 = 6$, $\bar{X}_3 = 5$. The mean of all 9 scores is $\bar{X}_T = (3 + 4 + 5 + 5 + 7 + 6 + 3 + 7 + 5)/9 = 5$. We find the variance of the three sample means about the mean of all sample scores, $\bar{X}_T$, in the usual way; since we want an unbiased estimator we divide by one less than the number of samples:

$$\frac{(4 - 5)^2 + (6 - 5)^2 + (5 - 5)^2}{2} = 1.$$

This quantity we have computed gives an estimate of $\sigma_{\bar{X}}^2 = \dfrac{\sigma^2}{n}$, so $\sigma^2 = n\sigma_{\bar{X}}^2 \approx$ 3(1) = 3, since each sample is of size 3. This first estimate of σ^2 depends on variability between samples and will be referred to as $\hat{\sigma}^2$ **between** (read "σ hat squared between"; the hat is put on to remind you that this is an estimate based on sample quantities):

$$\hat{\sigma}^2 \, between = 3.$$

It should be noted that we could have computed $\hat{\sigma}^2$ *between* directly by multiplying each squared deviation by the size of the sample involved:

$$\hat{\sigma}^2 \, between = \frac{3(4 - 5)^2 + 3(6 - 5)^2 + 3(5 - 5)^2}{2} = 3.$$

When samples are of different sizes, each squared deviation will be weighted by the size of the sample from which it comes. Thus, if:

$$n_1 = 9, \bar{X}_1 = 3, n_2 = 8, \bar{X}_2 = 6, n_3 = 7, \bar{X}_3 = 5, \bar{X}_T = 5$$

then

$$\hat{\sigma}^2 \, between = \frac{9(3 - 5)^2 + 8(6 - 5)^2 + 7(5 - 5)^2}{2} = 22.$$

The mathematical justification requires more background than you possess. Be of great faith, please!

Now we need another and independent estimate of σ^2. We learned earlier that s^2 is an unbiased estimate of σ^2. If the three samples come from the same population (if H_0 is true), then we appear to have three ways of estimating σ^2: we could use s_1^2 or s_2^2 or s_3^2. Instead, we use a weighted average of all three, assigning weights according to the degrees of freedom in each sample.

	Sample 1	Sample 2	Sample 3	Total
ΣX	12	18	15	45
ΣX^2	50	110	83	243
n	3	3	3	
D	2	2	2	
$s^2 = \dfrac{\Sigma X^2 - \dfrac{(\Sigma X)^2}{n}}{n - 1}$	1	1	4	

$$\sigma^2 \approx \frac{2(1) + 2(1) + 2(4)}{2 + 2 + 2} = 2$$

This estimate of σ^2 depends on variability within each sample, and will be called $\hat{\sigma}^2$ **within:**

$\hat{\sigma}^2 \ within = 2.$

$$F = \frac{\hat{\sigma}^2 \ between}{\hat{\sigma}^2 \ within} = \frac{3}{2} = 1.5$$

What about degrees of freedom? For $\hat{\sigma}^2$ *between,* the number of degrees of freedom (D_X) is the number of samples minus 1. In this sample, $D_X = 3 - 1 = 2$. For $\hat{\sigma}^2$ *within,* the number of degrees of freedom (D_Y) is the total number of scores (in all samples) minus the number of samples. In this example, $D_Y = 9 - 3 = 6$. For $\alpha = .05$, the critical value of $F(2, 6)$ is 5.14 for a right-tail test. The computed F value, 1.5, falls in the "fail to reject" region, so we cannot reject the hypothesis that the three populations from which the samples were taken have the same means. Since it was assumed that they have the same variances and that all are normally distributed, this means that they are identical populations.

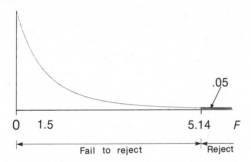

Now let's do the same problem again, and find computational formulas that will expedite the work when more or larger samples are used. First, some new notation is needed:

Let k = the number of samples.

n_T = the total number of scores in all samples.

The following steps are carried out (in Section 13.6), page 302, an explanation of why they work will be given):

1. List, in columns, the scores in each sample.
2. Add all the scores in the first sample, square the sum, and divide by the number of scores in the first sample.
3. Repeat Step 2 for each of the other samples.
4. Add the results of Steps 2 and 3; call this sum A.
5. Add all the scores in all samples; call this sum B.
6. Square each score and then add these squared scores for all samples; call this sum C.

7. Add the number of scores in each sample; call this sum n_T.

8. Compute $\hat{\sigma}^2\ between = \dfrac{A - B^2/n_T}{k - 1}$

   and $\hat{\sigma}^2\ within = \dfrac{C - A}{n_T - k}$

9. $F = \dfrac{\hat{\sigma}^2\ between}{\hat{\sigma}^2\ within}$ is compared with the critical value $F_\alpha(k - 1, n_T - k)$ found in the appropriate F table.

Here is the work involved in our example:

	Sample 1	Sample 2	Sample 3	
	3	5	3	
	4	6	5	
	5	7	7	Total
Sum:	12	18	15	$45 = B$
Sum squared:	144	324	225	
Number in sample:	3	3	3	$9 = n_T$
$\dfrac{\text{Sum squared}}{\text{Number in sample}}$	48	108	75	$231 = A$
Sum of squares:	$9 + 16 + 25 =$ 50	$25 + 36 + 49 =$ 110	$9 + 25 + 49 =$ 83	$243 = C$

	Numerator	Degrees of Freedom (Denominator)	Quotient
$\hat{\sigma}^2\ between$	$A - \dfrac{B^2}{n_T} = 231 - \dfrac{45^2}{9} = 6$	$k - 1 = 2$	$\dfrac{6}{2} = 3$
$\hat{\sigma}^2\ within$	$C - A = 243 - 231 = 12$	$n_T - k = 6$	$\dfrac{12}{6} = 2$

$$\hat{\sigma}^2\ between = \frac{231 - (45)^2/9}{3 - 1} = 3$$

$$\hat{\sigma}^2\ within = \frac{243 - 231}{9 - 3} = 2$$

$F = \dfrac{3}{2} = 1.5$; $F_{.05}(2, 6) = 5.14, > 1.5$, so we fail to reject H_0, as before.

Note, however, that even when H_0 is rejected this test does **not** disclose how many pairs of means differ significantly. This problem will be discussed in Section 13.8, page 307.

Example 1   Random samples of tomato seeds are treated with one of three different chemicals or left untreated, then planted in similar test plots. The weights of tomatoes produced by each sample are as follows; test whether $\mu_1 = \mu_2 = \mu_3 = \mu_4$ at a .01 level of significance.

	$\bar{X}_1$	$\bar{X}_2$	$\bar{X}_3$	$\bar{X}_4$	
	80	70	78	90	
	82	75	81	94	
	85	76			
	82				
Sum	329	221	159	184	$893 = B$
Sum squared	108,241	48,841	25,281	33,856	
n	4	3	2	2	$11 = n_T$
$\dfrac{\text{Sum squared}}{n}$	27,060.3	16,280.3	12,640.5	16,928	$72,909.1 = A$
Sum of squares	27,073	16,301	12,645	16,936	$72,955 = C$

$$\hat{\sigma}^2 \text{ between} = \frac{72,909.1 - 893^2/11}{3} = 137.9$$
$$\hat{\sigma}^2 \text{ within} = \frac{72,955 - 72,909}{11 - 4} = 6.56$$
$$\left.\rule{0pt}{40pt}\right\} F = \frac{137.9}{6.56} = 21.0$$

The critical value of F is $F_{.01}(3, 7) = 8.45$, so H_0 is rejected: there is a difference in at least 2 of the population means.

If a constant is subtracted from each score in a set, the variance is not changed (see page 74). Often work can be simplified if the same constant is subtracted from every score in every sample. If 80 is subtracted from each score in Example 1, we shall obtain the same F score.

Example 2   Repeat Example 1 after subtracting 80 from each score.

	$\bar{X}_1$	$\bar{X}_2$	$\bar{X}_3$	$\bar{X}_4$	
	0	−10	−2	10	
	2	−5	1	14	
	5	−4			
	2				
Sum	9	−19	−1	24	$13 = B$
Sum squared	81	361	1	576	
n	4	3	2	2	$11 = aT$
$\dfrac{\text{Sum squared}}{n}$	20.3	120.3	.5	296	$429.1 = A$
Sum of squares	33	141	5	296	$475 = C$

$$\hat{\sigma}^2 \; between = \frac{429.1 - 13/11}{3} = 137.9$$

$$\hat{\sigma}^2 \; within = \frac{475 - 429.1}{11 - 4} = 6.56$$

$$F = \frac{137.9}{6.56} = 21.0, \; \text{as before.}$$

The critical value is again 8.45, so H_0 is again rejected.

13.6 WHERE DO THE COMPUTATIONAL FORMULAS COME FROM?

Let's take a closer look at the three samples of students in the discussion on page 297. Subscripts 1, 2, and 3 will refer to the three samples: $\bar{X}_1$ is the mean of the first sample, n_3 is the number of scores in the third sample, and so on. In this notation,

$$A = \frac{(\Sigma X_1)^2}{n_1} + \frac{(\Sigma X_2)^2}{n_2} + \frac{(\Sigma X_3)^2}{n_3}$$

$$B = \Sigma X_1 + \Sigma X_2 + \Sigma X_3$$

$$C = \Sigma(X_1{}^2) + \Sigma(X_2{}^2) + \Sigma(X_3{}^2)$$

We found (page 298) that $\hat{\sigma}^2 \; between$ is the variance of the sample means about the mean $\bar{X}_T$ of all sample scores, with weights according to the size of each sample:

$$\sigma^2 \; between = \frac{n_1(\bar{X}_1 - \bar{X}_T)^2 + n_2(\bar{X}_2 - \bar{X}_T)^2 + n_3(\bar{X}_3 - \bar{X}_T)^2}{k - 1}$$

(In this example, $k = 3$.) Let's look at the first term in the numerator.

$$n_1(\bar{X}_1 - \bar{X}_T)^2 = n_1\bar{X}^2 - 2n_1\bar{X}_1\bar{X}_T + n_1\bar{X}_T{}^2 \qquad [\text{squaring } (\bar{X}_1 - \bar{X}_T)^2]$$

$$= n_1\left(\frac{\Sigma X_1}{n_1}\right)^2 - 2n_1\left(\frac{\Sigma X_1}{n_1}\right)\bar{X}_T + n_1\bar{X}_T{}^2 \left[\text{since } \bar{X}_1 = \frac{\Sigma X_1}{n_1}\right]$$

$$= \frac{(\Sigma X_1)^2}{n_1} - 2(\Sigma X_1)\bar{X}_T + n_1\bar{X}_T{}^2$$

The other two terms in the numerator of $\hat{\sigma}^2 \; between$, $n_2(\bar{X}_2 - \bar{X}_T)^2$ and $n_3(\bar{X}_3 - \bar{X}_T)^2$, can be expanded in a similar fashion. Adding all three, the numerator of $\hat{\sigma}^2 \; between$ becomes

$$\left[\frac{(\Sigma X_1)^2}{n_1} - 2(\Sigma X_1)\bar{X}_T + n_1\bar{X}_T{}^2\right] + \left[\frac{(\Sigma X_2)^2}{n_2} - 2(\Sigma X_2)\bar{X}_T + n_2\bar{X}_T{}^2\right]$$

$$+ \left[\frac{(\Sigma X_3)^2}{n_3} - 2(\Sigma X_3)\bar{X}_T + n_3\bar{X}_T{}^2\right]$$

$$= \underbrace{\left[\frac{(\Sigma X_1)^2}{n_1} + \frac{(\Sigma X_2)^2}{n_2} + \frac{(\Sigma X_3)^2}{n_3}\right]}_{= A} \underbrace{-2(\Sigma X_1 + \Sigma X_2 + \Sigma X_3)\bar{X}_T + (n_1 + n_2 + n_3)\bar{X}_T{}^2}_{= -\dfrac{B^2}{n_T}?}$$

If we can show that the last terms in the sum $= -\dfrac{B^2}{n_T}$, then we have succeeded in showing that σ^2 between $= \dfrac{A - B^2/n_T}{k-1}$. But this is easy to show, since $n_1 + n_2 + n_3 = n_T$ and $\bar{X}_T = \dfrac{\Sigma X_1 + \Sigma X_2 + \Sigma X_3}{n_1 + n_2 + n_3} = \dfrac{(\text{sum of all } X \text{ scores})}{n_T} = \dfrac{B}{n_T}$:

$$-2(\Sigma X_1 + \Sigma X_2 + \Sigma X_3)\bar{X}_T + (n_1 + n_2 + n_3)\bar{X}_T{}^2 = -2B\left(\frac{B}{n_T}\right) + (n_T)\left(\frac{B}{n_T}\right)^2$$

$$= -\frac{2B^2}{n_T} + \frac{B^2}{n_T} = -\frac{B^2}{n_T}$$

Therefore $\hat{\sigma}^2$ between $= \dfrac{A - B^2/n_T}{k-1}$.

Now what about $\hat{\sigma}^2$ within? It is the weighted average of the sample variances, with weights assigned according to the degrees of freedom in each sample:

$$\hat{\sigma}^2 \text{ within} = \frac{(n_1 - 1)s_1{}^2 + (n_2 - 1)s_2{}^2 + (n_3 - 1)s_3{}^2}{(n_1 - 1) + (n_2 - 1) + (n_3 - 1)}$$

Again, concentrate first on the first term in the numerator.

Using the computational formula for sample variance (see page 71):

$$(n_1 - 1)S_1{}^2 = (n_1 - 1)\left[\frac{\Sigma X_1{}^2 - (\Sigma X_1)^2/n_1}{n_1 - 1}\right] = \Sigma X_1{}^2 - \frac{(\Sigma X_1)^2}{n_1}$$

Weighting by degrees of freedom "cancels" the denominator. The same will be true for the other samples, so the numerator becomes

$$\Sigma X_1{}^2 - \frac{(\Sigma X_1)^2}{n_1} + \Sigma X_2{}^2 - \frac{(\Sigma X_2)^2}{n_2} + \Sigma X_3{}^2 - \frac{(\Sigma X_3)^3}{n_3}$$

After rearrangement, this is

$$(\Sigma X_1{}^2 + \Sigma X_2{}^2 - \Sigma X_3{}^2) - \left[\frac{(\Sigma X_1)^2}{n_1} + \frac{(\Sigma X_2)^2}{n_2} + \frac{(\Sigma X_3)^2}{n_3}\right] = C - A.$$

That takes care of the numerator. Since $(n_1 - 1) + (n_2 - 1) + (n_3 - 1) = n_T - k$,

$$\hat{\sigma}^2 \text{ within} = \frac{C - A}{n_T - k}.$$

13.7 EXERCISES

1. At a .05 level of significance, do the populations from which the following samples are taken have the same means? Assume they are normally distributed and have the same variance.

Sample 1	Sample 2	Sample 3	Sample 4
0	−1	1	3
1	2	4	5
2	5	7	8
3			10

For Exercises 2–4 use analysis of variance to test whether the samples come from populations with the same mean ($\alpha = .05$). State assumptions that must be made; H_0 and H_1; and whether or not H_0 is rejected.

2.

Sample 1	Sample 2	Sample 3	Sample 4
0	4	4	8
−2	0	6	6
2	2	2	4

3. Sample 1: 10, 12, 14, 16, 18, 20
Sample 2: 8, 10, 12, 14, 16
Hint: Subtract 12 from each score.

4. Sample 1: 10, 12, 14, 16, 18, 20
Sample 2: 6, 8, 10, 12, 14
Compare with Exercise 3; you already have some sums and sums of squares. Also compare the results.

5. A botanist plants random samples of each of five different strains of corn on five plots of land. The plots are of the same size and fertility and the same fertilizer is used on each. The yields per plot are as follows:

Strain 1	Strain 2	Strain 3	Strain 4	Strain 5
4	7	10	16	10
3	8	14	14	13
6	9	12	10	12
2	8	9	7	10
2	5	5	3	14

At a .05 level of significance, do the different strains produce the same yield?

6. Students are taught statistics by three different methods: (a) using TV tapes, (b) using audio tapes, (c) with a live instructor. A random sample of 5 students taught by each method shows the following grades on the final examination.

(a)	(b)	(c)
75	80	90
78	75	95
80	70	60
60	75	80
95	80	80

Is there a difference in grades of students taught by different methods? (.05 level of significance.) Hint: Subtract 80 from each score.

7. The data in this table show the relative performance in German of random samples chosen from those using a traditional language lab and those using computer-assisted instruction:

	Language Lab	Computer-Assisted
Sum	27	46
Sum of squares	109	230
Sample size	8	10

(a) Use a t test to test H_0: $\mu_X = \mu_Y$, H_1: $\mu_X \neq \mu_Y$ at a .05 level of significance; first find out whether the populations have the same variance.

(b) Use analysis of variance to test H_0: $\mu_1 = \mu_2$ at a .05 level of significance.

8. The tensile strength of three different brands of cotton thread is compared by measurements on three spools of each brand, chosen at random. Assume the tensile strength of all three brands is normally distributed, and they have the same variance. Test at a .01 level of significance, using the following data:

Brand A	Brand B	Brand C
98	105	110
100	103	98
102	101	106

9. An agricultural chemist is measuring the effectiveness of three different formulations of an insecticide. He tests random samples of each formulation under standard conditions on colonies of fire ants, and collects the following data:

Percentage Kill, using Formulation

1	2	3
40.2	50.4	71.2
35.1	60.2	60.3
50.3	53.8	75.4
	40.2	80.3

At a .05 level of significance, is there a difference in the mean percentage kill of the three formulations? Assume normality and equal variances.

10. The same agricultural chemist as in Exercise 17 is carrying out the same experiment with three new formulations of insecticide. This time he collects the following data:

Percentage Kill,
using Formulation

1	2	3
40.2	50.4	71.2
39.6	51.0	72.0
43.0	52.0	73.0
	50.6	70.0

Test again at a .05 level of significance.

11. Do carpenters, masons, and electricians in Madison, New Jersey, have the same weekly wages? An economist assumes that the pay of all three groups is normally distributed and has the same variance. She determines the weekly wages of small random samples, with the following results:

Carpenters: $340, 300, 320, 260, 290
Masons: $330, 370, 350
Electricians: $390, 370, 350, 370

(a) Is the hypothesis that they have the same weekly wages correct? ($\alpha = .01$).

(b) If H_0 is rejected, use Scheffé's test with $\alpha = .01$ to see which pairs of means differ.

ANSWERS

1. $F = \dfrac{(238 - 50^2/14)/3}{(308 - 238)/10} = 2.8$; $F_{.05}(3, 10) = 3.71$. H_0 is still accepted: The means are equal.

In Exercises 2–5, assume that all populations are normal and that those in each exercise have the same variance.

2. $F = \dfrac{(168 - 36^2/12)/3}{(200 - 168)/8} = \dfrac{20}{4} = 5.00$; $F_{.05}(3, 8) = 4.07$. H_0 is rejected; the population means are not all the same.

3. $F = \dfrac{(54 - 18^2/11)/1}{(164 - 54)/9} = \dfrac{24.55}{12.22} = 2.01$; $F_{.05}(1, 9) = 5.12$. Fail to reject H_0.

4. Subtracting 12 from each score: $F = \dfrac{(74 - 8^2/11)/1}{(184 - 74)/9} = 5.58$; $F_{.05}(1, 9) = 5.12$. H_0 is rejected; $\mu_X \neq \mu_Y$.

Sample 1 is identical in Exercises 3 and 4. Sample 2 has the same variance in both, but the mean is 2 less in Exercise 4. In one case H_0 is still accepted, in the other it is rejected.

5. (Subtract 10 from each score.) $F = \dfrac{213.0/4}{189.2/20} = \dfrac{53.3}{9.46} = 5.6$; $F_{.05}(4, 20) = 2.87$. H_0 is rejected; the yields are different.

6. $F = \dfrac{(113.8 - (-27)^2/15)/2}{(1529 - 113.8)/12} = 32.6/117.9, < 1$ so H_0 is not rejected. It is not even necessary to look up the critical value of F in the table, since it is certainly greater than 1.

7. (a) $H_0: \mu_X - \mu_Y = 0,\ H_1: \mu_X - \mu_Y \neq 0,\ \alpha = .05.$

$$\bar{X} = \frac{27}{8} = 3.38,\ s_X{}^2 = (109 - 27^2/8)/7 = 2.55$$

$$\bar{Y} = \frac{46}{10} = 4.60;\ s_Y{}^2 = (230 - 46^2/10)/9 = 2.04$$

$s_X{}^2/s_Y{}^2 = \dfrac{2.55}{2.04} = 1.25$; with $\alpha = .05$ and a two-tail test, $F_{.025}(9,\ 7) = 4.82$, so it can be
assumed that the populations have the same variance.

$$t = \frac{(3.38 - 4.6) - 0}{\sqrt{\dfrac{7(2.55) + 9(2.04)}{(8 - 1) + (10 - 1)}\left(\dfrac{1}{8} + \dfrac{1}{10}\right)}} = \frac{-1.22}{.71} = -1.72$$

$D = (8 - 1) + (10 - 1) = 16,\ \alpha = .05$, critical value of $t = 2.12$. H_0 is accepted; there is
no significant difference between the two methods of instruction.

(b) $F = \dfrac{(302.7 - 73^2/18)/1}{(339 - 302.7)/16} = \dfrac{6.67}{2.27} = 2.94;\ F_{.05}(1,\ 16) = 4.49.$ H_0 cannot be rejected.

Notice that the computed and critical values of F are, within roundoff, the
square of the computed and critical values of t. For $H_0: \mu_1 = \mu_2,\ H_1: \mu_1 \neq \mu_2$, the t test
and the analysis of variance test are exactly equivalent.

8. Suggestion: Subtract 100 from each score.
$F = \dfrac{(92.3 - 23^2/9)/2}{(183 - 92.3)/6} = 1.11;\ F_{.01}(2,\ 6) = 10.9$, so H_0 is not rejected. The samples come
from populations with the same mean.

9. (Subtract 50 from each score.) $F = \dfrac{(2,104.70 - 67.4^2/11)/2}{(2,651.6 - 2,104.7)/8} = \dfrac{845.9}{68.4} = 12.4;\ F_{.05}(2,\ 8) = $
4.46. H_0 is rejected; the three formulations do not have the same mean kill.

10. Subtracting 60: $F = \dfrac{(1,948.2 - (-47)^2/11)/2}{(1,961.2 - 1,948.2)/8} = 537.7$

Subtracting 50: $F = \dfrac{(2,108.2 - 63^2/11)/2}{(2,121.2 - 2,108.2)/8} = 537.7$

$F_{.05}(2,\ 8) = 4.46$. Reject H_0; at least two populations have different means.

11. Subtracting 350: $F = \dfrac{[13,120 - (-160)^2/12]/2}{(18,400 - 13,120)/9} = \dfrac{5,493.5}{587} = 9.4;\ F_{.01}(2,\ 9) = 8.02$, so H_0
is rejected.

13.8   WHICH MEANS ARE DIFFERENT?

Suppose that $H_0: \mu_1 = \mu_2 \ldots = \mu_k$ has been rejected after an F test. The experimenter then would like to know how to account for this: which pairs of means are
significantly different? A number of statisticians have derived tests which can be
applied after the analysis of variance test has shown a significant difference in the

means. The test to be described here was devised by Scheffé.* It can be used for samples of different sizes, and is not much affected by populations that are not normal or that do not have equal variances. This test is valid for **all pairs of samples at once,** and thus differs considerably from repeated use of t tests.

Scheffé's Test   Scheffé's test includes the following steps:

1.   Compute $S = \sqrt{(k - 1) F_\alpha(k - 1, n_T - k)(\hat{\sigma}^2 \ within)}$
2.   If samples are of the same size n, see whether $|\bar{X}_i - \bar{X}_j|$ is less or greater than $S \sqrt{\dfrac{2}{n}}$.

   If samples are of different sizes, see whether $|\bar{X}_i - \bar{X}_j|$ is less than or greater than $S \sqrt{\dfrac{1}{n_i} + \dfrac{1}{n_j}}$.

   Carry out this step for each pair of sample means $\bar{X}_i, \bar{X}_j$.
3.   If $|\bar{X}_i - \bar{X}_j|$ is larger, $H_0: \mu_i = \mu_j$ is rejected, and we conclude that that pair of means contributes significantly to the rejection of the analysis of variance test. If $|\bar{X}_i - \bar{X}_j|$ is smaller, however, then we are unable to reject the hypothesis that $\mu_i = \mu_j$.

   Note: α in Scheffé's test is often the same as for the original analysis of variance test, but it doesn't need to be.

Example 1   A botanist plants random samples of each of five different strains of corn on five plots of land. The plots are of the same size and fertility and the same fertilizer is used on each. The mean yields of corn are 3.4, 7.4, 10, 10, and 11.8 bushels per plot. At a .05 level of significance, $H_0: \mu_1 = \mu_2 = \mu_3 = \mu_4 = \mu_5$ is rejected, with $A = 2{,}027.8$, $C = 2{,}217$ (see Exercise 5, page 304). Which pairs of population means differ, at a .05 level of significance?

$$n_T = 25, \ k = 5, \ \hat{\sigma}^2 \ within = \frac{2{,}217 - 2027.8}{25 - 5} = 9.46; \ F_{.05}(4,20) = 2.87;$$

$$S \sqrt{\frac{2}{n}} = \sqrt{4(2.87)(9.46)} \sqrt{\frac{2}{5}} = \sqrt{108.6}\sqrt{.4} = 6.6$$

The only sample means which differ by more than 6.6 are strains 1 and 5, so the statement "$\mu_1 = \mu_5$" can be rejected. We are unable, however, to reject the statement that any other pairs of means are equal.

Example 2   Consider the carpenters, masons, and electricians of Exercise 11, page 306. Here an analysis of variance test shows that $H_0: \mu_1 = \mu_2 = \mu_3$ is rejected, if $\alpha = .01$. Apply Scheffé's test at the same level of significance, to see which pair or pairs of means differ. $\bar{X}_1 = 302, \bar{X}_2 = 350, \bar{X}_3 = 370, n_1 = 5, n_2 = 3, n_3 = 4$.

$$k - 1 = 3 - 1 = 2, F_{.01}(2, 9) = 8.02, \hat{\sigma}^2 \ within = 587, S = \sqrt{2(8.02)(587)}$$
$$= 97.0.$$

---

*Scheffé, H., *The Analysis of Variance* (New York: John Wiley, 1959).

Since all pairs of means are to be examined, it is convenient to use a table such as the following:

		Differences of Means		n_i	$\sqrt{\dfrac{1}{n_i}+\dfrac{1}{n_j}}$		$S\sqrt{\dfrac{1}{n_i}+\dfrac{1}{n_j}}$	
		$\bar{X}_2 = 350$	$\bar{X}_3 = 370$	n_i	n_2	n_3		
Carpenters	$\bar{X}_1 = 303$	-48	-68	5	.730	.671	70.8	65.1
Masons	$\bar{X}_2 = 350$		-20	3		.764		74.1
Electricians	$\bar{X}_3 = 370$			4				

The statement "$\mu_1 = \mu_3$" is rejected since $|-68| > 65.1$. We are unable to reject the statement "$\mu_1 = \mu_2$" (since $|-48| < 70.9$) or "$\mu_2 = \mu_3$" (since $|-20| < 74.1$).

13.9 EXERCISES

1. In Exercise 2, page 304, the hypothesis of equal population means was rejected after looking at four samples, each of size 3. The means of the four samples are 0, 2, 4, and 6, respectively, and $\hat{\sigma}^2\ within = 4.0$. Use Scheffé's test with $\alpha = .05$ to see which pairs of means are different.

2. In Exercise 4, page 304, H_0: $\mu_X = \mu_Y$ was rejected; $\bar{X} = 15$, $\bar{Y} = 10$, $n_1 = 6$, $n_2 = 5$, $\hat{\sigma}^2\ within = 12.22$. Use Scheffé's test with α to see if it tells you the same thing. ($\alpha = .05$.)

3. An analysis of variance test is made with the following data: Sample 1: 0, 1, 1, 2, 2; Sample 2: 2, 3, 3, 4, 4; Sample 3: 4, 5, 5, 6, 6; Sample 4: 6, 7, 7, 8, 8. H_0: $\mu_1 = \mu_2 = \mu_3 = \mu_4$ is rejected at a .05 level of significance. Use Scheffé's test to see which pairs of means are different, again with $\alpha = .05$. ($A = 452.8$, $C = 464$.)

4. The amount of cola given out by three soda machines is measured as follows:

Machine A: 6.1, 6.4, 6.2, 6.3 ounces.
Machine B: 5.7, 5.9, 5.8, 5.7, 5.3 ounces.
Machine C: 6.5, 6.4, 6.2 ounces.

H_0: $\mu_1 = \mu_2 = \mu_3$ is rejected with $\alpha = .01$. Use Scheffé's test, at a .01 level of significance, to see which means differ. ($\hat{\sigma}^2\ within = .034$.)

ANSWERS

1. $S\sqrt{\dfrac{2}{n}} = \sqrt{3(4.07)(4)}\sqrt{\dfrac{2}{3}} = 5.71$; $\mu_1 \neq \mu_4$ since $|6 - 0| > 5.71$. We are unable to conclude that any of the other pairs of means differ.

2. $F_{.05}(1, 9) = 5.12$; $S \sqrt{\left(\frac{1}{n_1}\right) + \left(\frac{1}{n_2}\right)} = \sqrt{1(5.12)(12.22)} \sqrt{\frac{1}{6} + \left(\frac{1}{5}\right)} = 4.79$. $\bar{X}_1 = 5$, $\bar{X}_2 = 0$; $\mu_X \neq \mu_Y$ since $|5 - 0| > 4.79$.

3. $\hat{\sigma}^2$ within $= \dfrac{(464 - 452.8)}{16} = .70$, $F_{.05}(3, 16) = 3.24$, $S \sqrt{\dfrac{2}{n}} = \sqrt{3(3.24)(.70)} \sqrt{\dfrac{2}{5}} = 1.65$.
$\bar{X}_1 = 1.2$, $\bar{X}_2 = 3.2$, $\bar{X}_3 = 5.2$, $\bar{X}_4 = 7.2$; each pair of means is found to differ.

4. $S = \sqrt{2(8.02)(.034)} = .74$

Mean	Differences of Means		$\sqrt{\dfrac{1}{n_i} + \dfrac{1}{n_j}}$			$S \sqrt{\dfrac{1}{n_i} + \dfrac{1}{n_j}}$	
	$\bar{X}_2$	$\bar{X}_3$	n_i	n_2	n_3	n_2	n_3
$\bar{X}_1 = 6.25$	.57	$-.12$	4	.67	.76	.50	.56
$\bar{X}_2 = 5.68$		$-.69$	5		.73		.54
$\bar{X}_3 = 6.37$			3				

$\mu_1 \neq \mu_2$, $\mu_2 \neq \mu_3$

13.10 VOCABULARY AND SYMBOLS

F score

$F(D_X, D_Y)$

$F_\alpha(D_X D_Y)$

F distribution

analysis of variance   ANOVA

$\hat{\sigma}^2$ between

$\hat{\sigma}^2$ within

k

n_T

A, B, C

Scheffé's test S

13.11 REVIEW EXERCISES

1. What is the probability that a random score from an F distribution with $D_X = 7$ and $D_Y = 15$ is greater than 3.29?

2. Five percent of the scores in an F distribution with $D_X = 3$ and $D_Y = 24$ are greater than F_0; find F_0.

3. Two samples, of sizes 17 and 8, have standard deviations of 5.0 and 2.5 cm., respectively. At a .025 level of significance, do the populations from which they are taken have the same variance?

4. What assumptions about (a) the populations and (b) the samples should be satisfied if you are to use an F test of the null hypothesis that $\sigma_X^2 = \sigma_Y^2$?

5. Is there more variation in systolic than in diastolic blood pressure? In a random sample of 21 patients, the mean systolic blood pressure is 123.2 with a standard deviation of 15.0, while the mean diastolic blood pressure is 80 with a standard deviation of 10.0. ($\alpha = .05$.)

6. The mean height of a random sample of 61 women is 63.8 inches with a standard deviation of 3.1 inches, while a random sample of 41 men have a mean

height of 69.4 inches with a standard deviation of 3.4 inches. At a .01 level of significance, is there more variability in the height of men than of women?

7.   Heights of the 32 adults at the Groundhog's Eve party (see Exercise 9, page 50) were divided into those born in Wisconsin (where the party was), those from other states, and those from outside the United States:

	Wisconsin	Other States	Outside the United States
Mean	69.6	68.2	63.3
n	8	18	6
$\Sigma(X - 70)$	-3	-32	-40
$\Sigma(X - 70)^2$	107	244	274

(a) State H_0 and H_1 for an analysis of variance.
(b) Look at the sample means, since that's what it's all about. Do they look different?
(c) Is the difference explained by random variation from sample to sample alone? ($\alpha = .01$)

8.   One of the questions asked at the Groundhog's Eve party was the usual transportation used to work or school. Heights of adults in the different categories were:

Walk: Mean = 67.9 inches, $n = 9$, $\Sigma X = -1$, $\Sigma X^2 = 65$
Bike: Mean = 70.0 inches, $n = 7$, $\Sigma X = 14$, $\Sigma X^2 = 170$
Bus: Mean = 63.9 inches, $n = 7$, $\Sigma X = -29$, $\Sigma X^2 = 137$
Car: Mean = 67.9 inches, $n = 8$, $\Sigma X = -1$, $\Sigma X^2 = 85$

(The sums and sums of squares were found after subtracting 68 from each height.) Do an analysis of variance. ($\alpha = .01$.)

9.   In studying the metabolism of cancer patients (see Exercise 5, page 235), Iris Weisman measured the serum phosphate concentrations of patients with different kinds of cancer.

Breast cancer: Mean = 3.47, $n = 39$, sum = 135.2, sum of squares = 481.12
Colon cancer: Mean = 3.22, $n = 40$, sum = 128.7, sum of squares = 426.95
Stomach cancer: Mean = 3.65, $n = 16$, sum = 58.4, sum of squares = 217.78

(a) Test whether patients with cancers of these three sites have the same mean serum phosphate concentrations ($\alpha = .05$).
(b) Use Scheffé's test with $\alpha = .05$ to see which means are different.

10.   Iris Weisman was particularly interested in the effect of treatment with 5FU on the serum phosphate concentrations in cancer patients. The change in concentration was calculated for each patient. In the following table, X stands for the drop in serum phosphate level after treatment:

	n	$\bar{X}$	ΣX	ΣX^2
Breast cancer	39	.30	11.6	20.42
Colon cancer	40	.38	15.3	16.37
Stomach cancer	16	.79	12.6	13.42
Total	95		39.5	50.21

(a) Is the effect the same for all three types of cancer? ($\alpha = .01$.)

(b) If H_0 is rejected, use Scheffé's test with $\alpha = .01$ to see which types of cancer do not have the same mean decrease in serum phosphate concentration.

11. In the study of "Audience Interest in Mass Media Messages about Lung Disease in Vermont" quoted earlier (see Exercise 3, page 194), one of the concepts highly rated was health services. Here are the ratings of older respondents by age group:

		$\bar{X}$	s	n	Sum	Sum of squares
					Subtracting 65.7:	
	Over 50	81.6	24.4	43	683.7	35,876
Age	35 to 50	65.7	36.1	50	0	63,857
	Under 35	65.6	34.5	57	−5.7	66,655
	Total			150	678.0	166,388

Does age group have an effect on the average interest rating of health services? ($\alpha = .05$)

12. Here are the prices, in $1,000 units, of some houses listed on the (un)-Real Estate page of the New York Times, February 18, 1979. Prices were picked at random.

Manhattan: 550, 750, 250, 290
New Jersey: 140, 300, 80, 135
Westchester: 298, 250, 245, 67

At a .05 level of significance, is the mean price of a house the same in all three locations? State your conclusions and any misgivings you may have about the problem.

13. In a study of ethnicity and blood pressure in Detroit whites (see Exercise 18, page 204), Harburg and others collected the following data on systolic blood pressure:

Ethnicity of Father	n	Mean	Standard Deviation	Sum	Sum of Squares
				Subtracting 124.8:	
Northern Europe	171	126.5	15.0	290.7	38,744
Central Europe	191	124.8	16.0	0.0	48,640
French	48	121.3	14.8	−168.0	10,883
Mediterranean	82	120.4	12.7	−360.8	14,652
Total	492			−238.1	112,919

(a) At a .05 level of significance, is there a difference in systolic blood pressure for Detroit whites with fathers from different ethnic backgrounds?

(b) If H_0 is rejected, use Scheffé's test to see which means differ.

ANSWERS

1. .025

3. $F = \dfrac{5^2}{2.5^2} = 4.0$, $F_{.025}(16, 7) = 4.55$. Fail to reject.

5. $F = \dfrac{15^2}{10^2} = 2.25$, $F_{.05}(20, 20) = 2.12$. Reject H_0. The results are questionable since the samples are not independent.

7. (a) $\mu_{\text{Wisc.}} = \mu_{\text{U.S.}} = \mu_{\text{foreign}}$ $(\mu_1 = \mu_2 = \mu_3)$; H_1: at least one pair of means differs.

(b) The mean height of those born outside the United States looks low.

(c) $F = \dfrac{[324.68 - (-75)^2/32]/2}{(625 - 324.68)/(32 - 3)} = \dfrac{74.44}{10.36} = 7.19$; $F_{.01}(2, 29) = 5.42$, so reject H_0. (But don't take this one seriously; the folks at that party weren't a random sample from any of the three populations.)

9. (a) $\dfrac{(1095.95 - 322.3^2/95)/2}{(1125.85 - 1095.95)/(95 - 3)} = 3.85$; $F_{.05}(2, 92) < 3.10$, reject H_0.

(b) $S = \sqrt{2(3.09)(.325)} = 1.42$

	Mean	Differences of Means $\bar{X}_2$	Differences of Means $\bar{X}_3$	n_i	$\sqrt{\dfrac{1}{n_i} + \dfrac{1}{n_j}}$		$S\sqrt{\dfrac{1}{n_i} + \dfrac{1}{n_j}}$	
Breast	3.47	.25	−.18	39	.225	.297	.320	.422
Colon	3.22		−.43	40		.296		.420
Stomach	3.65			16				

We conclude that $\mu_2 \neq \mu_3$.

11. $F = \dfrac{(1{,}0871.4 - 678^2/150)/2}{(166{,}388 - 1{,}0871.4)/147} = 3.69$; $F_{.05}(2, 147) < 3.07$; reject H_0.

13. (a) $F = \dfrac{[2{,}669.71 - (-238.1)^2/492]/3}{(112{,}919 - 2{,}669.71)/488} = \dfrac{851.5}{225.92} = 3.77$; $F_{.025}(3{,}488) < 3.23$, so reject H_0.

(b) $S = \sqrt{3(2.68)(225.95)} = 42.62$.

$\bar{X}_i$	Difference of Means $\bar{X}_2$	Difference of Means $\bar{X}_3$	Difference of Means $\bar{X}_4$	n_i	$\sqrt{\dfrac{1}{n_i} + \dfrac{1}{n_j}}$			$S\sqrt{\dfrac{1}{n_i} + \dfrac{1}{n_j}}$		
126.5	1.7	5.2	6.1	171	.105	.163	.134	4.5	6.9	5.7
124.8		3.5	4.4	191		.161	.132		6.9	10.4
121.3			.9	48			.182			7.8
120.4				82						

$\mu_1 \neq \mu_4$.

14

LINEAR REGRESSION AND CORRELATION

Many problems in statistics are concerned with the relation between two variables which can be paired together—height and weight of students, age and take-home pay of plumbers, IQ and score on a mechanical aptitude test, for example. There are two different questions that need to be answered: Are the variables related? If they are related, can knowledge of one be used to predict the other? (Is there a relation between a student's rank in his class at Lynn High School and his grade-point average at Boonton College? And if the Dean of Admissions at Boonton knows a particular student's rank at Lynn, can he predict how well the student will do at Boonton if admitted there?)

In this chapter you will learn how to plot the pairs of data, find the equation of the straight line which best fits the plotted points, and find a coefficient which measures how well the points fit the straight line. For this it will not be necessary to make any assumptions about the population from which the samples are taken. If the second variable is to be predicted when a value of the first is known, or if the closeness with which all paired scores in the populations fit a straight line is to be measured, then it is essential to make assumptions about the distributions in the population. These assumptions and inferences about the populations will be taken up in Chapter 15.

14.1 INTRODUCTION

Suppose you know the years of school completed by and the income of each of a number of people. In each case observations are paired: for the first individual both years of schooling (or X) and income (or Y) are known; the pair (X_1, Y_1) represents the first individual; the second individual provides the pair of observations (X_2, Y_2), and so on. We shall not be interested in a list of years of schooling and a separate list of incomes, but only in measurements that are paired together. (Can X scores be the number of births per year in the United States, 1930–1950, and Y scores be the number of United States college freshmen, 1948–1968? The answer is "yes" if each X score is paired with a Y score 18 years later.) The pair (X, Y) is called an **ordered pair** since order matters: if X represents years of school and Y is income, then (12, 10,000) implies 12 years of school and an income of $10,000, while (10,000, 12) implies 10,000 years of school and an income of $12.

A warning about notation: You are probably accustomed to x- and y-coordinates. Here we shall use X and Y instead, because capital letters have been used consistently for scores. Also the equation $y = mx + b$, where m is the slope and b is the y-intercept of the line for which this is the equation, is probably familiar to you. Here, however, the equation of a straight line will be $Y = a + bX$, so that you can refer to other statistics texts which are pretty uniform in this respect; now the slope is b and the Y-intercept is a.

A certain familiarity with straight lines is required. If you can answer the following questions, skip to Section 14.6, page 271. If not, refer to the section noted. The answers are at the top of page 325.

1. Can you graph the line $Y = 1 + 2X$? (If not, see Section 14.2.)
2. What is the slope of the line $Y = 4 + 2X$? What is the Y-intercept? (Section 14.3.)
3. What is the equation of the line through P: (1, 4) and Q: (2, 3)? (Section 14.4.)

14.2 GRAPH OF A STRAIGHT LINE

Equations of the form $Y = a + bX$ have graphs which are straight lines. (It is also true that any straight line which is not a vertical line has the equation $Y = a + bX$, and therefore this is called a linear equation.) Upon replacing X by 2 and Y by 4, the equation $Y = 2 + X$ becomes an identity: $4 = 2 + 2$. We say that the point P whose coordinates are (2, 4) lies on the line $Y = 2 + X$. Q: (0, 2) is another such point, since $2 = 2 + 0$. Since two points determine a straight line, we can now sketch the graph. To check arithmetic, it is a good idea to find a third point, such as R: (1, 3), whose coordinates satisfy the equation, and see that it does lie on the line determined by P and Q. It is customary and efficient to put these results in a table, as shown below. Choose any convenient X values, and find the corresponding Y values by substituting the chosen X values in the equation.

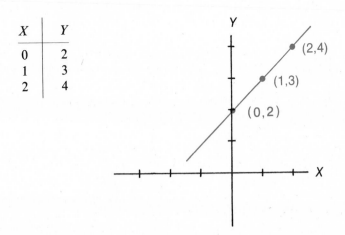

X	Y
0	2
1	3
2	4

Example 1   Graph the line whose equation is $Y = 3 - 2X$.
Choose three convenient X values: 0, 1, 4, say.

Answers   **1.**

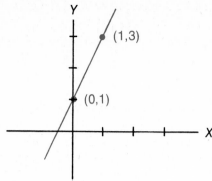

2. Slope = 2, Y-intercept = 4.

3. $Y = 5 - X$.

Next find the corresponding Y values. If $X = 0$, then $Y = 3 - 2(0) = 3$. Similarly, $Y = 1$ when $X = 1$; $Y = -5$ when $X = 4$. Then plot the points and draw the straight line.

X	Y
0	3
1	1
4	-5

Sometimes different scales must be used on the X and Y axes.

Example 2   Graph the line whose equation is $Y = 200 + 100X$.

X	Y
-2	0
0	200
1	300

Labels on the X and Y axes are sometimes omitted if the units are 0, 1, 2, . . . on both, but should be included if both are 0, 100, 200, . . . (for example), and must be included if they are different on the two axes, as in this example.

14.3   THE SLOPE OF A LINE

The slope of a line is a number that tells you not only whether Y values of points on the line are increasing or decreasing as X values increase but also how quickly they are changing. The slope of a vertical line is not defined; the slope of any horizontal line is 0. For any other line, we can take any two points on the line and find the ratio of the change in Y values to the change in X values:

$$\text{slope} = \frac{\text{change in } Y \text{ values}}{\text{change in } X \text{ values}}$$

Example 1   What is the slope of the line $Y = \dfrac{1}{2} + \dfrac{1}{2}X$?

Choose any two points P and Q on the line. Suppose their X coordinates are 3 and 7. Then P is at (3, 2) and Q is at (7, 4), and the slope of this line is $\dfrac{4-2}{7-3} = \dfrac{1}{2}$. It doesn't matter in which order you take the points:

$$\text{slope} = \frac{2-4}{3-7} = \frac{-2}{-4} = \frac{1}{2}.$$

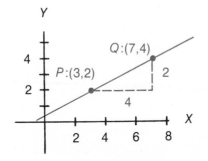

But if you take the Y-coordinate of Q first in the numerator, take the X-coordinate of Q first in the denominator $\left(\textbf{not } \dfrac{4-2}{3-7} = \dfrac{2}{-4} = -\dfrac{1}{2}\right)$. Nor does it matter which two points on the line you take. If instead of P and Q, you choose $R: (-1, 0)$ and $S: \left(4, \dfrac{5}{2}\right)$, the slope is $\dfrac{0-5/2}{-1-4} = \dfrac{-5/2}{-5} = \dfrac{1}{2}.$

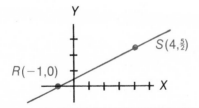

Why is the slope of a vertical line not defined? Choose, for example, the points $(2, 1)$ and $(2, 3)$ on the vertical line $X = 2$; slope $= \dfrac{3 - 1}{2 - 2} = \dfrac{2}{0} = ?$ The slope is not defined because division by 0 is not defined.

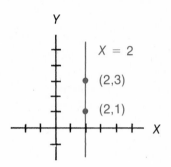

Choose two points on the horizontal line $Y = 2$ and check that its slope is 0.

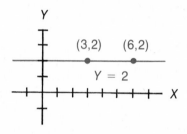

Example 2   What is the slope of the line whose equation is $Y = 2 + 3X$?
$(0, 2)$ and $(1, 5)$ are two points on the line.

$$\text{slope} = \frac{5 - 2}{1 - 0} = 3 \quad \left(\text{or } \frac{2 - 5}{0 - 1} = \frac{-3}{-1} = 3 \right).$$

X	Y
0	2
1	5

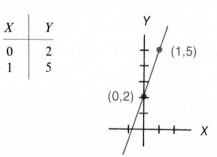

But 3 is the **coefficient of X** (the number that multiplies X) in the equation $Y = 2 + 3X$. Will this always be the case?

Example 3   What is the slope of the line $Y = a + bX$, where a and b are any constants? Assume for graphing that a and b are positive.

If $X = 0$, $Y = a$; if $X = 1$, $Y = a + b$. So two points on the line are P: $(0, a)$ and Q: $(1, a + b)$.

$$\text{slope} = \frac{(a + b) - a}{1 - 0} = \frac{b}{1} = b$$

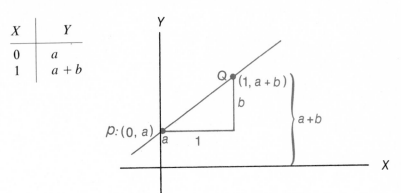

X	Y
0	a
1	$a + b$

The slope of the line $Y = a + bX$ is b, the coefficient of X.

In the examples shown so far in this section, a point moving along the line has been rising as its X-coordinate increases. But what if the point **falls** as it moves to the right?

Example 4   What is the slope of the line $Y = 3 - 2X$?
P: $(0, 3)$ and Q: $(1, 1)$ lie on the line.

$$\text{slope} = \frac{3 - 1}{0 - 1} = \frac{2}{-1} = -2 \ \text{ or } = \frac{1 - 3}{1 - 0} = \frac{-2}{1} = -2$$

Note that the slope is still the coefficient of X and is negative if the coefficient is negative.

Lines (a) through (c) below have positive slope; lines (d) through (f) have negative slope; (g) has slope 0; the slope of (h) is not defined. If you think of a (non-vertical) line as a road leading to the right, the slope is positive if the road goes uphill, 0 if it is level, and negative if the road goes downhill.

The Y-intercept of a line is the Y-coordinate of the point at which the line crosses the X axis; it is found by letting $X = 0$ in the equation of the line.

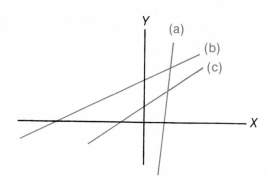 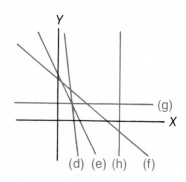

Equation	Slope	Y-intercept
$Y = 1 + 2X$	2	1
$Y = 3 - 2X$	-2	3
$Y = -4 + X$	1	-4
$Y = a + bX$	b	a

If the slope and Y-intercept of a line are known, it is easy to graph the line, so note the last line above with care: The coefficient of X is the slope of the line $Y = a + bX$ and the constant term is the Y-intercept.

Example 5   Graph the line $Y = 4 - 2X$.

The Y-intercept is 4; that is, the line crosses the Y axis at $(0, 4)$. The slope is -2; as X increases by 1, Y decreases by 2. The points P: $(0, 4)$ and Q: $(1, 2)$ determine the line.

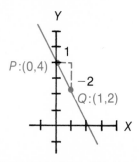

14.4 · THE EQUATION OF A LINE THROUGH TWO GIVEN POINTS

The graph of the equation $Y = a + bX$ is a straight line. To find the equation of a line through two given points, you must determine the values of a and b. Finding b is easy—it is the slope of the line and is determined by the given points. To find a

concentrate on one of the given points. It lies on the line, so its coordinates satisfy the equation of the line. Using this information, we can find a. Study the following examples.

Example 1   Find the equation of the line through (1, 4) and (2, 2).

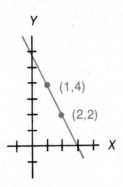

The equation is $Y = a + bX$. $a = ?$ $b = ?$

The slope of the line is $\dfrac{4 - 2}{1 - 2} = -2$, so $b = -2$ and the equation is $Y = a - 2X$. The point P: (1, 4) lies on the line, so these coordinates satisfy the equation. $4 = a - 2$ (1), so $a = 6$; $Y = 6 - 2X$ is the equation. (Check that the line passes through Q: (2, 2) since these coordinates also satisfy the equation $Y = 6 - 2X$.)

Example 2   Find (a) the equation and (b) the Y-intercept (that is, the point where the line crosses the Y axis) of the line through (3, 5) and (4, 7).

(a) Slope $= \dfrac{5 - 7}{3 - 4} = 2 = b$; $Y = a + 2X$.

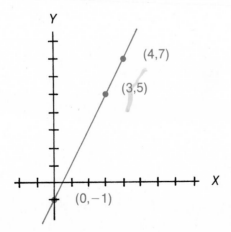

P: (3, 5) lies on the line, so $5 = a + 2(3)$ and therefore $a = -1$. The equation of the line is $Y = -1 + 2X$.

(b) When $X = 0$, $Y = -1$. The Y-intercept is $-1 = a$.

14.5 EXERCISES

1. (a) Plot the points (1, 3), (2, 5), and (3, 7) on the same graph.
   (b) What is the equation of the line on which all these points lie?
   (c) What is the Y-intercept of the line?

2. Estimate the slope and the Y-intercept of the lines shown in (a).

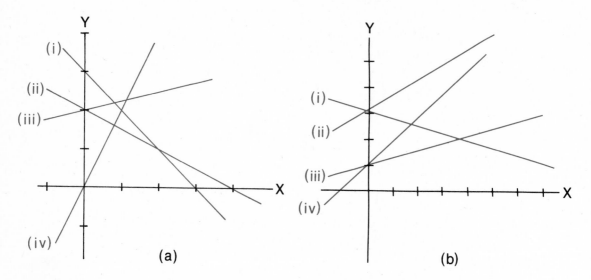

(a)

(b)

3. Match up the lines shown in 2 (a) with the following equations:
   (a) $Y = 2 + \dfrac{1}{4}X$
   (b) $Y = 3 - X$
   (c) $Y = 2X$
   (d) $Y = 2 - \dfrac{1}{2}X$

4. Match up the lines shown in (b) above with the following equations:
   (a) $Y = 30 + \dfrac{20X}{3}$
   (b) $Y = 10 + 10X$
   (c) $Y = 10 + \dfrac{10X}{3}$
   (d) $Y = 30 - \dfrac{10X}{3}$

5. (a) Graph the line $Y = 1 + 2X$ if you are particularly interested in X values between -1 and $+3$.
   (b) Graph the line $Y = 1 + 2X$ if you are particularly interested in X values between 100 and 104.

6. For each of the following, determine the slope of the line and the Y-intercept, and sketch the graph.

(a) $Y = -2 + 3X$.
(b) $Y = 5 - 2X$.
(c) $X = 4$.
(d) $X + 2Y = 4$.

7. Find the equation of the line through the given points, and sketch its graph.

(a) P: (0, 4), Q: (4, 6)
(b) P: (1, 3), Q: (2, 1)
(c) P: (3, 0), Q: (3, 4)
(d) P: (−2, 5), Q: (4, −3)

ANSWERS

1. (a)

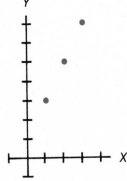

(b) $Y = 1 + 2X$
(c) 1 [since (0, 1) lines on the line].

2. (i) $b = -1$, $a = 3$; (ii) $b = -\dfrac{1}{2}$, $a = 2$; (iii) $b = \dfrac{1}{4}$, $a = 2$; (iv) $b = 2$, $a = 0$.

3. (a) (iii); (b) (i); (c) (iv); (d) (ii)

4. (a) (i); (b) (iv); (c) (ii); (d) (i).

5.

(a)

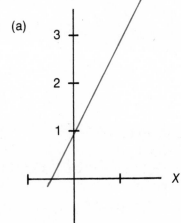

(b)

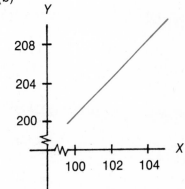

6.

	Slope	Y-intercept
(a)	3	-2
(b)	-2	5
(c)	Not defined	None
(d)	$-\dfrac{1}{2}$	$2\left(Y = 2 - \dfrac{1}{2}X\right)$

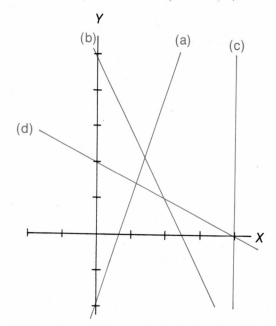

7. (a) $Y = \dfrac{1}{2}X + 4$; (b) $Y = -2X + 5$; (c) $X = 3$; (d) $Y = -\dfrac{4}{3}X + \dfrac{7}{3}$.

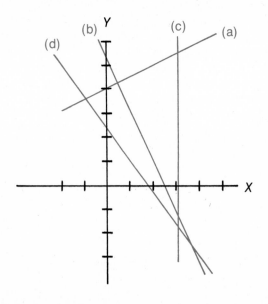

14.6 THE REGRESSION EQUATION

Suppose the number of hours spent in studying for a final exam in statistics and the number of questions answered correctly by each student are as follows:

Hours of study (X)	5	8	6	9	10	8	5	4	10	4	10	7	9
Correct answers (Y)	5	8	7	9	10	7	4	4	8	2	9	6	8

These data can be shown pictorially by plotting the corresponding pairs of scores as points on a graph. The resulting picture is called a **scatter diagram** or **scatter plot:**

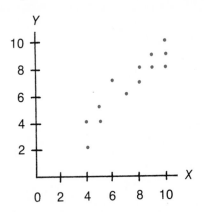

We shall learn how to draw the line which best fits a given set of points, and how to find its equation. The line is called a **regression line,** the equation a **regression equation.** The word "regression" is used because a man named Sir Francis Galton developed the techniques for finding the equation on the line while comparing the heights of parents and children. He discovered that short parents have children who are shorter than average, but not as short as their parents, while tall parents have children who are taller than average but not as tall as their parents—there is a "regression to the mean."

It will always be possible to find a straight line which best fits a given set of points. Sometimes it is not wise to do so, however. Later we shall study correlation, and then we shall have a measure of how close to a straight line the given points lie. Until then, always make a scatter diagram and see if a straight line seems reasonable as in (a) below; don't bother to find the regression line if the scatter diagram looks like (b) or (c):

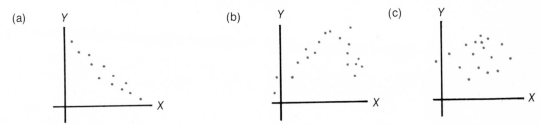

In more advanced texts, it is shown how to fit a regression curve of the form $Y = a + bX + cX^2$ to a scatter plot such as (b). We shall work only with straight lines, however.

The purpose of a regression equation is to use one variable to predict another. If you count cricket chirps per minute, can you tell the temperature? If she knows an entering freshman's standing in his high-school class, can the Dean of Admissions predict his grade-point average at the end of his freshman year? Can a company which manufactures small appliances give finger dexterity tests to job applicants and predict which ones will assemble parts most quickly? The variable whose value is known (cricket chirps per minute or standing in high-school class or score on finger dexterity test) is called the independent variable and is (usually) represented by X; the variable whose value is being predicted (temperature or grade-point average or number of parts assembled) is called the dependent variable and is (usually) represented by Y.

Suppose we are given the following pairs of X and Y scores:

X	Y
10	10
15	40
35	40

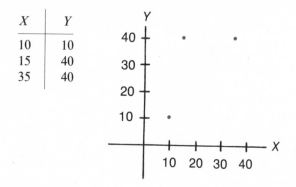

We shall try to find an equation $Y_c = a + bX$ so that we can predict a Y score by substituting in a value of X. For example, if $X = 20$, what corresponding Y score would be predicted? If a and b can be determined, the answer is $a + 20b$. The subscript c (for computed Y) is used to distinguish Y values computed or predicted from the regression equation from Y values given in the original pairs.

The difference between the given score Y and the predicted score Y_c is known as the **error of estimation.** The **regression line,** or **the line which best fits the given pairs of scores, is the line for which the sum of the squares of these errors of estimation is minimized.**

Example 1   Which of the following equations describes the line which best fits the three points (i.e., the three pairs of X and Y values) given above?

(a) $Y_c = 30$. This is obtained by letting $a = \bar{Y}$, $b = 0$.

(b) $Y_c = -50 + 6X$. This is the equation of the line determined by the first two points: (10, 10) and (15, 40).

(c) $Y_c = -2 + 1.2X$. This equation is satisfied by the first and third pairs: (10, 10) and (35, 40).

(d) $Y_c = 12.8 + .86X$. This is the true regression equation; the method by which it is derived will be shown shortly.

Answer   (a) $Y_c = 30$ (b) $Y_c = -50 + 6X$ (c) $Y_c = -2 + 1.2X$ (d) $Y_c = 12.8 + .86X$

X	Y	Y_c	$Y - Y_c$	$(Y - Y_c)^2$	Y_c	$Y - Y_c$	$(Y - Y_c)^2$	Y_c	$Y - Y_c$	$(Y - Y_c)^2$	Y_c	$Y - Y_c$	$(Y - Y_c)^2$
10	10	30	-20	400	10	0	0	10	0	0	21.4	-11.4	130.0
15	40	30	10	100	40	0	0	16	24	576	25.7	14.3	204.5
35	40	30	10	100	160	-120	14,400	40	0	0	42.9	2.9	8.4
				600			14,400			576			342.9

(a)   (b)   (c)   (d)

The sum of the squares of errors is the least for the equation $Y_c = 12.8 + .86X$; therefore, this is the best fit of the four equations given.

How is the regression equation determined? By more advanced mathematics, we can show that the values of b and a may be computed using these formulas:

$$b = \frac{n\Sigma XY - (\Sigma X)(\Sigma Y)}{n\Sigma X^2 - (\Sigma X)^2}$$

$$a = \frac{\Sigma Y}{n} - b\frac{\Sigma X}{n} = \bar{Y} - b\bar{X}$$

least squares method

Warnings

(1) n is the number of **pairs** of X and Y scores which are used in determining the regression line. In Example 2 below, $n = 3$, **not** 6.

(2) Be careful to distinguish between $(\Sigma X)^2$ and ΣX^2; $(\Sigma X)^2$ directs you to add first and then square the sum, while ΣX^2 directs you to square each X score and then add the squares.

(3) The denominator of b is always greater than 0. It is $n(n - 1)$ times the variance of the X scores, and the variance is always greater than 0 unless all X's are the same.

But be of good cheer: it is easy to apply these formulas.

Example 2   Find the regression line for the pairs of X and Y scores shown in the box below.

X^2	X	Y	XY
100	10	10	100
225	15	40	600
1225	35	40	1400
1550	60	90	2100

Notice that the X^2 column is written to the left of the X column, so that your eye doesn't have to travel far.

$$b = \frac{n\Sigma XY - (\Sigma X)(\Sigma Y)}{n\Sigma X^2 - (\Sigma X)^2} = \frac{3(2100) - 60(90)}{3(1550) - 60^2} = \frac{6300 - 5400}{4650 - 3600} = \frac{900}{1050} = .857$$

$$a = \frac{90}{3} - \frac{60}{3}(.857) = 30 - 20(.89) = 12.9$$

The equation of the regression line is, then, $Y_c = 12.9 + .86X$

Example 3   Find the regression line for the six pairs of X and Y scores shown in the box below:

X^2	X	Y	XY
4	2	0	0
9	3	3	9
16	4	4	16
25	5	4	20
36	6	6	36
49	7	11	77
139	27	28	158

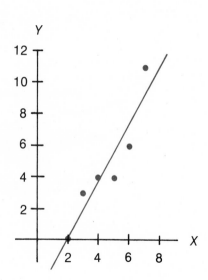

$$b = \frac{6(158) - 27(28)}{6(139) - 27^2} = \frac{948 - 756}{834 - 729} = \frac{192}{105} = 1.829$$

$$a = \frac{28}{6} - 1.829\left(\frac{27}{6}\right) = -3.564$$

$$Y_c = -3.56 + 1.83X$$

Again it should be emphasized that n is 6 here, not 12.

Sometimes $n\Sigma XY$ is approximately equal to $(\Sigma X)(\Sigma Y)$ and it is necessary to avoid rounding off before subtraction. Here is an example to illustrate this:

Example 4
(a) What is the equation of the regression line for the pairs (X, Y) shown in the box below?
(b) What Y is predicted when $X = 72$?

X^2	X	Y	XY
4,900	70	127	8,890
4,356	66	112	7,392
5,625	75	146	10,950
5,041	71	131	9,301
4,489	67	116	7,772
3,844	62	97	6,014
28,255	411	729	50,319

(a) $b = \dfrac{6(50,319) - 411(729)}{6(28,255) - 411^2} = \dfrac{301,914 - 299,619}{169,530 - 168,921} = \dfrac{2,295}{609} = 3.77$

$a = \dfrac{729}{6} - \dfrac{411}{6}(3.77) = -137$

$Y_c = -137 + 3.77X.$

(b) When $X = 72$, $Y_c = -137 + 3.77(72) = 134$. (This gives you practice in using the regression equation for prediction, but note that the prediction is meaningless without assumptions about the distribution of the Y population that will be taken up in Chapter 15. This is similar to the point estimate of a population mean; see the discussion on page 168.)

The values of a and b are determined to three significant figures, but the results would have been very different if all the numbers in the numerator and denominator of b had been rounded to three or even four figures before subtraction.

In the examples given so far, b has been positive. The slope of the regression line may be negative.

Example 5
What is the predicted value of Y when $X = 6$ for the data shown in the box below?

X^2	X	Y	XY
1	1	6	6
4	2	3	6
16	4	2	8
64	8	1	8
85	15	12	28

$$b = \frac{4(28) - 15(12)}{4(85) - 15^2} = \frac{112 - 180}{340 - 225} = \frac{-68}{115} = -.591$$

$$a = \frac{12}{4} - (-.591)\left(\frac{15}{4}\right) = 3 + 2.22 = 5.22$$

$$Y_c = 5.22 - .591X.$$

When $X = 6$, $Y_c = 5.22 - .591(6) = 1.67$.

A regression equation is used for prediction (in the last example, $Y_c = 1.67$ when $X = 6$). But a prediction is meaningless unless you can estimate how valid it is, and this can't be done unless certain assumptions about the population distribution can be made. This whole project will be discussed in the next chapter. For the moment, be aware of the following:

1. A regression equation $Y_c = a + bX$ can be found even when the data are far from linear; for example:

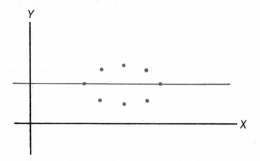

Finding the regression equation in such a case is a waste of time, however, so always make a scatter diagram first and see whether your data lie at least approximately on a straight line.

2. Decide ahead of time which variable is the predictor (X) and for which variable you wish to make a prediction (Y). You will not get the same equation if you interchange the two.

3. We have found only the best (in the sense of "least squares") linear equation $Y_c = a + bX$ to fit data; it may be preferable to use the least squares method but to fit an equation of another type: $Y_c = a + b \log X$ or $Y_c = a + bX + cX^2$, for example. But these other types are not considered in this book.

4. Finding a regression equation is possible when there is no cause and effect relationship between X and Y, and **never** proves that such a relationship exists. The number of cigarettes sold and the size of the police force in selected cities under 100,000 may show a linear relation, but smoking more cigarettes does not cause the hiring of more police; rather, both are related to the size of the city and thus increase as the population increases.

14.7 EXERCISES

1. By computing $\Sigma(Y - Y_c)^2$ for each, find which of the following equations best fits (in the sense of "least squares") the pairs of scores $(1, 2)$, $(2, 2)$,

(2, 3), (3, 4):

    (a) $Y_c = 1 + X$

    (b) $Y_c = 1 + .5X$

    (c) $Y_c = \dfrac{3}{4} + X$

2. On a multiple-choice statistics exam, the professor made a mistake in computing the regression equation for the points (1, 3), (1, 4), (2, 5), (3, 7), and offered the following choice of answers: (a) $Y_c = 1$, (b) $Y_c = 1 + X$, (c) $Y_c = 1 + 3X$, and (d) $Y_c = 3X$. Unfortunately, none of these is the true regression equation. Which of these four choices gives the best fit in the sense of least squares?

3. Draw a scatter plot for each of the following:

(a)

X	3	5	6	2	3	8	6	4	7	1	2	4
Y	5	1	1	6	4	6	2	2	4	9	5	3

(b)

X	32	36	30	36	34	35	30	32	38	36	37	34	33
Y	5	10	4	8	7	7	2	7	12	9	10	6	5

For which of these sets does it make sense to find a linear regression equation? Explain.

4. (a) Find the equation of the line through (1, 2) and (3, 4).

    (b) Find the regression equation through (1, 2) and (3, 4). Compare with (a).

5. Use the method of least squares to fit a straight line to the given data. Plot the data and the fitted line.

(a)

X	Y
1	2
2	2
2	3
3	4

(b)

X	Y
1	14
3	12
6	9
7	8
13	2

In each case, what is the predicted value of Y for $X = 4$?

6. In the following scores the X variable is IQ and Y is the grade on a statistics quiz.

X^2	X	Y	XY
10,000	100	10	1,000
12,100	110	12	1,320
14,400	120	15	1,800
16,900	130	18	2,340
19,600	140	20	2,800
73,000	600	75	9,260

(a) Make a scatter diagram.
(b) Find the equation of the straight line which best fits the given X and Y scores.
(c) What is the Y-intercept of the regression line? its slope?
(d) What is the predicted grade for a student whose IQ is 125? 105?
(e) Add the regression line to the scatter diagram.

7. In an earlier exercise (Exercise 8, page 169), you studied the relation between length and width of zircon crystals from the paper by geologist Jagannadham. Of course this relationship can also be studied by regression analysis. If X is the length, Y the width of a crystal, the 209 crystals in his sample have $\bar{X} = 54.48$ mm., $\bar{Y} = 22.94$ mm., $\Sigma X^2 = 654,478$, $\Sigma Y^2 = 117,676$, and $\Sigma XY = 267,698$.

(a) Fit the least squares regression line.
(b) Does it appear that width is proportional to length, except for random variation? (If A is proportional to B, then $A = kB$ where k is a constant.)

Use the following data in Exercises 8 to 12. The table on the left shows the United States Government Service (GS) scale of annual salaries (in dollars, rounded to the nearest \$100) for Step 1 of GS levels 1–18 on October 23, 1978. Within each grade there are steps, and the salaries for steps in GS 5 are shown in the table on the right.

GS level	Salary Step 1	GS 5 Step	Salary
1	\$ 6,600	1	\$10,510
2	7,400	2	10,860
3	8,400	3	11,210
4	9,400	4	11,560
5	10,500	5	11,910
6	11,700	6	12,260
7	13,000	7	12,610
8	14,400	8	12,960
9	15,900	9	13,310
10	17,500	10	13,660
11	19,300		
12	23,100		
13	27,400		
14	32,400		
15	38,200		
16	44,800		
17	52,400		
18	61,400		

8. (a) Find the equation of the regression line for first-step salary (Y) against GS level (X) for levels GS 1 to GS 10.
(b) Find the difference between the predicted and actual salaries for Steps 1, 3, and 10. Comment on how good the prediction is.

9. (a) Find the equation of the regression line for first-step salary (Y) against GS level (X) for levels GS 11 to GS 18.
   (b) Find the difference between the predicted and actual salaries for Steps 13 and 18. Comment on how good the prediction is.

10. (a) The equation of the regression line for first-step salary (Y) against GS level (X) using all 18 levels is $Y_c = -4{,}172 + 2{,}859X$. Find the difference between the predicted and actual salaries for Steps 1, 3, 10, 13, and 18.
    (b) Compare the results in (a) with those in Exercises 8(b) and 9(b). Can you explain why the results in (a) are so much worse? (A graph of the original data might help.)

11. Why can descriptive statistics be used but not statistical inference on the data given above?

12. Look at the data for steps in GS 5 carefully. Can you find a short-cut for finding the regression equation of salary against step? If not, compute it.

ANSWERS

1.

(a) $Y_c = 1 + X$ · (b) $Y_c = 1 + .5X$ · (c) $Y_c = \dfrac{3}{4} + X$

X	Y	Y_c	$(Y - Y_c)^2$	Y_c	$(Y - Y_c)^2$	Y_c	$(Y - Y_c)^2$
1	2	2	0	1.5	.25	1.75	$\dfrac{1}{16}$
2	2	3	1	2	0	2.75	$\dfrac{9}{16}$
2	3	3	0	2	1	2.75	$\dfrac{1}{16}$
3	4	4	$\dfrac{0}{1}$	2.5	$\dfrac{2.25}{3.50}$	3.75	$\dfrac{1}{16}$
							$\dfrac{12}{16} = .75$

Of the three, (c) gives the best fit, since .75 is less than 1 and also less than 3.50; (c) is the regression line.

2.

(a) $Y_c = 1$ · (b) $Y_c = 1 + X$ · (c) $Y_c = 1 + 3X$ · (d) $Y_c = 3X$

X	Y	Y_c	$(Y - Y_c)^2$	Y_c	$(Y - Y_c)^2$	Y_c	$(Y - Y_c)^2$	Y_c	$(Y - Y_c)^2$
1	3	1	4	2	1	4	1	3	0
1	4	1	9	2	4	4	0	3	1
2	5	1	16	3	4	7	4	6	1
3	7	1	36	4	9	10	9	9	4
			65		18		14		6

The best choice is (d), since this gives the smallest value (6) for $\Sigma(Y - Y_c)^2$.

3.

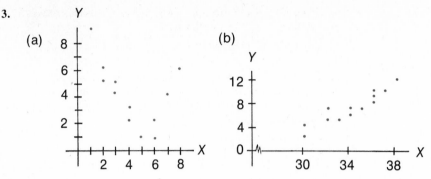

(a)

(b)

It makes sense for (b). The points in (a) will fit a parabola better than a straight line.

4. (a) $Y = 1 + X$

(b) $b = \dfrac{28 - 24}{2(10) - 4^2} = 1$, $a = \dfrac{6}{2} - 1\left(\dfrac{4}{2}\right) = 1$, $Y_c = 1 + X$.

5. (a) $b = \dfrac{4(24) - 8(11)}{4(18) - 8^2} = 1$, $a = \dfrac{11}{4} - 1\left(\dfrac{8}{4}\right) = .75$, $Y_c = .75 + X$; when $X = 4$, $Y_c = 4.75$.

(b) $b = \dfrac{5(186) - 30(45)}{5(264) - 30^2} = -1.00$, $a = \dfrac{45}{5} - (-1.00)\dfrac{30}{5} = 15.0$, $Y_c = 15.0 - 1.00X$;

when $X = 4$, $Y_c = 11.0$.

(a)

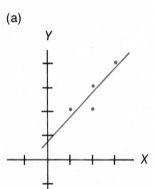

(b)

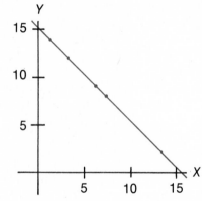

6. (b) $b = \dfrac{5(9,260) - 600(75)}{5(73,000) - 600^2} = .260$, $a = 15 - .26(120) = -16.2$, $Y_c = -16.2 + .26X$

(c) Y-intercept $= a = -16.2$, slope $= b = .260$.

(d) $Y_c = -16.2 + .26(125) = 16.3$ when $X = 125$, $Y_c = 11.1$ when $X = 105$.

(e)

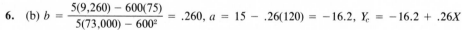

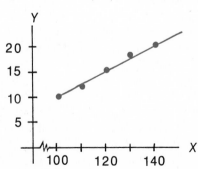

7. (a) Since $\bar{X} = 54.48$ and $n = 209$, $\Sigma X = 11,386.3$; similarly, $\Sigma Y = 4,794.5$. $b =$
$$\frac{209(267,698) - 11,386.3(4,794.5)}{209(654,478) - 11,386.3^2} = .190, a = 22.94 - .190(54.48) = 12.59, Y_c = 12.6 +$$
$.19X$.

(b) If width is proportional to length, then the Y-intercept is 0. In this sample, it is not. But one would have to learn how to test the null hypothesis that the Y-intercept for the population is 0, given that the Y-intercept for a sample of size 209 is 12.6, to answer this question properly.

8. (a) $Y_c = 4,827 + 1,210X$

(b)

X	Y	Y_c	$Y - Y_c$
1	6,600	6,036	564
3	8,400	8,456	−56
10	17,500	16,924	576

9. (a) $Y_c = -48,796 + 5,943X$

(b)

X	Y	Y_c	$Y - Y_c$
13	27,400	28,461	−1,061
18	61,400	58,175	3,225

10. (a)

X	Y	Y_c	$Y - Y_c$
1	6,600	−1,313	7,913
3	8,400	4,405	3,995
10	17,500	24,418	−6,918
13	27,400	32,995	−5,595
18	61,400	47,290	14,110

(b) The scatter plot fits a curve rather than a straight line. However, the curve can be approximated fairly well by two straight lines.

11. The data are not for a random sample from any population, so statistical inference is meaningless.

12. There is a constant increase of $350 between steps, so the data fit a straight line with slope of 350. Take any pair of coordinates, for example (1, 10,510): $10,510 = a + 350(1)$; $a = 10,160$; $Y_c = 10,160 + 350X$.

14.8 PEARSON CORRELATION COEFFICIENT: ROUGH ESTIMATES

A correlation coefficient, like a regression equation, is determined for X and Y scores that are paired. There are a number of different correlation coefficients; we shall concentrate on the Pearson correlation coefficient, r. This correlation coefficient will give a measure of the closeness to linearity of a relationship between two variables, and will determine the precision with which the regression equation discussed in Section 14.6 can be used for prediction. Note, however, that in correlation problems there is no necessity for considering one variable as independent and the other as dependent or predicted. For example, we may consider the relationship between weight and height of coal miners. Tall people tend to be heavier and short people lighter, but height could be used to predict weight or vice versa.

When all pairs of X and Y values are on the regression line, the correlation coefficient is either $+1$ or -1: it is $+1$ if the slope of the regression line is positive (in which case we say the correlation is **perfect and positive**); it is -1 if the slope of the regression line is negative (when the correlation is said to be **perfect and negative**).

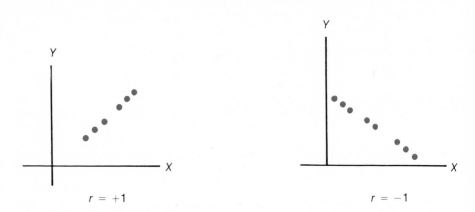

$$r = +1 \qquad\qquad\qquad r = -1$$

If some pairs of X and Y scores do not lie on the regression line, the correlation coefficient r is between -1 and $+1$. (Warning: If in determining a correlation coefficient you come up with results such as $r = 2.3$ or $r = -1.5$, check your work! Something is wrong.) The value of r can be estimated **very roughly** by enclosing the cloud of scores in a tight lining, and measuring the length L and width W of the lining. Then

$$r \approx \pm \left(1 - \frac{W}{L}\right)$$

where the sign of r is determined by the sign of the slope of L. Use any units of measurement you like, but it may be easiest to let the width W be 1 unit and estimate L as a multiple of W. Here are some examples:

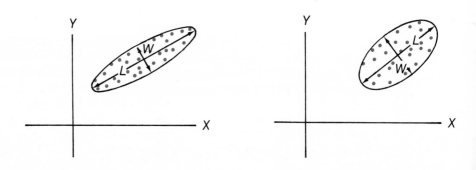

$$W = 1, L = 4, r \approx 1 - \frac{1}{4} = \frac{3}{4} \qquad W = 1, L = 2, r \approx 1 - \frac{1}{2} = \frac{1}{2}$$

$$W = 1, L = 1, r \approx 1 - \frac{1}{1} = 0 \qquad W = 1, L = 3, r \approx -\left(1 - \frac{1}{3}\right) = -\frac{2}{3}$$

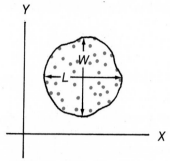
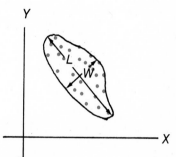

If only the points were fairly dense inside an elliptical cloud, this method would work pretty well.

Unfortunately, in practice, scores don't tend to behave nicely, and isolated points make it hard to determine exactly how the lining should be drawn. Inclusion of the point A would lower the estimate of r. It's probably better to exclude it, and then lower the estimate of r a little.

$$1 - \frac{W}{L} = 1 - \frac{1}{4} = \frac{3}{4}; r \approx .7.$$

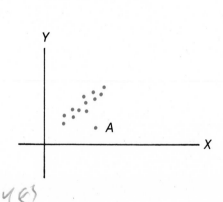
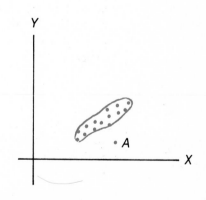

14.9   ERROR, EXPLAINED, AND TOTAL VARIATIONS

The fact that some observed points do not lie on a regression line—the line of "best fit"—does not deny the hypothesis that the line represents a relation between X and Y scores. If these scores are obtained by a psychologist, say, in carrying out an experiment, he expects experimental errors in his observations. The term "experimental errors" lumps together possible differences in factors not controlled by the experiment, as well as inaccuracies in measurement; there may also be round-off errors. The **standard error of estimate** (symbol: s_e) is a measure of these errors:

$$s_e = \sqrt{\frac{\Sigma(Y - Y_c)^2}{n - 2}}$$

Error Variation. The numerator $\Sigma(Y - Y_c)^2$ of the equation above is called the error variation. (In the remainder of this section we shall concentrate on variations rather than variances or standard deviations.*

Definition: Error variation $= \Sigma(Y - Y_c)^2$

Recall that in finding the regression line, the criterion of "best fit" was that the sum of the squares of the errors $\Sigma(Y - Y_c)^2$ should be made as small as possible. Thus if the Y_c's are computed using any other a and/or b in the regression equation, the resulting standard error of estimate (and error variation) would be larger.

Example 1

Find (a) the error variation, (b) the standard error of estimate for the pairs of X and Y scores below. The regression equation is $Y_c = -3.564 + 1.829X$ (determined in Example 3, page 328).

X	Y	Y_c	$Y - Y_c$	$(Y - Y_c)^2$
2	0	.094	−.094	.01
3	3	1.923	1.077	1.16
4	4	3.752	.248	.06
5	4	5.581	−1.581	2.50
6	6	7.410	−1.410	1.99
7	11	9.239	1.761	3.10
	28	28.00	.001	8.82

(a) Error variation $= \Sigma(Y - Y_c)^2 = 8.82$.

(b) $s_e = \sqrt{\dfrac{8.82}{4}} = 1.485$.

In computing the quantities in the table, observe that $\Sigma Y = \Sigma Y_c$ and $n(Y - Y_c) = 0$ (within round-off). Thus two totals for checking arithmetic are available.

Explained Variation. If knowing an X score is of no help whatsoever in predicting the corresponding Y score with a linear regression equation—that is, if there is no linear relationship between X and Y scores—then the best estimate of Y_c for any X is $Y_c = \bar{Y}$. In this case, $\Sigma(Y_c - \bar{Y})^2 = 0$. But if there is some linear relation between X and Y, $\Sigma(Y_c - \bar{Y})^2$ measures what is gained by using Y_c instead of $\bar{Y}$ for prediction; $\Sigma(Y_c - \bar{Y})^2$ is called the explained variation.

---

*But we shall return to the standard error of estimate in Chapter 15.

> **Definition:** Explained variation $= \Sigma(Y_c - \bar{Y})^2$

Example 2   For the same pairs of scores as in Example 1, compute the explained variation.

X	Y	Y_c	$Y_c - \bar{Y}$	$(Y_c - \bar{Y})^2$
2	0	.094	−4.573	20.91
3	3	1.923	−2.744	7.53
4	4	3.752	− .915	.84
5	4	5.581	.914	.84
6	6	7.410	2.743	7.53
7	$\dfrac{11}{28}$	9.239	$\dfrac{4.572}{- .002}$	$\dfrac{20.91}{58.56}$

Explained variation $= \Sigma(Y_c - \bar{Y})^2 = 58.56$.

Observe that the sum of the $Y_c - Y$ column is 0 (within round-off). This should always be the case; again, you have a useful check on arithmetic.

Total Variation. The variation of the Y's about their mean can also be computed. The quantity $\Sigma(Y - \bar{Y})^2$ is called the total variation.

Example 3   For the same pairs of X and Y scores used in Examples 1 and 2, compute the total variation.

X	Y	$Y - \bar{Y}$	$(Y - \bar{Y})^2$
2	0	−4.667	21.78
3	3	−1.667	2.78
4	4	− .667	.45
5	4	− .667	.45
6	6	1.333	1.78
7	$\dfrac{11}{28}$	$\dfrac{6.333}{- .002}$	$\dfrac{40.11}{67.35}$

$$\bar{Y} = \frac{28}{6} = 4.667$$

Total variation $= \Sigma(Y - \bar{Y})^2 = 67.35$.

Again there is a check on arithmetic: the sum of the $Y - \bar{Y}$ column must be 0 (except for round-off).

Study the figures below carefully so that you realize what distances are being squared to find the variations:

It can be shown that $\Sigma(Y - Y_c)^2 + \Sigma(Y_c - \bar{Y})^2 = \Sigma(Y - \bar{Y})^2$:

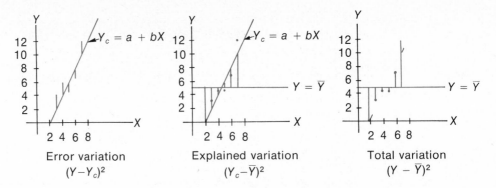

Error variation + Explained variation = Total variation

No higher mathematics is needed to show that this will always be the case, but the algebra involved is so long and tedious that a proof is not included here.

Example 4   Check that the sum of the error variation and the explained variation equals the total variation, within round-off, for the scores in Examples 1, 2, and 3.

	Variation	Formula
Error	8.82	$\Sigma(Y - Y_c)^2$
Explained	58.56	$\Sigma(Y_c - \bar{Y})^2$
Total	67.35	$\Sigma(Y - \bar{Y})^2$

$$8.82 + 58.56 = 67.38$$
$$\approx 67.35$$

Within round-off, the formula checks.

If all the data points lie on the regression line, each $Y - Y_c = 0$, so $\Sigma(Y - Y_c)^2 = 0$; the explained variation equals the total variation. If, on the other hand, knowing an X score is of no help in predicting a Y score, then the best choice for the equation of the regression line is $Y_c = \bar{Y}$; therefore the explained variation $\Sigma(Y_c - \bar{Y})^2$ equals 0.

The ratio of the explained variation to the total variation measures how well the linear regression line fits the given pairs of scores. It is called the **coefficient of determination,** and is denoted by r^2.

$$r^2 = \frac{\text{Explained variation}}{\text{Total variation}}$$

In the examples above,

$$r^2 = \frac{58.56}{67.35} = .869.$$

The explained variation is never negative (it is the sum of squares, which are never negative), and is never larger than the total variation (since the error variation, also the sum of squares, is never negative); so r^2 is always between 0 and 1. If the explained variation equals 0, $r^2 = 0$; this occurs when knowing the X value of a pair is of no help in predicting the Y value, and the best choice of Y_c, for any X, is $\bar{Y}$.

If r^2 is known, then $r = \pm\sqrt{r^2}$. The sign of r is **the same as the sign of b from the regression equation.** Now at last the coefficient for correlation r is defined:

$$r = (\text{sign of } b)\sqrt{r^2} = (\text{sign of } b)\sqrt{\frac{\text{Explained variation}}{\text{Total variation}}}$$

Since r^2 is between 0 and 1, r is between -1 and $+1$.

In the above examples, the equation of the regression line was $Y_c = -3.57 + 1.83X$; $b = +1.83$, so $r = +\sqrt{.869} = .93$.

An r close to 1 or to -1 indicates a good fit, because then the explained variation is a large part of the total variation and the error variation is a small part; an r close to 0 indicates a poor fit, since in this case the explained variation is a small part of the total variation and the error variation is a large part of it.

14.10   COMPUTATIONAL FORMULAS FOR r

If you understand the definition of the correlation coefficient r given in the last section, you should be quite clear on what r does: it measures how closely the (X, Y) pairs of data are linearly related, and it tells you whether the slope of a regression line is positive or negative. But to find r, it's necessary to find first the regression equation $Y_c = a + bX$, and sometimes you are interested only in r. Also, the computations for explained and total variation are staggering if n is large. Fortunately, there is a computational formula for r which is very much simpler:

$$r = \frac{n\Sigma XY - (\Sigma X)(\Sigma Y)}{\sqrt{n\Sigma X^2 - (\Sigma X)^2}\sqrt{n\Sigma Y^2 - (\Sigma Y)^2}}$$

Again, the derivation is long and tedious, but not difficult—that is, it uses algebraic manipulations but not advanced mathematics—and is omitted here.

Example 1   Use the computational formula to determine r for the scores in Examples 1–3 of the previous section.

X^2	X	Y	Y^2	XY
4	2	0	0	0
9	3	3	9	9
16	4	4	16	16
25	5	4	16	20
36	6	6	36	36
49	7	11	121	77
139	27	28	198	158

$$n = 6; \ r = \frac{6(158) - 27(28)}{\sqrt{6(139) - 27^2}\sqrt{6(198) - 28^2}} = \frac{948 - 756}{\sqrt{105}\sqrt{404}} = \frac{192}{206} = .93$$

as before.

| Example 2 | Find r for the pairs of X and Y scores below. |

X^2	X	Y	XY	Y^2
1	1	6	6	36
4	2	3	6	9
16	4	2	8	4
64	8	1	8	1
85	15	12	28	50

$$r = \frac{4(28) - 15(12)}{\sqrt{4(85) - 15^2}\sqrt{4(50) - 12^2}} = \frac{112 - 180}{\sqrt{340 - 225}\sqrt{200 - 144}}$$

$$= \frac{-68}{\sqrt{115(56)}} = \frac{-68}{80.2} = -.85$$

Since $n\Sigma X^2 - (\Sigma X)^2 = n(n - 1)\left[\dfrac{\Sigma X^2 - (\Sigma X)^2/n}{n - 1}\right]$ and $\dfrac{\Sigma X^2 - (\Sigma X)^2/n}{n - 1}$ is the variance of the X scores, the quantity under the square root sign must be positive. (If it were not, you couldn't find the square root anyway.) The same comment holds for $n\Sigma Y^2 - (\Sigma Y)^2$: check your arithmetic if it ever appears to be negative!

As with regression lines, n is the number of X scores or the number of Y scores, not the total number of scores.

Before reading on, try this next one yourself.

| Example 3 | Find r. |

X^2	X	Y	XY	Y^2
	1	4		
	2	5		
	3	6		

$$r = \frac{n\Sigma XY - (\Sigma X)(\Sigma Y)}{\sqrt{n\Sigma X^2 - (\Sigma X)^2}\sqrt{n\Sigma Y^2 - (\Sigma Y)^2}} = ?$$

Answer

$$r = \frac{3(32) - 6(15)}{\sqrt{3(14) - 6^2}\sqrt{3(77) - 15^2}} = \frac{6}{6} = 1$$

Are you surprised? The three points lie on the same line.

There are other formulas for r that sometimes prove useful. The similarity between the formula above for r and the equation for b in the regression equation $Y_c = a + bX$ should have struck you:

$$b = \frac{n\Sigma XY - \Sigma X \Sigma Y}{n\Sigma X^2 - (\Sigma X)^2}$$

$$r = \frac{n\Sigma XY - \Sigma X \Sigma Y}{n\Sigma X^2 - (\Sigma X)^2} \cdot \frac{\sqrt{n\Sigma X^2 - (\Sigma X)^2}}{\sqrt{n\Sigma Y^2 - (\Sigma Y)^2}} = b\left(\frac{\sqrt{n\Sigma X^2 - (\Sigma X)^2}}{\sqrt{n\Sigma Y^2 - (\Sigma Y)^2}}\right)$$

If the numerator and denominator of the fraction at the right are divided by $\sqrt{n(n-1)}$, these become, respectively, the standard deviations of the X and Y scores, s_X and s_Y.

Substituting, then,

$$r = b\left(\frac{s_X}{s_Y}\right)$$

Solving for b,

$$b = r\left(\frac{s_Y}{s_X}\right)$$

Ordinarily the first thing you do with sets of scores is to find their mean and standard deviation. Therefore you can find either r or b and then use one of these formulas to find the other.

Example 4 In Example 5, page 329, the regression equation for the following scores was found to be $Y_c = -.59X + 5.22$. Determine r.

X	Y	X^2	Y^2
1	6	1	36
2	3	4	9
4	2	16	4
8	1	64	1
15	12	85	50

$$s_X = \sqrt{\frac{85 - 15^2/4}{3}} = 3.10$$

$$s_Y = \sqrt{\frac{50 - 12^2/4}{3}} = 2.16$$

$$b = -.591$$

$$r = -.591\left(\frac{3.10}{2.16}\right) = -.85$$

14.11 SOME WARNINGS

1. Even though two variables show a high correlation (whether positive or negative), this does not show any cause-and-effect relationship between them.

There is a high correlation between boys' length of pants and ability to read in grades K–6: kindergarteners wear shorts and read poorly, while sixth graders wear long pants and read pretty well. Would you save money on teachers' salaries by passing a law that all boys must wear long pants? In 100 cities chosen at random, a study shows a positive correlation between the number of churches and the number of bars. Would you conclude that religion causes drinking or vice versa? And if you wanted to encourage church attendance, would you increase the number of liquor licenses?

This is, of course, the source of the controversy about smoking and lung cancer. Even the tobacco companies admit there is a positive correlation between the two. Proving there is not a common cause of both—urban atmosphere or nervous tension, for example—is a more difficult matter.

2. Sometimes a sample will show a correlation of two attributes, whereas subsamples do not show the same correlation.

Some people think there is a negative correlation between grades received and amount of time spent studying by college students. (It is probably not true that bright students study less than dummies—in fact, in challenging courses they are likely to study more.) But even if the correlation were negative, this wouldn't mean that you would get better grades by studying less! In a particular course, bright students might study less but get higher grades, while dull students were getting lower grades even though studying more.

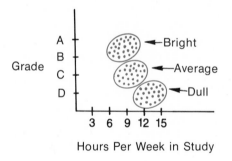

In general, if a population consists of several groups with different means, an apparent correlation may be produced or lost, as these diagrams indicate:

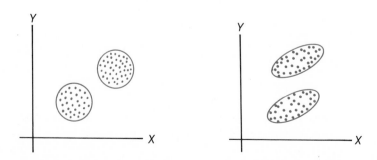

Your best remedies to these pitfalls are (a) always draw a scatter diagram, and (b) take a careful look at the source of your sample.

3.   Be careful in interpreting r, once you have computed it.

An apparently helpful but very dubious statement is the following: "If r is greater than .7, there is a very strong association between X and Y scores; if r is between .3 and .7 there is some association; if r is between 0 and .3, there is negligible association." Of course the same statement should be made for negative values of r: $r = +0.8$ and $r = -0.8$ show exactly the same degree of correlation but in opposite directions.

Why is this statement so dubious? First, because r measures only linear relationships. In the scatter diagram shown below, r will be close to 0 even though there is a very close relationship between X and Y, since it is not a linear relationship. Usually you can tell whether this is the reason for a low correlation by looking at the scatter diagram.

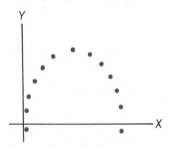

Secondly, recall that r^2 measures the proportion of the variation in Y scores explained by a linear relation between Y and X. If $r = .3$, then $r^2 = .09$; only 9% of the variation in Y scores is explained. If $r = .7$, then 49% is explained, and so on for other r values.

Finally, when can we say "There is a high correlation between these variables"? The answer depends on how r is to be used. A sociologist studying delinquency in teen-agers might correlate number of arrests with a variety of other scores: attitudes toward authority, education of parents, IQ, participation in religious services, and so on. Any correlation greater than .2 might indicate a factor worthy of further study. On the other hand, if the College Entrance Examination Board found a correlation as low as .7 between the scores of students who take two different verbal aptitude tests, I'm sure they would be very alarmed. In general, that which is a low correlation to a physicist or a chemist is a very high correlation to a sociologist.

4.   So far, no assumptions have been made about the distributions of the populations from which the X and Y scores are drawn. Even without these assumptions, we can determine the regression equation and the correlation coefficient for a sample, as part of descriptive statistics. As soon as we are to make inferences about the populations, however, it will be absolutely essential to make assumptions about the populations. In the next chapter we shall do so.

14.12 EXERCISES

1. List three sets of paired scores between which you would expect to find a high positive correlation.

2. List three sets of paired scores between which you would expect to find no correlation.

3. List three sets of paired scores between which you would expect to find a high negative correlation.

4. For the following sets of scores, plot a scatter diagram, estimate r from the scatter diagram, and compute r.

(a) X	Y	(b) X	Y	(c) X	Y
1	1	1	2	1	7
3	3	2	5	3	5
5	4	3	3	2	6
7	8	4	5	1	6
		5	2	7	1
				2	5
				6	3
				6	2

5. For which of the following problems would you find the equation for the regression line and for which would you find r?
  (a) Is there a linear relationship between the exchange rate of pounds sterling and the price of Scotch in the United States?
  (b) What is the relationship between the length of a boy's pants and his grade in school?
  (c) Do factory workers who are heavy drinkers have more absenteeism?
  (d) Can the attendance at a baseball game be predicted if the afternoon temperature forecast at 8 A.M. is known?
  (e) How many people will attend a baseball game if the afternoon temperature forecast at 8 A.M. is 70°F?
  (f) What effect does the age of dishwashers have on the number of repairs needed?
  (g) Can a student's grade-point average upon graduation be estimated if his freshman grades are known?

6. Estimate r:

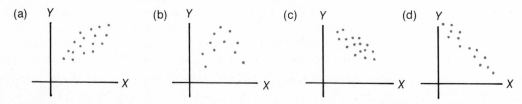

7. For the following scores, compute:
    (a) the regression equation.
    (b) the error variation $\Sigma(Y - Y_c)^2$.
    (c) the explained variation $\Sigma(Y_c - \bar{Y})^2$.
    (d) the total variation $\Sigma(Y - \bar{Y})^2$.
    (e) r^2, using the results of (c) and (e).
    (f) r.
    Finally,
    (g) draw a scatter diagram.
    Does the result of (f) surprise you?

X	-1	0	1	-1	1
Y	1	0	1	-1	-1

8. Follow the same directions as in Exercise 7 for the following scores:

X	-1	0	2	-2	1
Y	1	0	-2	2	-1

9. (a) The regression equation for the points $(2, 9)$, $(4, 6)$, $(6, 0)$ is $Y_c = 14 - 2.25X$. Compute (i) $\Sigma(Y - Y_c)^2$, (ii) $\Sigma(Y - \bar{Y})^2$, (iii) $\dfrac{\Sigma(Y - Y_c)^2}{\Sigma(Y - \bar{Y})^2}$, (iv) r.

   (b) If each Y value in (a) is multiplied by 10, the new regression equation is $Y_c = 140 - 22.5X$. Compute the new values of (i) $\Sigma(Y - Y_c)^2$, (ii) $\Sigma(Y - \bar{Y})^2$, (iii) $\dfrac{\Sigma(Y - Y_c)^2}{\Sigma(Y - \bar{Y})^2}$, (iv) r.

   (c) compare the results of (a) and (b).

10. An account executive for an advertising firm presents the following figures to a client:

(X) Percentage of sales spent on advertising	(Y) Sales (thousands of dollars)
1	100
2	107
3	115
4	125
5	135

   (a) Find the regression line which fits these data.
   (b) Find $\Sigma(Y - Y_c)^2$.
   (c) Find $\Sigma(Y - \bar{Y})^2$.
   (d) Use (b) and (c) to find r.
   (e) Use the computational formula to find r.

11. The numbers of years in school (X) and the salaries (Y) in thousands of dollars of eight working women chosen at random in Denver are as follows:

X	12	16	19	10	13	16	12	12
Y	10	9	20	8	6	18	9	7

(a) Make a scatter diagram.

(b) What salary would you predict for a woman with a master's degree? (She has had 17 years of school.)

(c) Find the coefficient of correlation between the number of years in school and the salary earned by these women.

12. A doctor collects the following information on mean calories per day and weight lost per week:

Calories:	2,000	1,200	1,500	500	800
Weight lost (lb):	1	3	2	5	4

(a) Find the correlation between calories per day and weight loss per week.

(b) What weight loss would you predict for a dieter who is restricted to 700 calories per day?

13. Ministers' salaries (in thousands of dollars) and sales of alcohol (in billions of dollars) are as follows for the years 1970–1974:

	1970	1971	1972	1973	1974
Salaries	8.3	9.0	10.0	10.5	11.1
Alcohol consumption	7.9	8.3	8.7	9.2	9.6

(a) Find r.

(b) Can you explain why the coefficient of correlation is so high?

14. Don Smith follows several stocks daily and compares their activity to that of the market as a whole. X values are average gain or loss in points each day in his selected stocks compared to Y values that represent point gain or loss in the market for an average share. For one work week, the data read:

	X	Y
Monday	+1 1/2	−1/8
Tuesday	−1/4	−3/4
Wednesday	+1/4	−3/4
Thursday	+1 1/8	−1/2
Friday	−1/2	−3/8

(a) Compute r.

(b) Did Don Smith pick good stocks?

15. The equation of the regression line through the points (1, 2), (2, 2), (2, 3), and (3, 4) is $Y_c = .75 + X$. Find the correlation between X and Y scores:

(a) using $r = b\left(\dfrac{s_X}{s_Y}\right)$.

(b) using $r = \dfrac{n\Sigma XY - \Sigma X\Sigma Y}{\sqrt{n\Sigma X^2 - (\Sigma X)^2}\sqrt{n\Sigma Y^2 - (\Sigma Y)^2}}$.

ANSWERS

4.

(a)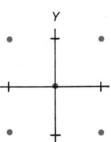

(b)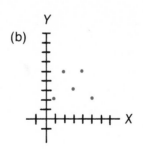

(c)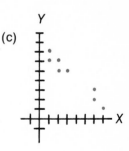

(a) $r \approx .9$; (b) $r \approx 0$; (c) $r \approx -.9$.

5. Regression for (b), (e), (f); correlation for (a), (c), (d), (g).

6. (a) $r \approx .5$; (b) $r \approx 0$; (c) $r \approx -.6$; (d) $r \approx -.9$

7. (a) $b = \dfrac{5(0) - 0(0)}{5(5) - 0^2} = 0$; $a = 0 - 0\left(\dfrac{0}{5}\right) = 0$, $Y_c = 0$.

(b) 4. (c) 0. (d) 4. (e) $\dfrac{0}{4} = 0$. (f) 0.

(g)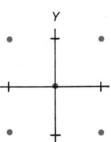

8. (g)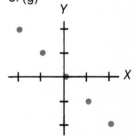

8. (a) $b = \dfrac{5(-10) - 0(0)}{5(10) - 0^2} = -1$, $a = \dfrac{0}{5} - (-1)\dfrac{0}{5} = 0$, $Y_c = -X$.

(b) 0; (c) 10; (d) 10; (e) $\dfrac{10}{10} = 1$; (f) $b = -1$, so $\sqrt{r} = -\sqrt{1} = -1$; (g) The points lie on a straight line whose slope is negative.

9. (a) (i) $\Sigma(Y - Y_c)^2 = (-.5)^2 + 1^2 + (-.5)^2 = 1.50$, (ii) $4^2 + 1^2 + (-5)^2 = 42$, (iii) $\dfrac{1.50}{42} = .04$, (iv) .98.

(b) (i) $(-5)^2 + 10^2 + (-5)^2 = 150$, (ii) $40^2 + 10^2 + (-50)^2 = 4{,}200$, (iii) $\dfrac{150}{4{,}200} = .04$, (iv) $-.98$.

(c) When Y values are multiplied by 10, $\Sigma(Y - Y_c)^2$ and $\Sigma(Y - \bar{Y})^2$ are both multiplied by 100; their quotient is unchanged, and therefore r is not changed.

10. (a) $b = \dfrac{5(1{,}834) - 15(582)}{5(55) - 15^2} = 8.80$, $a = 90$, $Y_c = 90 + 8.80X$.

(b) Y_c column: 98.8, 107.6, 116.4, 125.2, 134.0.

$\Sigma(Y - Y_c)^2 = 1.44 + .36 + 1.96 + .04 + 1.00 = 4.80$.

(c) $\bar{Y} = 116.4$; $\Sigma(Y - \bar{Y})^2 = 269.0 + 88.36 + 1.96 + 73.96 + 345.96 = 779.2$.

(d) Explained variation $= 779.2 - 4.80 = 774.4$; $r^2 = \dfrac{774.4}{779.2} = .9938$, $r = .997$.

(e) $r = \dfrac{440}{\sqrt{50}\sqrt{3,896}} = .997$.

11. (a)

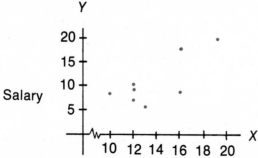

Number of Years in School

(b) $b = \dfrac{8(1,282) - 110(87)}{8(1,574) - 110^2} = 1.394$; $a = \dfrac{87}{8} - 1.394\left(\dfrac{110}{8}\right) = -8.29$.

$Y_c = -8.29 + 1.39X$; if $X = 17$, $Y_c = 15.3$; her predicted salary is \$15,300.

(c) $r = \dfrac{686}{\sqrt{492}\sqrt{1,511}} = .80$.

12. (a) $r = \dfrac{5(14,300) - 6,000(15)}{\sqrt{5(55) - 15^2}\sqrt{5(8,580,000) - 6,000^2}} = -1.00$.

(b) $b = \dfrac{-18,500}{6,900,000} = -.00268$, $a = \dfrac{15}{5} - (-.00268)\left(\dfrac{6,000}{5}\right) = 6.22$, $Y_c = 6.2 - .0027X$;

when $X = 700$, $Y_c = 4.3$.

13. (a) $r = \dfrac{5(430.43) - 48.9(43.7)}{\sqrt{25.54}\sqrt{9.26}} = .99$.

(b) Perhaps salaries increased because of inflation, and alcohol consumption because of increased population as well as inflation.

14. (a) $r = \dfrac{5(-.5625) - 2.125(-2.5)}{\sqrt{5(3.89) - 2.13^2}\sqrt{5(1.55) - (-2.5)^2}} = .55$.

(b) Yes, since $\bar{X} = .4$, $> \bar{Y} = -.50$. But note that the samples are very small and are not random.

15. (a) $r = 1\left(\dfrac{.82}{.96}\right) = .85$; (b) $r = \dfrac{4(24) - 8(11)}{\sqrt{4(18) - 8^2}\sqrt{4(33) - 11^2}} = .85$.

14.13 THE RELATION BETWEEN REGRESSION AND CORRELATION

Correlation measures the strength of a linear relation between two variables, while the regression line actually describes the relation. There are other differences.

In a regression equation, one variable (X) is the independent variable and is the predictor, while the other variable (Y) is to be predicted. In correlation, on the other hand, there is no distinction between the two variables, and either one may be designated as X. (It should be emphasized again that in neither case is a cause-and-effect relationship assumed or shown.)

A psychologist is more apt to use correlation: She collects data on a great many variables, and wants to know whether there is a linear relationship between any of them. An economist, on the other hand, is much more apt to use regression: She has more exact data, and more prior theory to tell her a linear relationship does exist. Correlation is used when the relation between variables is first being studied, and the regression equation is determined in a later part of the investigation.

In correlation, both variables may take on any values in a wide range. This is sometimes, but not always, true for prediction by a regression line. If, for example, the effect on freezing point of diluting a chemical with water is to be studied, one might add to a fixed amount of the chemical no water, then 10 cc., 20 cc., . . . of water. The X value is not chosen randomly, but is given fixed values by the experimenter.

14.14   VOCABULARY AND SYMBOLS

graph of a straight line	predicted value of Y
slope of a straight line	Pearson correlation coefficient r
equation of a straight line	standard error of estimate s_e
linear relationship	error variation
regression line	explained variation
error of estimation	total variation
"least squares" fit	coefficient of determination r^2
$Y_c = a + bX$	

14.15   REVIEW EXERCISES

1. (a) What is the slope of the line through (4, 6) and (−2, 10)?
    (b) Find the equation of this line.
    (c) What is its Y-intercept?

2. (a) What is the purpose of (i) a regression line? (ii) a correlation coefficient?
    (b) Give an example of the need for each, different from those discussed in this text.

3. (a) Which of these lines best fits, in the sense of least squares—that is, in minimizing $\Sigma(Y - Y_c)^2$—the points (−1, 0), (0, −1), (1, −1), (1, −2), (−2, 0): (i) $Y_c = -1 - X$, (ii) $Y_c = -1 - \frac{1}{2}X$, (iii) $Y_c = -1$?
    (b) Find the equation of the regression line.

4. Upon graduation, the scores of six chemistry majors on the mathemati-

cal aptitude test (MAT) were compared with their cumulative grade-point averages (GPA):

MAT	GPA
600	2.8
720	3.8
640	3.6
590	2.9
620	3.2
680	3.6

(a) Make a scatter diagram.
(b) Find the equation of the straight line which best fits the given values.
(c) Give the Y-intercept of the regression line and its slope.
(d) What is the predicted GPA for a student whose math score was 700?

5.

X	0	1	5
Y	3	4	5

For these scores, $\bar{X} = 2$, $\bar{Y} = 4$. Each of the following lines passes through $(\bar{X}, \bar{Y}) = (2, 4)$: (a) $Y_c = 4$; (b) $Y_c = 2X$; (c) $Y_c = 4 - 2b + bX$ (if $Y_c = a + bX$ passes through (2, 4), then $a = 4 - 2b$).

For each line given, show that $\Sigma(Y - Y_c) = 0$.

6. The heights and weights of five men are as follows:

Height (inches)	64	68	70	72	74
Weight (pounds)	160	170	180	190	195

(a) What weight would you predict for a man 69 inches tall?
(b) What height would you predict for a man who weighs 180 pounds?

7. An economist gives the following estimates of sales price and demand for a product:

Price (in dollars)	Demand (in tons)
1	9
2	7
3	6
4	3
5	1

(a) What demand would he predict if the sales price is $1.50?
(b) What price should be set for the product if the demand is to be 8?

8. The following data are given in a paper by Morissey, Tessler, and

Parrin, "Being Seen But Not Admitted; A Note on Some Neglected Aspects of State Hospital Administration":*

Trends in Census Reduction, Monthly Admissions, and Monthly "Seen but not Admitted"

Worcester State Hospital, 1969–1977

Fiscal Year*	Average daily census	Average monthly admissions†	Average monthly "seen but not admitted"	Ratio of admissions to "seen but not admitted"
1969	993	152	‡	‡
1970	932	125	32	3.9 : 1
1971	738	118	39	3.0 : 1
1972	708	110	37	3.0 : 1
1973	579	94	47	2.0 : 1
1974	587	95	38	2.5 : 1
1975	480	99	52	1.9 : 1
1976	452	104	57	1.8 : 1
1977	462	125	55	2.3 : 1

*Year ending June 30.
† Figures are based on total admissions. Due to the way hospital records are maintained, the numbers of voluntary as opposed to involuntary admissions during this time period could not be differentiated. This results in a conservative estimate of the actual nonadmission rate for voluntary applicants.
‡ Data unavailable.

Using each of the four variables in turn as Y, do a plot of Y on $X =$ year, draw what you think is the best-fitting regression line lightly in pencil (with a ruler!) by eye, calculate the regression line, and draw it.

9. The median heights of boys and girls aged 1 to 6 are as follows:

	Height	
Age	Boys	Girls
1	32.6	31.9
2	35.7	35.4
3	38.7	38.3
4	41.7	41.0
5	44.5	44.0
6	46.5	46.5

Find the coefficient of correlation between (a) heights of boys and girls, (b) age and median height of boys.

10. (a) Plot earnings per share against time, using one symbol or color for American Express and another for McGraw-Hill for the following data:

*American Journal of Orthopsychiatry, 49, 1979, pp. 153–156.

Year ending December 31:		1973	1974	1975	1976	1977	1978
Earnings per share	American Express Company		2.09	2.29	2.70	3.65	4.31
	McGraw-Hill Book Company	1.11	1.22	1.35	1.64	2.98	

    (b) Calculate the regression line for each company's earnings per share against time (X). Be sure to ease your work by subtracting some constant like 1976 from each year, but then be careful in drawing your lines on the graph.

    (c) Find the correlation coefficient between earnings per share of the two companies. What year(s) should you leave out?

    11. The approximate federal and New York State taxes are shown for gross incomes in thousands reported in joint returns for federal income taxes:

Income	21	25	34	40	53	98	136	189	250
Taxes(%)	36	41	46	51	56	65	69	72	74

    (a) Plot income (X) and taxes (Y).

    (b) What does the slope b of the regression line express?

    (c) Compute and draw the regression line.

    (d) Compute r.

    12. C. Smoszyk's study of a way to help relieve the anxiety of cancer patients (see Exercise 9, page 84) showed scores of anxiety state before and after treatment, and the change:

Patient		A	B	C	D	E	F	G	H	I	J
Anxiety state	Before treatment	23	48	69	58	48	44	57	20	26	34
	After treatment	20	26	67	57	33	24	44	21	26	27
	Change	−3	−22	−2	−1	−15	−20	−13	1	0	−7

    Find the correlation (a) between the patients' anxiety state scores before and after treatment, and (b) between the anxiety state score before treatment and the change. Illustrate each correlation by drawing the scatter diagram.

    13. In Borough Championships on February 17, the times in the 60-yard dash were 6.5, 6.8, 6.8, 6.8, and 6.9 seconds for the Brooklyn team and 6.7, 6.8, 6.9, 6.9, and 6.9 seconds for the Queens team. A statistics student determined the coefficient of correlation to be .88. Interpret.

    14. Anxiety reduction in 15 patients treated with the drug CDX were measured on the Hamilton Anxiety Rating Scale (HARS) and with the Hopkins Symptoms Check List (HSCL):

Subject	1	2	3	4	5	6	7	8	9	10	11	12	13	14	15
HARS	−5	−2	−7	−1	−5	−5	3	3	−12	−12	−20	−7	−4	−18	−11
HSCL	−6	−9	−10	−7	0	0	9	0	−1	−4	−9	−9	−6	−9	−1

    (a) Draw a scatter plot.

    (b) Estimate r.

    (c) Calculate r.

15. Robert W. Jackman* devised a method of predicting the amount of coup d'état activity in a (new) country from variables such as electoral turnout, presence of a dominant ethnic group, number of political parties, etc. To test his model, he correlated the prediction scores (X) with the amount of coup d'état activity experienced by 29 African nations from 1960 (or independence if later) through 1975 (Y). (Y is the sum of 5 points for each successful coup, 3 for each attempted coup, and 1 for each plotted one.)

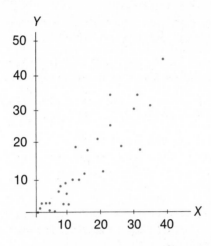

(a) From the plot shown, estimate the correlation coefficient.

(b) If you were prime minister of one of these countries, would you watch Jackman's index? Comment critically.

ANSWERS

1. (a) $\dfrac{6-10}{4-(-2)} = -\dfrac{2}{3}$; (b) $Y = \dfrac{26}{3} - \dfrac{2}{3}X$; (c) $\dfrac{26}{3}$.

3. (a) (ii), since the sum of squares is $\dfrac{3}{4}$ for it, compared to 2 for (i) and 3 for (iii).

(b) $b = \dfrac{5(-3)-(-1)(-4)}{5(7)-(-1)^2} = -.56$; $a = \dfrac{-4}{5} - (-.56)\left(\dfrac{-1}{5}\right) = -.91$; $Y_c = -.91 - .56X$.

5.

		(a)		(b)		(c)	
X	Y	Y_c	$Y - Y_c$	Y_c	$Y - Y_c$	Y_c	$Y - Y_c$
0	3	4	−1	0	3	$4-2b$	$-1+2b$
1	4	4	0	2	2	$4-b$	b
5	5	4	1	10	−5	$4+3b$	$1-3b$
			0		0		0

This exercise demonstrates that $\Sigma(Y - Y_c) = 0$ for any line which passes through

*American Political Science Review, 72, 1978, pp. 1262–1275.

$(\bar{X}, \bar{Y})$. Check that the equation of the regression line is $Y_c = \dfrac{23}{7} + \dfrac{5}{14}X$, and that this line also passes through $(\bar{X}, \bar{Y})$.

7. (a) X = price, Y = demand; $Y_c = 11.2 - 2.00X$; demand = 8.2.
   (b) X = demand, Y = price; $Y_c = 5.55 - .49X$; sales price should be $1.63.

9. (a) (Subtract 40 from each height.) $r = \dfrac{6(143.88) - (-.3)(-2.9)}{\sqrt{6(140.33) - (-.3)^2}\sqrt{6(148.91) - (-2.9)^2}} =$ 1.00.

   (b) $r = \dfrac{6(48.40) - 21(-.3)z}{\sqrt{6(140.33) - (-.3)^2}\sqrt{6(91) - 21^2}} = 1.00.$

11. (a)

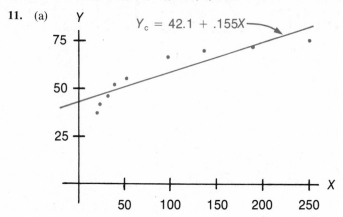

$Y_c = 42.1 + .155X$

(b) How progressive taxes are.

(c) $b = \dfrac{9(56,215) - 846(510)}{9(132,952) - 846^2} = \dfrac{74,475}{480,852} = .155;\ a = 56.7 - (.155)\left(\dfrac{846}{9}\right) = 42.1;\ Y_c =$ 42.1 + .155X.

(d) $r = \dfrac{74,475}{\sqrt{480,852}\sqrt{9(30,476) - 510^2}} = .90.$

13. Nonsense. There is no reason to pair Brooklyn runners and Queens runners.

15. (a) $r \approx +\left(1 - \dfrac{1}{10}\right) = .9.$

   (b) First I'd want to read the man's article carefully and look at any other information readily available on the subject. I'd consider how good his definitions and his identification of coups are.

15

LINEAR REGRESSION AND CORRELATION: STATISTICAL INFERENCE

In the last chapter, you learned how to find the Pearson correlation coefficient r and the linear regression equation for a sample of paired X and Y values. It was not necessary to make any assumptions about the distribution of either X or Y values in the populations. Now it is time to answer questions about the relationship of X and Y scores in the populations, and about the predicted value of Y for a given X in the population. For these, it is absolutely essential to make some assumptions about the population distributions. There is an intimate relation between regression and correlation. But as you learn to make statistical inferences about regression and correlation in the population, you must also study the differences between them, in particular the different assumptions about the population distributions.

15.1 INFERENCES ABOUT REGRESSION: ASSUMPTIONS ABOUT THE POPULATION

Suppose that you have made a scatter diagram for n pairs of X and Y scores and decided that they do approximately fit a straight line, and that you have then determined the regression equation $Y_c = a + bX$. Now you are concerned about the population from which the X and Y pairs were taken. Three questions need to be answered: First, is there a linear relationship between **all** X and Y pairs? Second, if there is, can you estimate the mean of all Y scores for any X in the population? Last, if in the future another X score is chosen from the population, can you predict what the corresponding Y score will be?

The following assumptions must be made about the population:

1. If a psychologist, in carrying out an experiment, could measure **all** Y scores in the population for each X score, he would expect experimental errors in his observations, due to inaccuracies in measurement, round-off errors, or differences in factors not controlled by the experiment. But it is assumed that the **mean** of the Y's for any given X falls on a straight line $Y = A + BX$. This means that there is a true linear relationship between pairs of scores, and that A and B for the population equation are approximated by a and b of the sample regression equation $Y_c = a + bX$.

2. For each X, the corresponding values of Y in the population are assumed

to be normally distributed, and the different normal distributions for different values of X all have the same variance. This diagram should help you picture these first two assumptions.

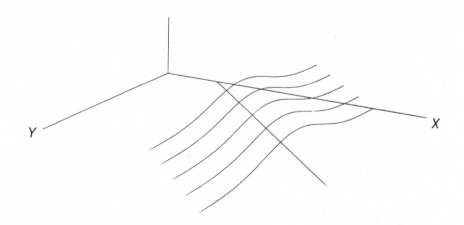

3. X scores used in determining a regression line commonly do not form a random sample from the population, and no assumption about the X population need be made; rather, particular X scores may be chosen by the experimenter. For example, the chemist studying the boiling point of solutions may first add 5 g. of chemical to 100 cc. of water, then 10, 15, 20, . . . g. A sociologist might study social adjustment scores of children with 0, 1, 2, 3, 4, . . . siblings. (When we study the accuracy of prediction of Y for a given X score, we shall find that it is greater for X values near the mean $\overline{X}$, so you may not want to have a wide spread in X values unless your sample size is large. But do not confuse this practical restriction on X with an assumption about the distribution of X scores in the population.)

15.2   TESTING H_0: $B = 0$

Is there indeed a linear regression equation $Y = A + BX$ for the population that will help in prediction? If there is not, then choosing $\overline{Y}$ for any X is the best that can be done. But this means the best population equation is $Y = \overline{Y}$—that is, $B = 0$. Therefore the hypothesis to be tested is H_0: $B = 0$. Only if this hypothesis is rejected is it worth going on to consider how the sample regression equation can be used for prediction of Y in the population, and with what accuracy this can be done.

What is the alternative hypothesis? Sometimes you will suspect from the nature of the experiment that the slope is indeed positive (or negative); in this case, use H_1: $B > 0$ (or $B < 0$) and a one-tail test; otherwise, use H_1: $B \neq 0$ and a two-tail test.

Before testing the null hypothesis, we must, as always, look at the sampling distribution of slopes. Suppose that, from a population which meets all the assumptions given in the preceding section, we take all different samples of size n and compute $\dfrac{b - B}{s_b}$ for each of these, where:

$$s_b = \frac{s_e}{\sqrt{\Sigma(X - \bar{X})^2}}$$

and s_e is your old acquaintance, the standard error of estimate $\sqrt{\dfrac{\Sigma(Y - Y_c)^2}{n - 2}}$. (See page 338.) The sampling distribution we end up with is a t distribution with $n - 2$ degrees of freedom.

Fortunately, computational formulas for $\Sigma(Y - Y_c)^2$ and $\Sigma(X - \bar{X})^2$ can be found:

$$\Sigma(Y - Y_c)^2 = \Sigma Y^2 - a\Sigma Y - b\Sigma XY$$

$$\text{and } \Sigma(X - \bar{X})^2 = \Sigma X^2 - \frac{(\Sigma X)^2}{n}$$

(The last one is the numerator of the formula for variance, with which you have long been familiar.) With a little algebra, the computational formula for s_b is seen to be

$$s_b = \sqrt{\frac{n(\Sigma Y^2 - a\Sigma Y - b\Sigma XY)}{(n - 2)[n\Sigma X^2 - (\Sigma X)^2]}}$$

Note that $n\Sigma X^2 - (\Sigma X)^2$ is the denominator of b.

To test H_0: $B = 0$

Compute $t = \dfrac{b - B}{s_b}$, with $s_b = \sqrt{\dfrac{n(\Sigma Y^2 - a\Sigma Y - b\Sigma XY)}{(n - 2)[n\Sigma X^2 - (\Sigma X)^2]}}$

Compare with critical value(s) of t for given α, $D = n - 2$.

Example 1   At a .05 level of significance, test H_0: $B = 0$, H_1: $B \neq 0$, for the (X, Y) scores shown below.

X^2	X	Y	XY	Y^2
4	2	0	0	0
9	3	3	9	9
16	4	4	16	16
25	5	4	20	16
36	6	6	36	36
49	7	11	77	121
139	27	28	158	198

$$b = \frac{6(158) - 27(28)}{6(139) - 27^2} = \frac{192}{105} = 1.8286^*$$

$$a = \frac{28}{6} - 1.8286\left(\frac{27}{6}\right) = -3.5619^*$$

$$\Sigma Y^2 - a\Sigma Y - b\Sigma XY = 198 - (-3.5619)(28) - 1.8286(158) = 8.82.$$

Now turn to page 338 and see that the error variation computed there for the same scores by using $\Sigma(Y - Y_c)^2$ has the same value, within round-off.

$$s_b = \sqrt{\frac{6(8.82)}{4(105)}} = .355.$$

$$t = \frac{1.83 - 0}{.355} = 5.16.$$

For $a = .05, n = 6, D = 4$, the critical values of t for a two-tail test are ± 2.78. Since 5.16 falls outside these limits, H_0 is rejected and H_1 accepted; the slope of the regression line in the population from which this sample was taken is not 0.

15.3 CONFIDENCE AND PREDICTION INTERVALS FOR THE REGRESSION EQUATION

You have determined the regression equation $Y_c = a + bX$ for a sample of n pairs of X and Y scores. Then you tested $H_0: B = 0$ and rejected it, so you can assume there is a meaningful linear relation between all pairs of X and Y scores in the population. Can you then estimate the mean Y score for any X in the population? And, if so, what is the accuracy of your estimate? Intuition perhaps suggests that we should in some fashion find two straight lines parallel to the regression equation and make a claim such as "a 95% confidence interval for the mean Y corresponding to X_0 is $Y_c \pm$ (a constant):

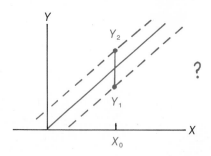

---

*To prevent round-off problems, b and a are carried to four decimal places in demonstrating that indeed $\Sigma(Y - Y_c)^2 = \Sigma Y^2 - a\Sigma Y - b\Sigma XY$.

This implies that the confidence interval is the same for all values of X. Unfortunately, this is not the case: It is smaller for values of X scores near the sample mean $\bar{X}$. Let us see why this is so:

The sample regression line $Y_c = a + bX$ always passes through the point whose coordinates are $(\bar{X}, \bar{Y})$: Since $a = \bar{Y} - b\bar{X}$, $Y_c = a + bX = \bar{Y} - b\bar{X} + bX$, and thus $Y_c = \bar{Y}$ if $X = \bar{X}$. Then we should expect the population regression line to pass through (μ_X, μ_Y). There are several sources of error. First, the slope of the population regression equation may not be the same as that of the sample regression equation, so the population line is in some region pivoted around the point (μ_X, μ_Y).

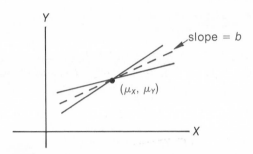

Secondly, we can determine confidence intervals for μ_X and μ_Y, but we do not know them exactly; this is another source of difficulty in placing the population regression line. The result is that the confidence interval for this mean Y score will be narrower when X_0 is close to $\bar{X}$ than when it is far away.

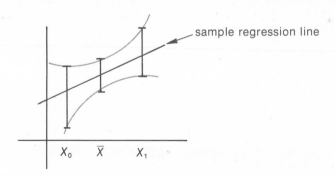

Suppose X_0 is a particular X score, and remember the assumption that, for a given X_0, the corresponding Y scores in the population are normally distributed. Two questions can now be asked:

1. What is the mean of the Y scores corresponding to X_0?
2. If an observation is made in the future with $X = X_0$, can we give limits within which, with given probability, the observed Y value will fall?

Consider the first question. It may seem plausible to use the sample regression line and choose $a + bX_0$ as our estimate of the mean of the corresponding Y scores. But here we have the same problem as in finding μ when $\bar{X}$ is known for a sample: A point estimate is of no use unless we have some notion of its accuracy, and it becomes an interval estimate when we do (see the discussion on page 168). Let μ_{Y/X_0} be the mean value of Y corresponding to X_0. A confidence interval for μ_{Y/X_0} is found by using a t-distribution with $n - 2$ degrees of freedom, with

$$t = \frac{Y_c - \mu_{Y/X_0}}{s_{Y/X_0}}$$

where

$$s_{Y/X_0} = s_e\sqrt{\frac{1}{n} + \frac{n(X_0 - \bar{X})^2}{n\Sigma X^2 - (\Sigma X)^2}} = \sqrt{\frac{\Sigma Y^2 - a\Sigma Y - b\Sigma XY}{n - 2}}\sqrt{\frac{1}{n} + \frac{n(X_0 - \bar{X})^2}{n\Sigma X^2 - (\Sigma X)^2}}$$

s_e reflects the spread of Y values about the sample regression line, common for all values of X (see Assumption 2, Section 15.1); the square root reflects the increasing spread of the Y scores as X_0 gets farther from $\bar{X}$. (Note that this factor has its smallest value when $X_0 = \bar{X}$ and the last term under the square root becomes zero.)

Example 1

For the scores given below, the regression equation is

$$Y_c = -3.56 + 1.83X, \ s_e = \sqrt{\frac{8.82}{4}} = 1.485$$

and $n\Sigma X^2 - (\Sigma X)^2 = 105$. (See Example 1, page 341.)

Find a 95% confidence interval for μ_{Y/X_0} when X_0 is (a) 2.0, (b) 4.5.

X	Y
2	0
3	3
4	4
5	4
6	6
7	11

(a) At a 95% confidence level, with $n = 6$ and $D = 4$, $t = \pm 2.78$; $\bar{X} = 4.5$. When $X = 2$, $Y_c = -3.56 + 1.83(2) = -.100$.

$$s_{Y/2} = 1.485\sqrt{\frac{1}{6} + \frac{6(2 - 4.5)^2}{105}} = 1.075.$$

The probability is .95 that paired random samples from these populations will have a computed Y_c value for $X = 2$ within 2.78 t units of the mean of all Y values for $X = 2$, $\mu_{Y/2}$. But 2.78 t units $= 2.78(1.075) = 2.99$ Y units. The paired samples I am considering were chosen at random, and $Y_c = -.100$ when $X = 2$. Therefore I have 95% confidence that $\mu_{Y/2}$ is between $-.100 - 2.99$ and $-.100 + 2.99$ (between -3.1 and $+2.9$).

> To find a confidence interval for μ_{Y/X_0}:
> 1. Determine t for the given confidence coefficient with $D = n - 2$.
> 2. Find s_{Y/X_0}.
> 3. Find Y_c when $X = X_0$: $Y_c = a + bX_0$.
> 4. Confidence limits for μ_{Y/X_0} are $Y_c - ts_{Y/X_0}$ and $Y_c + ts_{Y/X_0}$.

When $X = 2$, a 95% confidence interval for the mean of corresponding Y scores is -3.1 to 2.9.

(b) Again $t = \pm 2.78$, $\bar{X} = 4.5$; $s_{Y/X_0} = 1.485\left(\sqrt{\dfrac{1}{6}}\right) = .606$; when $X = 4.5$,

$Y_c = -3.56 + 1.83(4.5) = 4.68$.

$\mu_{Y/4.5} = 4.68 \pm 2.78(.606) = 3.0$ or 6.4.

When $X = 4.5$, a 95% confidence interval for the mean of corresponding Y scores is 3.0 to 6.4. Note that when $X = \bar{X}$, the width of the confidence interval is $6.4 - 3.0 = 3.4$, while it is $2.9 - (-3.1) = 6.0$ when $X = 2.5$.

Example 2

For the X and Y pairs of scores shown below,
(a) Find the equation of the regression line.
(b) Find 90% confidence intervals for μ_{Y/X_0} when X_0 is (i) 1, (ii) 2, (iii) 2.5, (iv) 3, (v) 4.
(c) Sketch the sample regression line and the limits of the confidence intervals for μ_{Y/X_0}.

X^2	X	Y	XY	Y^2	Y_c
1	1	0	0	0	.3
4	2	2	4	4	1.1
9	3	1	3	1	1.9
16	4	3	12	9	2.7
30	10	6	19	14	

Answers

(a) $b = \dfrac{4(19) - 10(6)}{4(30) - 10^2} = \dfrac{16}{20} = .80$

$a = \dfrac{6}{4} - .8\left(\dfrac{10}{4}\right) = -.50$

$Y_c = -.50 + .80X$.

(b) At a 90% confidence level, with $n = 4$ and $D = 2$, $t = \pm 2.92$.

$\mu_{Y/X_0} = Y_c \pm 2.92 s_{Y/X_0}$

$s_{Y/X_0} = \sqrt{\dfrac{14 - (-.50)6 - .80(19)}{2}}\sqrt{\dfrac{1}{4} + \dfrac{4(X_0 - 2.5)^2}{20}}$

$= .95\sqrt{.25 + .20(X_0 - 2.5)^2}$

	X_0	Y_c	s_{Y/X_0}	Confidence interval for μ_{Y/X_0}	Width of interval
(i)	1	.3	.79	−2.0 to 2.6	4.6
(ii)	2	1.1	.52	−0.4 to 2.6	3.0
(iii)	2.5	1.5	.48	0.1 to 2.9	2.8
(iv)	3	1.9	.52	0.4 to 3.4	3.0
(v)	4	2.7	.79	0.4 to 5.0	4.6

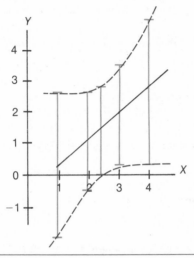

Now let's look at the second question asked on page 361: If an observation is made in the future with $X = X_0$, can we give limits within which the observed Y value will fall, with given probability? The computations here will be very similar to those for the first question, but the point of view is very different. We are no longer trying to estimate a population parameter (as in answering the first question); we are simply predicting a single observation to be made in the future.

The notation Y_p will be used for the Y score in a population predicted for the particular value X_0. A prediction interval for Y_p is found by using the t distribution:

$$t = \frac{Y_c - Y_p}{s_{Y_p}}$$

with $n - 2$ degrees of freedom, where

$$s_{Y_p} = s_e \sqrt{1 + \frac{1}{n} + \frac{(X_0 - \bar{X})^2}{\Sigma(X - \bar{X})^2}}\,.$$

$$= \sqrt{\frac{\Sigma Y^2 - a\Sigma Y - b\Sigma XY}{n - 2}} \sqrt{1 + \frac{1}{n} + \frac{n(X_0 - \bar{X})^2}{n\Sigma X^2 - (\Sigma X)^2}}$$

This is almost the same formula as for s_{Y/X_0}; it differs only in 1 added under the square root sign. But adding 1 under the square root increases s_{Y/X_0} and therefore the prediction interval for Y_p is wider than the confidence interval for μ_{Y/X_0}. This should be expected, since you should expect more variability in one score picked from a normal distribution than in the mean of many scores.

The prediction interval is

$$Y_c - ts_{Y_p} \text{ to } Y_c + ts_{Y_p}$$

where

$$Y_c = a + bX$$

Example 3   For the scores in Example 1, predict the interval within which the Y score corresponding to (a) $X_0 = 2$, (b) $X_0 = 4.5$, will fall, with probability .95 that your interval is correct.

(a) As before, $s_e = 1.485$, $n\Sigma X^2 - (\Sigma X)^2 = 105$, $t = \pm2.78$, and $Y_c = -.100$;

when $X_0 = 2$, $s_{Y_p} = 1.485\sqrt{1 + \dfrac{1}{6} + \dfrac{6(2 - 4.5)^2}{105}} = 1.83.$

$$Y_c - ts_{Y_p} = -.100 - 2.78(1.83) = -5.2$$

$$Y_c + ts_{Y_p} = -.100 + 2.78(1.83) = 5.0$$

A 95% prediction interval for Y when $X = 2$ is -5.2 to 5.0.

(b) When $X_0 = 4.5$, $s_{Y_p} = 1.485\sqrt{1 + \dfrac{1}{6}} = 1.60$ and $Y_c = 4.68$.

$$Y_c - ts_{Y_p} = 4.68 - 2.78(1.60) = .2$$

$$Y_c + ts_{Y_p} = 4.68 + 2.78(1.60) = 9.1$$

A 95% prediction interval for Y when $X = 4.5(=\bar{X})$ is .2 to 9.1.
Compare the confidence and prediction intervals for Y:

X_0	Confidence interval for μ_{Y/X_0}	Width	Prediction interval for Y_p	Width
2	−2.5 to 2.3	4.8	−5.2 to 5.0	10.2
4.5(=$\bar{X}$)	3.0 to 6.4	3.4	.2 to 9.1	8.9

Example 4   For the scores in Example 2, (a) find a 90% prediction interval for Y when X_0 is (i) 1, (ii) 2, (iii) 2.5, (iv) 3, (v) 4; (b) sketch the regression line and the limits of the prediction intervals for Y_p.

(a) As in Example 2, $t = \pm2.92$ and $s_e = .95$. $Y_p = Y_c \pm 2.92s_{Y_p}$, and

$$s_{Y_p} = s_e\sqrt{1 + \dfrac{1}{4} + \dfrac{4(X_0 - 2.5)^2}{20}}$$

	X_0	s_{Y_p}	Y_c	Prediction interval for Y_p	Width of interval
(i)	1	1.24	.3	−3.3 to 3.9	7.2
(ii)	2	1.08	1.1	−2.1 to 4.3	6.4
(iii)	2.5	1.06	1.5	−1.6 to 4.6	6.2
(iv)	3	1.08	1.9	−1.3 to 5.1	6.4
(v)	4	1.24	2.7	−.9 to 6.3	7.2

(b)

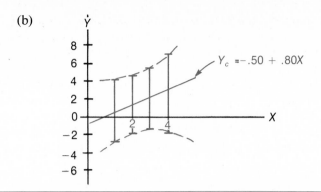

Both the confidence intervals for μ_{Y/X_0} and the prediction intervals for Y_p obtained in these examples have been very wide. How can they be narrowed? One way is to increase the sample size, since n and $n - 2$ occur in denominators. A second way is to decrease the spread of the X scores, since $\Sigma(X - \bar{X})^2$ also occurs in a denominator. If you have a random sample, nothing can be done about this. But a regression equation is commonly used in a planned experiment where X scores can be controlled. Do not be too ambitious about the spread of your X scores unless you are able to have a large sample size.

doesn't make sense?

15.4 EXERCISES

1. The regression equation for the following pairs of scores is $Y_c = 5.21 - .59X$. At a .05 level of significance, find out whether or not $B = 0$.

X	1	2	4	8
Y	6	3	2	1

2. Upon graduation, the scores of six chemistry majors on the Mathematical Aptitude Test (MAT) were compared with their cumulative grade-point average (GPA):

(X) MAT	600	720	640	590	620	680
(Y) GPA	2.8	3.8	3.6	2.9	3.2	3.6

The regression equation is $Y_c = -1.51 + .0075X$, $\Sigma XY = 12,863$, $\Sigma X = 3,850$, $\Sigma X^2 = 2,482,900$, $\Sigma Y = 19.9$, and $\Sigma Y^2 = 66.85$. (See Exercise 4, page 351.) With $\alpha = .01$, test $H_0: B = 0$, $H_1: B \neq 0$.

3.

X	0	0	0	0	1	1	1	1	2	2	2	2	3	3	3	3
Y	-1	1	0	2	4	3	1	2	6	4	5	6	7	8	6	8

(a) Test $H_0: B = 0$, $H_1: B \neq 0$ $(\alpha = .05)$.
(b) Test $H_0: B = 2.5$, $H_1: B \neq 2.5$ $(\alpha = .05)$.
(c) Find a 95% confidence interval for B.

4. Refer to Exercise 3. Suppose that a new set of 64 pairs of scores is made up by repeating each of the 16 pairs in Exercise 3 four times. Find a 95% confidence interval for B, and compare your answer with the result of Exercise 3(c).

Note: ΣX, ΣY, ΣX^2, ΣY^2, ΣXY, and n are each multiplied by 4; a and b are not changed.

5. Suppose that a new set of 1,600 pairs of scores is made up by repeating each of the 16 pairs in Exercise 3 100 times. Find a 95% confidence interval for B, and compare your answer with the results of Exercises 3(c) and 4.

6.

X	12	16	19	10	13	16	12	12
Y	10	9	20	8	6	18	9	7

For these scores, the regression equation is $Y_c = -8.29 + 1.39X$, $\Sigma XY = 1,282$, $\Sigma X = 110$, $\Sigma X^2 = 1,574$, $\Sigma Y = 87$, and $\Sigma Y^2 = 1,135$.

(a) Test (with $\alpha = .05$) $H_0: B = 0$, $H_1: B \neq 0$.
(b) Find a 95% confidence interval for B.

7.

X	−6	−5	−4	−3	−2	−1	0	1	2	3	4	5	6
Y	8	6	6	4	4	2	2	0	0	−2	−2	−4	−4

$\Sigma X = 0$, $\Sigma X^2 = 182$, $\Sigma Y = 20$, $\Sigma Y^2 = 216$, $\Sigma XY = -182$, $n = 13$.

(a) Find the equation of the regression line.
(b) Test $H_0: B = 0$, $H_1: B < 0$ ($\alpha = .01$).
(c) Test $H_0: B = -1.1$, $H_1: B \neq -1.1$ ($\alpha = .01$).
(d) Find a 99% confidence interval for B.

8. Using the data given in Exercise 7, (a) give a 95% confidence interval for the mean of all Y scores in the population if (i) $X = 0.0$, (ii) $X = 1.0$, (iii) $X = 1.5$; (b) give a 95% prediction interval for Y if (i) $X = 0.0$, (ii) $X = 1.0$; (iii) $X = 1.5$.

9. A sociologist studying upward mobility gives 80 children a test which measures their social attitudes (Y). He thinks that one factor which can be used to predict the test score is the number of children in the family (X). He discovers that $\Sigma X = 200$, $\Sigma X^2 = 600$, $\Sigma Y = 440$, $\Sigma Y^2 = 4,000$, $\Sigma XY = 1,300$, $a = .500$, $b = 2.00$.

(a) Find a 95% confidence interval for the mean test score if there are two children in the family.
(b) In addition to the 80 children, another child is to take the test; he has one brother and no sisters. What test score would the sociologist predict for this child, with .95 probability of being correct?

10. A farmhand sorts apples into 2, 3, 4, or 5 grades. The length of time he takes for sorting depends on the number of grades as follows:

Number of grades	2	2	2	2	2	3	3	3	3	3	4	4	4	4	4	5	5	5	5	5
Time (seconds)	3	1	2	1	1	2	3	2	1	2	2	3	3	1	4	3	4	4	5	2

(a) With .90 probability, what is the predicted time for 6 grades?
(b) Can you explain why it may not be wise to extrapolate prediction for X values outside the range in the sample?

ANSWERS

1. $s_b = \sqrt{\dfrac{4[50 - 5.21(12) + .59(28)]}{2[4(85) - 15^2]}} = .26$, $t = \dfrac{-.59 - 0}{.26} = -2.27$; $\alpha = .05$, $D = 2$, two-tail test, $t = \pm 4.3$. H_0 is accepted: $B = 0$.

2. $s_b = \sqrt{\dfrac{6[66.85 - (-1.51)(19.9) - .00752(12,863)]}{4[6(2,482,900) - 3,850^2]}} = .0018$; $t = \dfrac{.00752 - 0}{.0018} = 4.18$; critical $t = \pm 4.60$. H_0 is not rejected; $B = 0$.

3. (a) $\Sigma X = 24$, $\Sigma X^2 = 56$, $\Sigma Y = 62$, $\Sigma Y^2 = 362$, $\Sigma XY = 139$, $n = 16$, $b = 2.30$, $a = .43$,
   $s_b = \sqrt{\dfrac{16[362 - .43(62) - 2.30(139)]}{14[16(56) - 24^2]}} = .236$, $t = \dfrac{2.30 - 0}{.236} = 9.75$; for $a = .05$, two-tail test, $D = 14$, the critical values of t are ± 2.14. H_0 is rejected.
   (b) $t = \dfrac{2.30 - 2.50}{.236} = -.85$; we fail to reject H_0.
   (c) $B = 2.30 \pm 2.14(.236) = 1.79$ or 2.81. A 95% confidence interval for B is 1.8 to 2.8.

4. $s_b = \sqrt{\dfrac{64[4(362) - .43(4)(62) - 2.30(4)(139)]}{62[64(4)(56) - 4^2(24^2)]}} = .112$; $B = 2.30 \pm 2.00(.112) = 2.08$ or 2.52. A 95% confidence interval for B is 2.08 to 2.52. When n is multiplied by 4, the width of the confidence interval is cut in half.

5. $s_b = \sqrt{\dfrac{1,600[100(362) - .43(100)(62) - 2.30(100)(139)]}{1,598[1,600(100)(56) - 100^2(24^2)]}} = .0221$; $B = 2.30 \pm 2.00(.0221)$
   $= 2.26$ or 2.34; a 95% confidence interval for B is 2.26 to 2.34.
   The width of the interval is .1, for 64 scores it was .5, for 16 scores it was 1.0. One hundred times as many scores gives a confidence interval one-tenth as wide.

6. (a) $s_b = \sqrt{\dfrac{8[1,135 - (-8.29)(87) - 1.39(1,282)]}{6[8(1,574) - 110^2]}} = .449$, $t = \dfrac{1.39 - 0}{.449} = 3.10$. $D = 6$,
   critical $t = \pm 2.45$; reject H_0. $B \neq 0$.
   (b) $B = 1.39 \pm 2.45(.449) = .29$ or 2.49; a 95% confidence interval for B is .29 to 2.49.

7. (a) $b = \dfrac{13(-182) - 0(20)}{13(182) - 0^2} = -1.00$, $a = \dfrac{20}{13} - 0 = 1.54$, $Y_c = 1.54 - 1.00X$.
   (b) $s_b = \sqrt{\dfrac{13[216 - 1.54(20) - (-1.00)(-182)]}{11[13(182) - 0^2]}} = .040$, $t = \dfrac{-1.00 - 0}{.040} = -25$; the critical value of t is -2.72, so H_0 is rejected.
   (c) $t = \dfrac{-1.00 - (-1.10)}{.040} = 2.50$; the critical values of t are ± 3.11, so we fail to reject H_0.
   (d) $B = -1.00 \pm 3.10(.040) = -1.12$ or $-.88$; a 95% confidence interval for B is -1.12 to $-.88$.

8. $s_{Y/X_0} = \sqrt{\dfrac{216 - 1.54(20) - (-1.00)(-182)}{11}} \sqrt{\dfrac{1}{13} + \dfrac{X_0^2}{182}} = .539 \sqrt{\dfrac{1}{13} + \dfrac{X_0^2}{182}}$.
   $s_{Y_p} = .539 \sqrt{1 + \dfrac{1}{13} + \dfrac{X_0^2}{182}}$

	X_0	s_{Y/X_0}	Y_c	μ_{Y/X_0}	Width	s_{Y_p}	Y_p	Width
(i)	0.0	.149	1.54	1.21 to 1.87	.66	.559	.31 to 2.77	2.46
(ii)	1.0	.155	.54	.20 to .88	.68	.561	−.69 to 1.77	2.46
(iii)	1.5	.161	.04	−.31 to .39	.70	.563	−1.20 to 1.28	2.48

9. (a) $s_e = \sqrt{\dfrac{[4{,}000 - .500(440) - 2.00(1{,}300)]}{78}} = 3.89$, $\bar{X} = 2.5$; if $X_0 = 2$, $Y_c = .50 +$

$2.00(2) = 4.50$, $s_{Y/2} = 3.89\sqrt{\dfrac{1}{80} + \dfrac{80(2 - 2.5)^2}{8{,}000}} = .476$, $D = 78$, $t = \pm 1.99$, $\mu_{Y/2} =$

$4.50 \pm 1.99(.476) = 3.56$ or 5.45. A 95% confidence interval for the mean test score for families with two children is 3.56 to 5.45.

(b) $s_{Y_p} = .389\sqrt{1 + \dfrac{1}{80} + \dfrac{80(2 - 2.5)^2}{8{,}000}} = .392$; $Y_p = 4.50 \pm 1.99(.392) = 3.72$ to 5.28.

10. (a) $b = \dfrac{20(188) - 70(49)}{20(270) - 70^2} = \dfrac{330}{500} = .66$, $a = \dfrac{49}{20} - .660\left(\dfrac{70}{20}\right) = .140$, $Y_c = 4.10$ when

$X = 6$. $s_{Y_p} = \sqrt{\dfrac{[147 - .14(49) - .66(188)]}{18}}\sqrt{1 + \dfrac{1}{20} + \dfrac{20(6 - 3.5)^2}{500}} = 1.077$. $Y_p =$

$4.10 \pm 1.73(1.077) = 2.24$ or 5.96. The probability is .90 that the time predicted for sorting 6 grades is 2.2 to 6.0 seconds.

(b) The relation between X and Y may be linear only for a certain range of X values. Also the prediction interval for Y gets wider as X gets farther from the mean of 3.5.

15.5   THE SAMPLING DISTRIBUTION OF r

If a population consists of the following pairs of scores:

X: 1   1   3   5   5
Y: 1   5   3   1   5

the population coefficient of correlation is 0. But if a sample of size 3 consists of the first, third, and fifth pairs, then the sample correlation coefficient is $+1$. This is, admittedly, a manufactured and unusual example, but it should make it clear that it is necessary to test hypotheses about the population correlation after the sample r has been determined. We must look at the sampling distribution of r.

We now need another Greek letter: ρ is the Greek r (**rho,** pronounced to rhyme with **no**—the h is silent); ρ is the symbol for the population coefficient of correlation.

If all possible samples of given size n are taken from a population, and the correlation between pairs of X and Y scores is computed for each sample, what is the sampling distribution of the r's? The question cannot be answered unless it is assumed that **the population is normally distributed in both X and Y scores.** In more detail, this means that for each X the corresponding Y's must be normally distributed, and for each Y the corresponding X's must be normally distributed. The accompanying figure, representing three dimensions, should help you see what the distribution looks like. A distribution such as this is called a **bivariate normal distribution.**

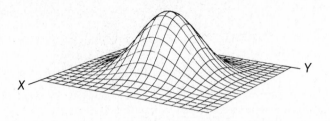

Unfortunately, the sampling distribution of r depends not only on the sample size n but also on the population correlation. When $\rho = 0$, the sampling distribution of r is symmetric about 0. The distribution is very skewed, however, if ρ is close to $+1$ or to -1. If $\rho = .80$, for example, samples taken from the population will have different r's. However, no r can be greater than $+1$, so the sample r's can not be more than .20 above the population correlation ρ, but may be considerably more below it. Therefore, the sampling distribution of r is negatively skewed if ρ is close to 1. Similarly, it will be positively skewed if ρ is close to -1.

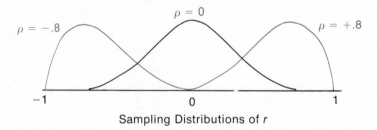

Sampling Distributions of r

For this reason, it will be necessary to use different methods for testing the null hypothesis $\rho = A$, depending upon whether $A = 0$ or is different from zero.

15.6 TESTING H_0: $\rho = 0$

Suppose r is computed for each sample of size n in a bivariate normal population with $\rho = 0$. The distribution of r scores is neither normal nor a t distribution. But if for each sample the quantity $\dfrac{r\sqrt{n-2}}{\sqrt{1-r^2}}$ is computed, then these quantities have a t distribution with $n - 2$ degrees of freedom.

Test H_0: $\rho = 0$ by computing $t = \dfrac{r\sqrt{n-2}}{\sqrt{1-r^2}}$ for a sample; compare with the critical values of t for $n - 2$ degrees of freedom and the given α.

Use a one-tail test if H_1: $\rho < 0$ or H_1: $\rho > 0$; use a two-tail test if H_1: $\rho \neq 0$.

.Example 1

A study of 66 students chosen at random from Johnson Junior High School shows there is a correlation of .3 between grades in art and in mathematics. Is there a significant correlation between grades in these two subjects for all students in the school? (.05 level of significance.) Assume that the grades of all students in both art and mathematics are normally distributed.

H_0: $\rho = 0$, H_1: $\rho \neq 0$; $a = .05$, $D = 66 - 2 = 64$, two-tail test, so $t = \pm 1.96$ (use the z table since $n > 40$); $r = .3$, $n = 66$.

$$t = \frac{r\sqrt{n-2}}{\sqrt{1-r^2}}$$

$$t = \frac{.3\sqrt{66-2}}{\sqrt{1-.3^2}} = \frac{2.4}{\sqrt{.91}} = 2.52$$

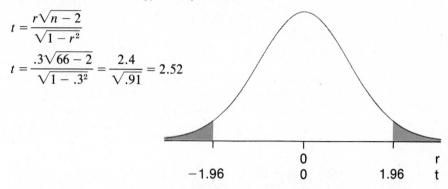

$$\begin{array}{ccc} & 0 & r \\ -1.96 & 0 & 1.96 \quad t \end{array}$$

Since 2.52 falls in the rejection region, H_0 is rejected: There is a significant correlation between grades in the two subjects.

Increase the size of your samples if you suspect that one variable or both are only approximately normal. If you think either one is **not** normally distributed, don't try to test a hypothesis about ρ.

Example 2

Eighteen students chosen at random from third grades in Morristown public schools show a correlation of $-.4$ between family income and hours per week spent watching TV. At a .01 level of significance, is there a significant negative correlation for all Morristown third graders?

H_0: $\rho = 0$, H_1: $\rho < 0$; one-tail test, $\alpha = .01$; $r = 0.4$, $n = 18$.

$$t = \frac{r\sqrt{n-2}}{\sqrt{1-r^2}} = \frac{-.4\sqrt{18-2}}{\sqrt{1-.4^2}} = \frac{-1.6}{.917} = -1.75$$

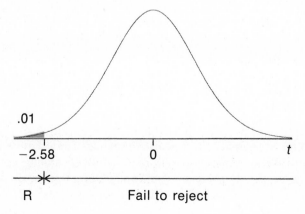

.01

$$\begin{array}{cc} -2.58 & 0 \qquad t \end{array}$$

R   Fail to reject

The critical value of t for $n - 2 = 16$ degrees of freedom with $\alpha = .01$ is -2.58.

We continue to accept H_0: There is no correlation between family income and time spent watching TV among Morristown third graders. This conclusion is not very trustworthy, however, for both family income and hours per week watching TV are probably skewed to the right rather than normal distributions.

15.7 TESTING H_0: $\rho = A$, $A \neq 0$

If $\rho \neq 0$, the sampling distribution of correlation coefficients r is skewed. But a famous statistician, R. A. Fisher, devised a transformation of r scores into Z scores so that the Z distribution is approximately normal for large n. (Do not confuse Z scores with standard z scores.) Fortunately, you do not need to carry out computations for this transformation; Table 8, Appendix C, gives corresponding r and Z values.

Example 1   What is the Z value corresponding to $r = .15$? to $r = -.70$?
If $r = .15$, then $Z = 1.51$.
If $r = -.70$, then $Z = -.867$.

If the Z score is computed for each pair of samples of size n in the populations, the distribution is approximately normal for large n ($n > 30$, say). Its mean μ_Z is the Z value corresponding to ρ, and its standard deviation is

$$\sigma_Z = \frac{1}{\sqrt{n - 3}}$$

A hypothesis is then tested with the usual z test for a normal distribution:

$$z = \frac{Z - \mu_Z}{\sigma_Z}$$

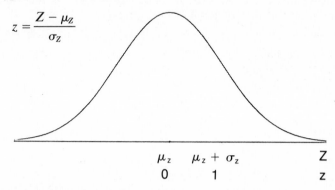

Example 2   Studies in past years have shown there is a correlation of .40 between grades in introductory economics and history courses at Drew University.

This year, a random sample of 39 students taking both courses shows a correlation of .34. Is there a significantly different correlation this year in the grades of all students taking both introductory courses? (.05 level of significance.)

H_0: $\rho = .40$, H_1: $\rho \neq .40$; $\alpha = .05$.
If $r = .34$, then $Z = .354$.
If $\rho = .40$, then $\mu_Z = .424$; the transformation from ρ to μ_Z has the same form as the transformation from r to Z.

$$\sigma_Z = \frac{1}{\sqrt{n-3}} = \frac{1}{\sqrt{36}} = .167$$

$$z = \frac{Z - \mu_Z}{\sigma_Z} = \frac{.354 - .424}{.167} = -.42$$

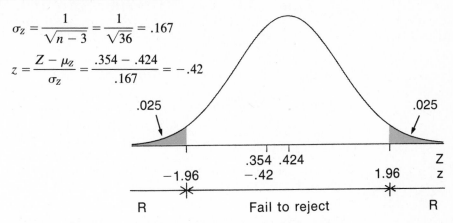

For a two-tail test with $\alpha = .05$, $z = \pm 1.96$; $-.42$ falls in the "fail to reject" region: The correlation is still .40 for grades of all students taking introductory economics and history courses, and has not changed.

The Fisher r-to-Z transformation is also used to find confidence intervals for ρ.

Example 3   Find a 95% confidence interval for ρ if a sample of 103 paired X and Y scores has a correlation of .58.
With a .95 confidence coefficient, $z = \pm 1.96$.
If $r = .58$, then $Z = .663$; $n = 103$ so $\sigma_Z = \frac{1}{\sqrt{100}} = .10$.

± 1.96 z units is $\pm 1.96(.10) = .196$ Z units, so I have 95% confidence that μ_Z is between $.663 - .196$ and $.663 + .196$ (between .467 and .859) Z units.
Table 8, Appendix C, is again used—this time looking up .467 and .859 in the Z column, and finding the corresponding ρ in the r column.
If $\mu_Z = .467$, then $\rho = .44$.
If $\mu_Z = .859$, then $\rho = .70$.
A 95% confidence interval for ρ is .44 to .70.

To find a confidence interval for ρ:
1. Find z for the given confidence coefficient.
2. Determine Z corresponding to the sample r.
3. Determine $\sigma_Z = \frac{1}{\sqrt{n-3}}$.
4. Confidence limits for μ_Z are $Z - z\sigma_Z$ and $Z + z\sigma_Z$.
5. Determine confidence limits for ρ by using Table 8 again.

Compare the results of the same problem if the sample size is considerably smaller:

Example 4   Find a 95% confidence interval for ρ if $r = .58, n = 39$. As in Example 1, $z = \pm 1.96$ and $Z = .663$.

$$\sigma_Z = \frac{1}{\sqrt{39-3}} = \frac{1}{\sqrt{36}} = .167.$$

$$\mu_Z = Z \pm z\sigma_Z = .663 \pm 1.96(.167) = .336 \text{ or } .990.$$

If $\mu_Z = .336$, then $\rho = .32$.
If $\mu_Z = .990$, then $\rho = .76$.
A 95% confidence interval for ρ is .32 to .76. A smaller n gives a wider confidence interval.

15.8 EXERCISES

1.   Fifty-one grade school students are given a reading test and the height of each child is measured. The computed value of r between score on the test and height is .80.
   (a) Test the hypothesis that there is no correlation between reading score and height for all grade school students. ($\alpha = .05$.)
   (b) Test the hypothesis that the correlation is at least .85 for all grade school students. ($\alpha = .05$.)
   (c) Can you explain the apparently high correlation?
   (d) Can you explain why the tests suggested in (a) and (b) should not be used?

2.   The correlation between paired scores is .30. Test $H_0: \rho = 0, H_1: \rho \neq 0$ if (a) $n = 11$, (b) $n = 102$. ($\alpha = .05$)

3.   The correlation between paired scores is .30. Test $H_0: \rho = .50, H_1: \rho < .50$ if (a) $n = 11$, (b) $n = 102$. ($\alpha = .05$)

4.   A sociologist studying upward mobility finds there is a correlation of .25 between number of children in the family and score on a social attitudes test for 80 children. (See Exercise 9, page 367.)
   (a) Test whether or not there is a positive correlation. ($\alpha = .01$)
   (b) in previous studies he had concluded that $\rho = .34$. Does this new study force him to change that conclusion? ($\alpha = .05$)

5.   There is a correlation of .75 between location of direction of the sound and frequency of the sound when experiments are carried out in 84 experiments when the frequency is between 50 and 300 Hertz. Find a 95% confidence interval for the population correlation.

6.   The correlation between the IQ's of 187 identical twins reared together is .87. Does this support the theory that, for genetic reasons, identical twins have exactly the same IQ, at a .05 level of significance? (Test $H_0: \rho = 1.00$. Why should it be a one-tail test?)

7. The correlation between age and systolic blood pressure among 228 men chosen at random is .20. Find a 95% confidence interval for this correlation among all men.

8. The correlation between the area and population of 10 counties in Texas, chosen at random, is −.2. At a .10 level of significance: (a) is there a positive correlation between area and population for all counties in Texas?; (b) is the correlation −.3 or less for all counties? (c) Find a 90% confidence interval for ρ. Does this explain the apparent contradiction in (a) and (b)?

9. Last year, the correlation between rainfall in inches and production of wheat per acre in North Dakota was .61. This year the correlation has been .54 on 139 randomly chosen acres in North Dakota. Has there been a significant difference this year? ($\alpha = .01$.)

ANSWERS

1. (a) $t = \dfrac{.80\sqrt{49}}{\sqrt{1 - .80^2}} = 9.33$; $\alpha = .05$, $D = 49$, critical values of t are ± 1.96; H_0 is rejected.

(b) H_0: $\rho = .85$, H_1: $\rho < .85$, $\sigma_z = \dfrac{1}{\sqrt{48}}$, $z = \dfrac{1.10 - 1.26}{.144} = -1.11$; the critical value of z is -1.65, so H_0 is accepted.

(c) Children's reading ability and height both increase as they grow older.

(d) Heights of children in any one grade may approximate a normal distribution, but for all grades they probably are not; the same is true of reading ability.

2. (a) $t = \dfrac{.30\sqrt{9}}{\sqrt{1 - .30^2}} = .94$; if $D = 9$, $\alpha = .05$, then $t = 2.26$; H_0 is still accepted.

(b) $t = \dfrac{.30\sqrt{100}}{\sqrt{1 - .30^2}} = 3.14$; if $D = 100$, $\alpha = .05$, then $t(= z) = 1.96$; H_0 is rejected.

3. The critical value of z is -1.64. (a) $z = \dfrac{.310 - .549}{1/\sqrt{8}} = -.68$; H_0 cannot be rejected.

(b) $z = \dfrac{.310 - .549}{1/\sqrt{99}} = -2.38$; H_0 is rejected.

4. (a) H_0: $\rho = 0$, H_1: $\rho > 0$; $t = \dfrac{.25\sqrt{78}}{\sqrt{1 - .25^2}} = 2.28$. Critical $t(= z) = 1.64$; reject H_0. There is a positive correlation.

(b) If $r = .25$, $Z = .255$; when $\rho = .34$, $\mu_z = .354$; $z = \dfrac{.255 - .354}{1/\sqrt{77}} = -.87$. Critical $z = -1.64$; he does not reject the conclusion that $\rho = .34$.

5. $r = .75$, $Z = .973$; $\mu_z = .973 \pm 1.96\dfrac{1}{\sqrt{81}} = .755$ or 1.191. A 95% confidence interval for ρ is .64 to .83.

6. The μ to μ_z transformation is not defined when $\rho = 1.00$, so H_0 can't be tested. Using H_0: $\rho = .99$, H_1: $\rho < .99$, we find $z = \dfrac{1.333 - 2.647}{1/\sqrt{184}} = -17.8$, so H_0 is rejected.

7. $\mu_z = .203 \pm 1.96\left(\dfrac{1}{\sqrt{225}}\right) = .072$ or $.334$; A 95% confidence interval for ρ is $.07$ to $.32$.

8. (a) H_0: $\rho = 0$, H_1: $\rho < 0$, $t = \dfrac{-.20\sqrt{8}}{\sqrt{1-(-.20)^2}} = -.58$, $D = 8$, critical $t = -1.40$; H_0 cannot be rejected on this evidence.

   (b) H_0: $\rho = -.3$, H_1: $\rho > -.30$, $z = \dfrac{-.203-(-.310)}{1/\sqrt{7}} = .28$; the critical value of z is 1.28. Again, H_0 cannot be rejected.

   (c) $\mu_z = -.203 \pm 1.64\left(\dfrac{1}{\sqrt{7}}\right) = -.82$ or $+.42$; a 90% confidence interval for ρ is $-.68$ to $+.40$. The confidence interval is very wide because n is small. This explains why H_0 cannot be rejected in either (a) or (b).

9. $z = \dfrac{.604-.709}{1/\sqrt{136}} = -1.22$; critical $z = \pm 2.58$. H_0 is still accepted.

15.9 VOCABULARY AND SYMBOLS

$Y = A + BX$	sampling distribution of r
sampling distribution of slopes	normal distribution in two variables
s_{Y/X_0}	bivariate normal distribution
Y_p	Z distribution $\quad \mu_Z,\ \sigma_Z$
$\sigma_{Y/X}$	Fisher r-to-Z transformation
ρ	

15.10 REVIEW EXERCISES

1. What assumptions about the distribution of all X and Y scores must be made if you are to test (a) H_0: $B = 0$, (b) $\rho = .40$?

2. Find a 95% confidence interval for the correlation between the birth weight of a baby and the average number of cigarettes smoked daily by the mother during pregnancy if this correlation is $-.46$ in a sample of 1,138 births.

3.

X	-1	-1	-1	0	0	0	0	0	1	1	1	2	2	2	3	3	3
Y	4	4	3	2	1	3	2	2	1	0	1	0	0	1	-1	-1	-2

   (a) Find the regression equation.
   (b) Test H_0: $B = 0$, H_1: $B \neq 0$ at a .05 level of significance.
   (c) Find a 90% confidence interval for the mean Y score corresponding to $X_0 = 1$.
   (d) Find, with .90 probability that you are correct, the Y score you would predict for an X score of 1.

4. In Exercise 7, page 332 you found that the regression equation between length (X) and width (Y) of 209 zircon crystals is $Y_c = 12.6 + .190X$, with $\Sigma X = 11,386$, $\Sigma X^2 = 654,478$, $\Sigma Y = 4,795$, $\Sigma Y^2 = 117,676$, and $\Sigma XY = 267,698$. Is

the slope of the population regression line different from 0, at a .01 level of significance?

5. Refer to Exercise 4. Find a 90% confidence interval for the mean width of zircon crystals whose length is 50 mm.

6. Refer to Exercise 4. Find, with .90 probability that you are correct, the width you would predict if you chose another zircon crystal at random and found its length to be 50.0 mm.

7. The regression equation for IQ (X) and grade on a statistics quiz (Y) for five students is $Y_c = -16.2 + .26X$ with $\Sigma X = 600$, $\Sigma X^2 = 73,000$, $\Sigma Y = 75$, $\Sigma Y^2 = 1,193$, and $\Sigma XY = 9,260$. (See Exercise 6, page 331.) Test to see whether or not the population regression line has positive slope, with $\alpha = .05$.

8. C. Smoszyk's study of 10 cancer patients (see Exercise 12, page 354) showed a correlation of .85 between anxiety scores before and after treatment.
   (a) At a .05 level of significance, is there any correlation between anxiety state scores before and after treatment for all such cancer patients?
   (b) If another researcher claimed that $\rho = .80$ before and after similar treatment of similar patients, would Smoszyk's study support or question this claim? $(\alpha = .05.)$

9. You also used Smoszyk's data to find a correlation of $-.30$ between anxiety state before treatment in 10 cancer patients and the change in their anxiety scores during treatment.
   (a) Test H_0: $\rho = 0$, H_1: $\rho \neq 0$. $(\alpha = .01)$
   (b) Test H_0: $\rho = -.40$, H_1: $\rho > -.40$. $(\alpha = .10)$

10. Anxiety reduction in 15 patients treated with the drug CDX showed a correlation of .46 between measurements on the Hamilton Anxiety Rating Scale (HARC) and the Hopkins Symptoms Check List (HSCL) for 15 patients (see Exercise 14, page 354).
   (a) At a .01 level of significance, is there any correlation for all such patients?
   (b) Is the correlation at least .60 for all such patients? $(\alpha = .10.)$

11. (a) Explain how a sampling distribution of correlations for samples of size 100 is obtained.
   (b) Why does it seem likely that the sampling distribution of correlations is different if $\rho = 0$ or $\rho = -.90$, and therefore different methods are used to test H_0: $\rho = 0$ and H_0: $\rho = -.90$.

12. If the correlation for 36 experimenters is .82 between the ability to recall nonsense syllables and time elapsed since presentation of the symbol, can the hypothesis that the population correlation is less than .70 be rejected, at a .05 level of significance?

13. The correlation between traffic flow and number of vehicular accidents at 548 street crossings is .34.

(a) Find a 95% confidence interval for this correlation at all street crossings.

(b) Which of the assumptions underlying this type of correlation analysis are probably violated?

14. Peter Huth, biochemist at the Madison, Wisconsin, Veterans Administration Hospital, wrote a thesis on the forces that move carnitine across the cell walls of rats' kidneys sliced thin. In plain diffusion the concentration of the substance on the inside (Y) will be proportional to the concentration outside the cell (X). Using one of two types of the enzyme, he collected the following data (concentrations are in nanomoles per milliliter):

X	80	80	80	160	160	160	640	640	640	1,000	1,000	1,000	2,000	2,000	2,000
Y	72	63	55	179	156	258	423	595	464	1,320	960	1,010	805	741	781

X	2,000	2,000	2,000	2,000	2,000	2,000	5,000	5,000	5,000	5,000	5,000	5,000
Y	1,650	1,350	667	2,240	2,620	2,180	5,950	1,980	5,420	3,380	2,930	4,320

(a) Plot the 27 points.

(b) He would like to test $H_0: A = 0$, $H_1: A \neq 0$. Why?

(c) Assuming that the assumptions necessary for testing $H_0: A = 0$ are the same as for testing $H_0: B = 0$, which are violated by these data?

ANSWERS

1. (a) The Y scores corresponding to each X score are normally distributed, and every normal distribution has the same variance; the means of the Y's for each X fall on a straight line $Y = AX + B$. (No requirement is put on the X's, and X values can even be chosen in advance by the experimenter.)

(b) X and Y scores form a bivariate normal distribution—that is, the X scores corresponding to each Y are normally distributed, and the Y scores corresponding to each X are also normally distributed.

3. (a) $b = \dfrac{17(-19) - 15(20)}{17(45) - 15^2} = \dfrac{-623}{540} = -1.15;$ $\quad a = \dfrac{20}{17} - (-1.15)\left(\dfrac{15}{17}\right) = 2.19;$ $\quad Y_c = 2.2 - 1.2X.$

(b) $s_b = \sqrt{\dfrac{17[72 - 2.19(20) - (-1.15)(-19)]}{15(540)}} = \sqrt{\dfrac{17(6.35)}{15(540)}} = .115,$ $\quad t = \dfrac{-1.15 - 0}{.115} = -10;$ $D = 15$, critical $t = \pm 2.13$ so H_0 is rejected; $B \neq 0$.

(c) $s_{Y/1} = \sqrt{\dfrac{6.35}{15}}\sqrt{\dfrac{1}{17} + \dfrac{17(1 - 20/17)^2}{540}} = .159;$ when $X = 1$, $Y_c = 1.04$. $\mu_{Y/1} = 1.04 \pm 1.75(.159) = .76$ or 1.32. A 90% confidence interval for the mean Y score corresponding to $X_0 = 1$ is .76 to 1.3.

(d) $s_{Y_p} = .651\sqrt{1 + \dfrac{1}{17} + \dfrac{17(1 - 20/17)^2}{540}} = .670;$ $\quad Y_p = 1.04 \pm 1.75(.670) = -.13$ or 2.21.

5. $s_{Y/50} = \sqrt{\dfrac{117,676 - 12.6(4,795) - .190(267,698)}{207}}\sqrt{\dfrac{1}{209} + \dfrac{209(50 - 54.48)^2}{207(7,144,906)}} = .386;$

$\mu_{Y/50} = 22.1 \pm 1.64(.386) = 21.47$ or 22.73.

7. $H_0: B = 0$, $H_1: B > 0$. $s_b = \sqrt{\dfrac{5[1,193 - (-16.2)(75) - .26(9,260)]}{3[5(73,000) - 600^2]}} = .0115$, $z = \dfrac{.26 - 0}{.0115} = 22.5$; critical $t = 2.35$; do not reject H_0.

9. (a) $t = \dfrac{-.30\sqrt{8}}{\sqrt{1 - (-.30)^2}} = -.89$; $D = 8$, critical $t = \pm3.36$, H_0 is not rejected.

   (b) $z = \dfrac{-.310 - (-.424)}{1/\sqrt{7}} = .30$, critical $z = 1.28$; H_0 is still accepted.

11. (a) Take a sample of size 100 and find the correlation r between pairs of scores in the sample; this r is one of the scores in the sampling distribution. Repeat this for all possible samples of size 100.
   (b) Since ρ is always between -1.00 and $+1.00$, the sampling distribution of correlations is symmetric if $\rho = 0$ but is skewed to the right if $\rho = -.90$.

13. (a) $\mu_z = .354 \pm 1.96\left(\dfrac{1}{\sqrt{545}}\right) = .270$ or $.438$; a 95% confidence interval for ρ is .26 to .41.
   (b) There are (almost certainly) no accidents at most street corners, so that the distribution of accidents is skewed to the right, not normal. Traffic flow is probably skewed to the right too, with a few street corners with very heavy traffic and many with comparatively little.

16 NONPARAMETRIC TESTS

A nonparametric test is one in which (a) it is not necessary to assume the population is normally distributed (or any other very strong assumption about the population distribution; for this reason nonparametric tests are also called ''distribution-free tests'') or (b) categorical or ranked data are used. You will find the computations quite simple; both the mean and the standard deviation are meaningless for ranked data.

16.1 INTRODUCTION

For many of the tests you have learned to use, it has been necessary to assume that the population is normally distributed. If it is normal, the error made in making inferences about the population from sample data can be estimated. But if it is not normal, the error may be large and cannot be estimated. It is therefore valuable to know some tests in which such a strong assumption as normality doesn't need to be made about the population distribution; these are called **nonparametric** tests. You have already studied one of them: the χ^2 test. There are many others, but only three of them are included here.

The word ''nonparametric'' is confusing. It was originally applied when **no** assumption needed to be made about the population distribution (as in the χ^2 tests of Chapter 12). Almost always, however, some assumption needs to be made; in some of the tests in this chapter, for example, the population distribution is assumed to be continuous so that comparative rankings can be given to scores in different samples (the third score in sample 1 is higher than the third score in sample 2). So ''nonparametric'' is now used to mean that a less stringent requirement than normality is made for the population distribution. Also, the meaning of the phrase ''nonparametric test'' has been extended to include any test using categorical or ranked data. Thus, a comparison of the heights of two samples of women is made with a nonparametric test if, instead of measuring each person's height, all the women are simply arranged in order of height in a row (the two tallest come from sample 1, the third tallest comes from sample 2, and so on). Then, because the data are ranked and not metric, a nonparametric test is used—even if it is known that the heights in both populations from which the samples were chosen are normally distributed.

Nonparametric tests have many advantages: You avoid the error caused by assuming a population is normally distributed when it is not, the computations that need to be made are often very simple, and the data may be easier to collect (almost certainly so when categorical or ranked rather than metric). Why are nonparametric tests not always used, then? The answer is that you don't get something for nothing. If the population distribution is normal so that, say, a t test or a nonparametric test may be chosen, the former will generally give a smaller value of β than the latter for a given, fixed value of α; if this is the case, sample size will have to be larger for the nonparametric test if the same limits on α and β are to be attained as with a t test. For the same sample size, then, the t test results usually will be more reliable. Also, the null hypotheses are sometimes more general; rejection may imply "two population distributions are different," but you don't know whether they have different means, different variances, or different distributions (for example, one normal and the other not).

16.2 THE SIGN TEST

The sign test is used in a "before-after" type of experiment, or with matched pairs; it is thus an alternative to a t test for dependent samples (See Sec. 11.10, pages 255–257). In a sign test the experimenter counts how many of the differences are positive and how many are negative, and then determines whether the number of plus (or minus) signs is larger than one would ordinarily expect as a result of random or chance variation. H_0 states that a positive difference ($+$ sign) and a negative difference ($-$ sign) are equally likely, or $P = 1/2$, where P is the probability of a particular change (say, positive) if one individual is chosen at random. Thus the sign test uses a binomial distribution with H_0: $P = 1/2$. A difference of 0 (neither positive nor negative change visible) will simply not be counted, and n is reduced accordingly.

Example 1   A treatment of anxiety by chlordiazepoxide reports the following changes in anxiety for 15 patients, measured on the HSCL scale (see Exercises 2–6, page 258): (A negative change show less anxiety after the treatment.)

$$-6, -9, -10, -7, 0, 0, 9, 0, -1, -4, -9, -9, -6, -9, -1$$

Does the drug treatment diminish anxiety? ($\alpha = .05$.)

H_0: $P = .5$ says that the treatment makes no difference, thus positive and negative changes (due to random variation) are equally likely. (Let $P = $ the probability of an increase ($+$), since there are fewer of those.) H_1: $P < .5$ says that an increase is less likely than a decrease. There is only one plus but there are 11 minus signs. The three 0's are omitted, so n is reduced from 15 to 12. But if P does equal .5, the probability of 1 or less plus signs is $.000 + .003$ by Table 2 with $n = 12$, $P = .5$. $.003$ is smaller than $\alpha (= .05)$; the null hypothesis is rejected.

Remember that the normal distribution can be used as an approximation to the binomial distribution if n is sufficiently large, with a correction of .5 for continuity. In this example (even though n is small), $z = \dfrac{1 + .5 - 12(.5)}{\sqrt{12(.5)(.5)}} = -2.60$, and $\Pr(X < 1.5) = \Pr(z < -2.60) = .5 - .495 = .005$. The .5 is added or subtracted, whichever makes z closer to 0 and the tail probability larger; note that .005 errs on the large, i.e., "conservative" side.

The binomial assumptions (see page 124) must be satisfied if the sign test is to be valid; this means that the sample consists of differences which are independent. In the example above, the patients must not influence each other.

Example 2   To study the temperature pattern in a microwave oven, Carol Dahl heated 10 slices of meat loaf and measured the temperature on the left side, in the center, and on the right side of the oven:

Slice	1	2	3	4	5	6	7	8	9	10	
Left	185	188	175	196	200	200	180	195	200	197	
Center	165	165	182	165	165	165	184	180	185	184	°F.
Right	165	192	180	200	176	200	190	195	195	190	

(a) Is the oven a different temperature on the left than in the center? ($\alpha = .05$.)

(b) Is the oven hotter on the right than in the center? ($\alpha = .05$.)

(a) H_0: $P = .5$, H_1: $P \neq .5$. $n = 10$, $X = 2$. $\Pr(X \leq 2 \text{ or } X \geq 8) = 2(.001 + .010 + .044) = .110, > .05$. The null hypothesis is not rejected; there is not enough evidence to say that slices of meat loaf heated in the oven are a different temperature when heated on the left side or in the center.

(b) H_0: $P = .5$, H_1: $P > .5$, $n = 9$, $X = 8$ (since the right side is hotter for 8 slices, cooler for 1). $\Pr(X \geq 8) = .018 + .002 = .020$. The null hypothesis is rejected.

Remember that the normal distribution can be used as an approximation to the binomial distribution if n is sufficiently large, but that a correction of .5 for continuity must be used (see page 150). Recall also that the mean of a binomial distribution with $P = 1/2$ is $nP = n/2$ and the standard deviation is $\sqrt{nPQ} = \sqrt{n}/2$. So to test H_0: $P = 1/2$ for large n, you will compute $z = \dfrac{X \pm .5 - n/2}{\sqrt{n}/2}$; the .5 term in the numerator is the correction for continuity, and the minus sign is chosen if $X > n/2$, the plus sign if $X < n/2$. (This choice of the sign reflects conservatism, since H_0 is less likely to be rejected with this choice.)

Example 3   In a test of 60 drivers, it is found that 32 have a slower reaction time, 10 show no difference, and 18 have faster reactions after drinking 1 ounce of alcohol.

H_0: $P = .5$.

H_1: $P > .5$ (this means P is the proportion of drivers with a slower reaction time).

One-tail test, $\alpha = .05$, so $z = 1.64$.

$$z = \frac{X \pm 1/2 - n/2}{\sqrt{n}/2}$$

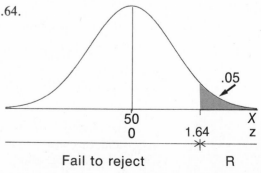

.05

| 50 | | X |
| 0 | 1.64 | Z |

Fail to reject   R

$X = 32 =$ number of drivers with slower reaction time

$n = 32 + 18 = 50$ (Note that the drivers with the same reaction time are not included; $n \neq 60$.)

$$z = \frac{32 - 1/2 - 50/2}{\sqrt{50/2}} = \frac{6.50}{3.54} = 1.84$$

H_0 is rejected; at a 5% level of significance, drivers have a slower reaction time after drinking 1 ounce of alcohol.

In using the sign test, you do not use any information you have about actual scores (and therefore about the size of the differences between scores). It is not surprising, therefore, that a t test for differences of matched pairs (see page 255) is preferred if the data are metric and if the required assumption that the differences are normally distributed is met. This is always the case: if data are metric, use of the actual data (rather than simply the rankings) will give a more reliable result **if the assumptions about the population distribution that are required for use of the test are met.**

16.3 THE MANN-WHITNEY TEST

This test (also called the Wilcoxon test or the rank-sum test) is used to test whether two independent random samples are taken from identical populations. It is assumed that the populations have identical shape and variability; do they also have the same median?

It must be possible to lump together the scores in the two samples and rank them. A test statistic U is computed, based on a comparison of the sums of ranks of the separate samples in the pooled set. Table 10, Appendix C, will be used for small samples (each less than 20), and a z test for large samples.

Example 1

Five runners from Brooklyn completed 300 yards in 33.2, 33.4, 33.7, 34.2, and 34.3 seconds, repectively; the times for five runners from Queens were 33.5, 33.6, 33.9, 34.2, and 34.5 seconds. Do the two samples constitute evidence that runners from one county tend to be faster at the 300-yard dash than runners from the other? ($\alpha = .05$.)

Even though both sample sizes equal 5, this is not a paired-samples problem as in the last section. Rather, each runner from Brooklyn is thought of as competing with each of the five runners from Queens, for a total of $5(5) = 25$ contests.

33.2 seconds (Brooklyn) is faster than each of the five times recorded for Queens; score 5 for U_X. The same is true for 33.4 seconds (Brooklyn); score 5 again. 33.7 seconds is shorter than 3 of the Queens times (longer than 2 of them); score 3. 34.2 seconds is shorter than one Queens time, and tied with one; score 1 for the win and 1/2 for the tie. And the last Brooklyn time, 34.3, scores 1 (ahead of 34.5 for Queens). In sum, $U_X = 5 + 5 + 3 + 1\frac{1}{2} + 1 = 15\frac{1}{2}$, out of 25 contests.

If you tally the wins for Queens runners, you get $U_Y = 3 + 3 + 2 + 1\frac{1}{2} + 0 = 9\frac{1}{2}$.

$U_X + U_Y = 15\frac{1}{2} + 9\frac{1}{2} = 25 = n_X(n_Y)$, where n_X and n_Y are the sizes of the two samples.

H_0 states the population medians are equal, and any difference between the samples $\left(15\frac{1}{2} \text{ wins and } 9\frac{1}{2} \text{ wins}\right)$ is due to chance variation. To test this at the 5% significance level, two-tail, use Table 9 in Appendix C. In the row for "larger sample size" = 5 and column for "smaller sample size" = 5, also, you find a critical value for U of 2. That's a lower critical value for the smaller U. That is, if one of the U's is 2 or smaller, then you can reject H_0. Your smaller rank sum, $9\frac{1}{2}$, isn't less than or equal to 2. H_0 is not rejected. No conclusion can be drawn about the populations of Brooklyn and Queens runners.

One thing is questionable about this example: the Mann-Whitney probabilities are computed on the assumption of two independent, random samples. Presumably the contestants in these races were selected to be the fastest runners in the population. So don't take this example too seriously, though it does illustrate the calculations for a Mann-Whitney test.

Note that the symbols X and Y are used differently here than in regression (the last two chapters) and rank correlation (the next topic). Here they denote values of one variable (running time in Example 1) in two different samples or populations; in correlation or regression they refer to values of two different attributes or variables pertaining to individuals in one sample or population. Switching back and forth will keep you on your toes, figuring out what you are looking at in any one problem.

If you are sure of your arithmetic, you may shorten the computation of U by finding the sum of ranks for only one of the samples, and then the check formula suggested above is used to find the sum of ranks for the other sample. ($U_Y = n_X n_Y - U_X$, for example.)

What if the sample sizes are larger than those given in Table 9, Appendix C? The sampling distribution of U approximates a normal distribution for samples of size about 20 or larger (and the approximation gets better as sample sizes increase), with mean

$$\mu_U = \frac{n_X n_Y}{2}$$

and with standard error

$$\sigma_U = \sqrt{\frac{n_X n_Y (n_X + n_Y + 1)}{12}}$$

Example 2   Two random samples, chosen independently, are ranked as follows:

X: 1, 2, 4, 5, 10, 11, 12, 14, 15, 16, 20, 21, 24, 27, 28, 29, 34, 35, 36
Y: 3, 6, 7, 8, 9, 13, 17, 18, 19, 22, 23, 25, 26, 30, 31, 32, 33, 37, 38, 39, 40, 41, 42, 43

Test, at a .05 level of significance, whether these are random samples from the same population.

H_0: the two samples come from the same population; H_1: they don't.

$n_X = 19$, $n_Y = 24$, $U_X = 302$, $U_Y = 154$. Check: $302 + 154 = 19(24)$.

$$\mu_U = \frac{19(24)}{2} = 228, \quad \sigma_U = \sqrt{\frac{19(24)(44)}{12}} = 40.9.$$

The z score for U_X is $z = \dfrac{U_X - \mu_U}{\sigma_U} = \dfrac{302 - 228}{40.9} = 1.81.$

The z score for U_Y is $\dfrac{154 - 228}{40.9} = -1.81$. But this is the negative of the z score for U_X! This will always be the case, so it's only necessary to compute one of them.

With $\alpha = .05$, using a two-tail test, the critical values of z are ± 1.96. H_0 cannot be rejected; there is no evidence that these samples come from different populations. (But if you hope that they do come from the same population, then you should take a large α in order to decrease β.)

Ranked data are used in a Mann-Whitney test, but the source is usually known metric data and means and standard deviations can be computed. Perhaps a t test of difference of means could be used (see Section 11.8). What are the relative advantages of t and Mann-Whitney tests? For both tests, it is assumed that the two sampled populations have the same shape and dispersion; for the t test, it is also necessary to assume that the shape is that of the normal curve. If there is any cause to doubt this (and the doubt is usually greatest for small samples), then the Mann-Whitney test should be used, since both test whether the measures of center of the two populations are the same. For a given α, the probability of rejecting H_0 when it is false (this is called the **power** of the test, and is measured by $1 - \beta$) is slightly higher with a t test. Sample size must be increased slightly if the same power is to be achieved using a Mann-Whitney test.

16.4 EXERCISES

1. A foolhardy* agricultural chemist who wants to compare Kepone and Myrex as insecticides for fire ants goes to a large pasture in which there are many nests and randomly chooses 36, which are then paired by size. He puts the same dosage of corn treated with Kepone or Myrex on one of each pair. A week later he returns to the pasture and notes that, in 8 pairs, the nest treated with Myrex has fewer ants; in 6 pairs, all fire ants have died; and in 4 pairs, Kepone has done better. Use the sign test to see whether the two chemicals differ in their effects on fire ants (.10 level of significance).

---

* Fire ants sting, and can kill a horse; both Kepone and Myrex are not biodegradable and have been banned.

2. Chuck Porter (Department of Entomology, University of Wisconsin, 1977) collected the following data on the monthly rainfall in cm. at Providencia, Amazon Jungle, Colombia, in 1970 and 1971 while working on his dissertation:

	Jan.	Feb.	Mar.	Apr.	May	June	July	Aug.	Sept.	Oct.	Nov.	Dec.
1970	16.3	11.2	21.1	48.3	45.2	70.1	48.3	82.6	57.9	37.3	38.4	36.8
1971	36.3	27.2	50.5	24.6	53.3	43.2	53.8	57.4	69.6	60.4	54.6	11.2

(a) Test whether 1971 was consistently wetter than 1970, to an extent greater than could be readily explained by chance or random variation. ($\alpha = .05$.)

(b) What is the computed probability if you are questioning whether either year is significantly wetter than the other?

3. Connie Smoszyk, working on her Master's thesis at the University of Wisconsin in 1976, gave a talking-it-out treatment to help relieve the anxiety of patients who had learned they had cancer, with the following results:

	Patient	a	b	c	d	e	f	g	h	i	j
Anxiety	Pre treatment	23	48	69	58	48	44	57	20	26	34
state	Post treatment	20	26	67	57	33	24	44	21	26	27

(a) Use the sign test to find out if the treatment decreased anxiety, at a .05 level of significance.

(b) Compare with the results of a t test. (See Exercise 1, page 257.)

4. Tony Rotatori experimented with a treatment to help obese, retarded adults to lose weight. The loss in weight during treatment was -10, -10, -10, $+1$, -6, -5, -4, $+1$, -4, $+3$ pounds, respectively, for 10 men.

(a) Test the null hypothesis of no effect by the sign test, at the 1% significance level.

(b) Compare your conclusion with what you got using a t test (Exercise 14, page 248.

(c) If you applied the sign test to the *percent* change in weight, how do you think this would affect your result?

5. Rotatori found the changes for a control group not receiving the treatment were (in pounds) $+5$, -3, -6, 0, $+4$, $+9$, $+2$, $+8$, respectively. Compare this sample with the changes for treated patients listed in Exercise 4, using Mann-Whitney, at a .05 level of significance.

6. Some children in Tucson were taught reading with the Kooda phonetics method and some with the MacLeod recognition method. The results of tests at the end of the first grade were as follows:

Kooda: 70, 95, 92, 74, 97, 73, 87, 88, 78, 85, 90, 76, 80, 82
MacLeod: 78, 94, 65, 74, 88, 90, 68, 80, 82, 72, 84, 85, 74, 85

Decide whether there is a difference in the results of the two methods ($\alpha = .05$), using a Mann-Whitney test.

7. A random sample of married men is chosen, and the men are asked whether they are satisfied or discontented with their marriages; then each man is asked how long he has been married. The results of this survey are as follows:

	Number of years married
Satisfied	1, 3, 3, 2, 7, 10
Not satisfied	16, 4, 9, 20, 16, 7, 4, 6, 18

At a .05 level of significance, is there a difference in the two groups?

8. Six students are chosen at random from those studying both Economics 101 and Psychology 24, the former with Professor A and the latter with Professor B. The students rate the professors as follows on teaching ability, on a 1–6 scale in which 6 is "excellent":

	Student					
	A	B	C	D	E	F
Professor A	5	4	6	4	4	5
Professor B	4	3	5	2	4	3

There is no reason to assume that ratings given by all students are normally distributed. Is there a significant difference in the ratings of the two professors? ($\alpha = .05$.)

9. Fifty American scientists are chosen at random, and for each the number of published scientific articles is counted. Of the 50, 27 are the oldest child in the family. Based on the following rankings, is there a difference in the number of publications of first-born and other children who become scientists? ($\alpha = .01$.)

	Rank (based on number of publications)
Oldest child	2, 2, 2, 4, 7, 7, 7, 11, 11, 11, 16, 16, 16, 16, 16, 22, 22, 22, 28, 31, 33, 33, 33, 38, 38, 38, 44
Not oldest child	5, 9, 13, 19, 20, 25, 25, 25, 27, 29.5, 29.5, 35, 36, 41, 41, 41, 43, 47.5, 47.5, 47.5, 47.5, 47.5, 47.5

10. Goodland Tire Company believes that a chemical additive to the rubber used in its tires will improve wear. Random samples, each of 25 tires, are tested under the same conditions with the following results:

	Wear (in thousandths of an inch)
Without additive	23, 14, 12, 11, 13, 18, 15, 16, 13, 10, 15, 20, 14, 9, 7, 20, 14, 18, 24, 19, 13, 16, 15, 20, 15
With additive	14, 19, 8, 10, 16, 12, 11, 9, 18, 12, 7, 6, 8, 11, 17, 13, 13, 15, 10, 8, 14, 7, 5, 13, 14

(a) Use a t test to see whether there is significantly less wear on tires made with rubber to which the chemical has been added ($\alpha = .05$). Assume that variances are the same for wear measurements on both types of tires, and that the measurements are normally distributed for both. $\bar{X} = 15.4$, $s_X^2 = 17.4$, $\bar{Y} = 11.6$, $s_Y^2 = 14.5$.

(b) Use a Mann-Whitney test to decide whether the chemical additive has improved the wear of tires, at a .05 level of significance.

ANSWERS

1. P = proportion in which Myrex does better. H_0: $P = .5$, H_1: $P > .5$, $n = 12$. (Ignore the 6 in which both were effective.) $X = 8$. Pr(8 or more in 12 trials with $P = .5$) = $.121 + .054 + .016 + .003 = .154$; this is greater than $.10$, so H_0 cannot be rejected; there is no difference.

2. (a) H_0: No difference except chance variation; $P = \frac{1}{2}$. H_1: $P > \frac{1}{2}$. 1971 rainfall is greater in 8 months. $\text{Pr}\left(X > 8 \text{ if } n = 12 \text{ and } P = \frac{1}{2}\right) = .121 + .054 + .016 + .003 + 0 = .194$, $> .05$. There is no difference.
   (b) $2(.194) = .39$

3. H_0: $P = \frac{1}{2}$, H_1: $P > \frac{1}{2}$ (P is the proportion of patients whose anxiety is lessened.). 8 less, 1 more, 1 no change, $n = 9$. $\text{Pr}\left(X = 8 \text{ or more in 9 trials if } P = \frac{1}{2}\right) = .018 + .002 = .02$, $< .05$, so H_0 is rejected; anxiety has been lessened by the treatment.

4. (a) H_0: $\text{Pr}(+) = \text{Pr}(-) = \frac{1}{2}$. H_1: $\text{Pr}(+) < \frac{1}{2}$. With 3 plus signs and $n = 10$, the significance probability = $.001 + .010 + .044 + .117 > .05$. H_0 cannot be rejected; there is not enough evidence to say that the treatment has weight-reducing value.
   (b) H_0; $\mu_d = 0$, H_1: $\mu_d < 0$ was not rejected using a t test; the results of the t and sign tests are the same.
   (c) The result would be the same, since the change in pounds and the percent change would have the same sign for each man.

5. X: $-10, -10, -10, -6, -5, -4, -4, 1, 1, 3$.
   Y: $-6, -3, 0, 2, 4, 5, 8, 9$.

   $$U_X = 0 + 0 + 0 + \frac{1}{2} + 1 + 2(1) + 2(3) + 4 = 13\frac{1}{2}$$

   $$U_Y = 3\frac{1}{2} + 7 + 7 + 9 + 10 + 10 + 10 + 10 = 66\frac{1}{2}$$

   $$U_X + U_Y = 80 = 8(10)$$

   Since $13\frac{1}{2}$ is less than 17, H_0 is rejected; there is a significant difference between the samples.

6. H_0: the population medians are the same for Kooda (X) and MacLeod (Y); H_1: they are different. $n_X = n_Y$, $U_Y = 4.5 + 12 + 0 + 2.5 + 9.5 + 10.5 + 0 + 5.5 + 6.5 + 1 + 7 + 7.5 + 2.5 + 7.5 = 76.5$, $U_X = 14^2 - 76.5 = 119.5$. Critical $U = 55$, so H_0 is accepted; Kooda and MacLeod tests have the same median.

7. $n_X = 6$, $n_Y = 9$, $U_Y = 8.5$, $U_X = 6(9) - 8.5 = 45.5$. The critical value of U is 10, > 8.5, so H_0 is rejected; the two groups differ.

8. $n_X = n_Y = 6$, $U_Y = 7$, $U_X = 6(6) - 7 = 29$. The critical value of U is 5, which is less than 7 and 29, so we fail to reject H_0.

9. $n_X = 27$, $n_Y = 23$, $U_Y = 23 + 20 + 17 + 2(12) + 3(9) + 9 + 2(8) + 2(4) + 4(1) + 6(0) = 148$; $\mu_U = \dfrac{27(23)}{2} = 310.5$, $\sigma_U = \sqrt{\dfrac{(27)(23)(51)}{12}} = 51.4$, $z = \dfrac{148 - 310.5}{51.4} = -3.16$. Critical $z = \pm 2.33$. H_0 is rejected; there is a difference.

10. (a) H_0: $\mu_X - \mu_Y = 0$, H_1: $\mu_X - \mu_Y > 0$ (X is without additive.) $\sigma_{(\bar{X}-\bar{Y})} = \sqrt{\dfrac{24(17.4) + 24(14.5)}{48}\left(\dfrac{2}{25}\right)} = 1.13$, $t = \dfrac{15.4 - 11.6}{1.13} = 3.36$. $D = 48$, critical $t(=z) = 1.64$; H_0 is rejected.

(b)

Score:		5	6	7	8	9	10	11	12	13	14	15	16	17	18	19	20	23	24
Frequency:	X	0	0	1	0	1	1	1	1	3	3	4	2	0	2	1	3	1	1
	Y	1	1	2	3	1	2	2	2	3	3	1	1	1	1	1	0	0	0

$U_X = 158$, $\mu_U = \dfrac{25(25)}{2} = 312.5$, $\sigma_U = \sqrt{\dfrac{25(25)(51)}{12}} = 51.5$, $z = \dfrac{158 - 312.5}{51.5} = -3.00$; the critical value of z is -1.64, so H_0 is rejected; the additive has improved tire wear.

16.5 SPEARMAN RANK CORRELATION COEFFICIENT

This coefficient, r_S, measures the correlation between two paired samples of ranked data. For example, two judges may each rank a random sample of n students according to their ability in writing. How well do the judges agree? Or the abilities of one individual in writing, accounting, mechanical skills, and so on may be ranked at the beginning and end of a training course. Does the individual still do best at writing, worst at accounting? As with the Pearson r, if the rankings agree (each member of the sample is ranked the same by both judges, or has the same rank in attributes at different times) we expect a high positive correlation; $r_S = +1$. If one judge thinks worst of the composition which the other judge rates best, and vice versa, then there is a high negative correlation.

The Spearman rank correlation coefficient is the Pearson correlation r applied to the **ranks** in two paired samples (not to the original scores). The customary formula for r could, of course, be used, but an even simpler computation for r_S may be substituted:

> 1. List the n pairs of ranks: X, Y.
> 2. Find the differences d between the two sets of ranks.
> 3. Square these differences and add the squares ($= \Sigma d^2$).
> 4. Compute r_S:
>
> $$r_S = 1 - \frac{6\Sigma d^2}{n(n^2 - 1)}$$

The Spearman rank correlation coefficient is useful (1) when only ranks are known or (2) when metric scores are known but it's easier to determine r_S than r. But there is one great drawback to use of the Spearman coefficient: even if r_S is close to 1 (or -1), we then know that as X increases, Y increases (or decreases), but we do not know by how much. If the Pearson r is close to $1\,(-1)$, on the other hand, we know there is a linear relationship between X and Y. Whenever X increases by a certain amount, Y increases (or decreases) by a constant amount; the regression line will fit the points (X, Y) almost exactly.

Example 1   Six paintings were ranked as follows by two judges; compute r_S: (a) using the four steps above, (b) using the formula for the Pearson r.

Painting	X First judge	Y Second judge	(a) d	d^2	(b) XY	X^2	Y^2
A	2	2	0	0	4	4	4
B	1	3	−2	4	3	1	9
C	4	4	0	0	16	16	16
D	5	6	−1	1	30	25	36
E	6	5	1	1	30	36	25
F	3	1	2	4	3	9	1
	(21)	(21)		10	86	91	91

(a) $\Sigma d^2 = 10$, $n = 6$.

$$r_S = 1 - \frac{6(10)}{6(6^2 - 1)} = .71$$

(b) $r = \dfrac{6(86) - 21(21)}{\sqrt{6(91) - 21^2}\sqrt{6(91) - 21^2}} = \dfrac{75}{105} = .71 = r_S.$

What if there are ties in either the X or the Y list? In this case, replace the equal ranks by the mean of the ranks for which they stand. Thus, if a sample of 5 shows 2 scores tied for rank 3 (that is, the ranked scores appear as 1, 2, 3, 3, 5), then both of the 3's are replaced by 3.5, the mean of 3 and 4.

Example 2   Compute r_S for the following rankings of 6 individuals on 2 tests:

Individual	A	B	C	D	E	F
Test X	1	2	2	4	5	6
Test Y	3	2	1	4	6	5

There is a tie for second place among the X rankings, so each of those 2's is replaced by 2.5.

X	Y	d	d^2
1	3	-2	4
2.5	2	.5	.25
2.5	1	1.5	2.25
4	4	0	0
5	6	-1	1.
6	5	1	1.
			8.5

$n = 6$

$$r_S = 1 - \frac{6(8.5)}{6(6^2 - 1)} = 1 - .24 = .76$$

Example 3   Eight factory workers chosen at random in the Beacon Light Company plant are ranked as follows on tests of finger dexterity and mechanical ability. Is there a correlation between them?

Worker	Finger dexterity	Mechanical ability	X	Y	d	d^2
A	1	2	1	3	-2	4
B	2	1	2	1	1	1
C	3	8	4	8	-4	16
D	3	2	4	3	1	1
E	3	2	4	3	1	1
F	6	6	6	6.5	$-$ .5	.25
G	7	6	7	6.5	.5	.25
H	8	5	8	5	3	9
						32.50

$n = 8$

The X column is the Finger Dexterity column adjusted for the triple tie for third place; the Y column is the Mechanical Ability column adjusted for ties for second and sixth places.

$$r_S = 1 - \frac{6(32.50)}{8(8^2 - 1)} = 1 - .39 = .61$$

What about testing hypotheses about the Spearman rank correlation coefficient for the population, ρ_S? If the null hypothesis is that $\rho_S = 0$, a t test can be made of the same type as when testing H_0: $\rho = 0$:

$$t = \frac{r_S \sqrt{n - 2}}{\sqrt{1 - r_S^2}}$$

with $n - 2$ degrees of freedom. However, n should be 10 or larger for this test.

Example 4 Eighteen workers chosen at random in the ABC factory show a Spearman rank correlation of .60 between finger dexterity and mechanical ability. For all workers in this factory, is there a significant correlation? (.05 level of significance.)

H_0: $\rho_S = 0$, H_1: $\rho_S \neq 0$; $\alpha = .05$.

$$r_S = .60, \ n = 18, \ \text{so} \ t = \frac{.6\sqrt{18 - 2}}{\sqrt{1 - .6^2}} = 3.00$$

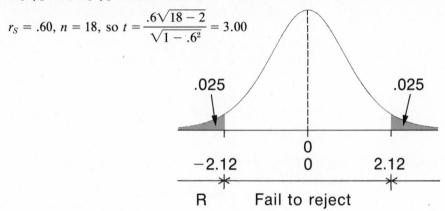

For a two-tailed test with 16 degrees of freedom and $\alpha = .05$, $t = \pm 2.12$. H_0 is rejected: There is a significant correlation.

16.6 EXERCISES

1. Ten dressmakers chosen at random in the Princess Dress Company are ranked as follows for craftsmanship and speed. Find the correlation between these two rankings.

	A	B	C	D	E	F	G	H	I	J
Craftsmanship	10	2	4	5	9	5	1	3	7	8
Speed	10	5	2	4	8	6	2	1	8	7

2. For the following paired scores, (a) determine r, (b) determine r_S, and (c) draw a scatter plot, and then (d) explain the difference between r and r_S.

X: 　3　5　0　2　4　1
Y: 　3　25　0　2　16　1

3. Giovanni Chianti's seven movies were ranked by the film department students of Henry Fisk College on character development and on suspenseful ending. Find the correlation between their rankings.

	Movie						
	A	B	C	D	E	F	G
Character development	1	2	2	4	5	6	6
Suspenseful ending	3	4	2	1	7	5	6

4. Facult~ ~chools were ~

arises bet~

meth~

~n do you rely on ~

~ticky issues, (b) smooth~

cing the issue by power play, ~

~hen the same faculty members were ~

s ~ ~etween faculty and administration. Scores ~

ag ~ ~nked from 1 for the most used to 6 for the least use~

me~ ~ at two of the medical schools:

	~flict~ ~etween ~ ~artments						Conflicts between faculty and administration					
Met~	~	~	c	d	e	f	a	b	c	d	e	f
Schoo	~	2	3	6	1	5	6	2	4	5	1	3
School	3	4	2	5	1	6	4	2	5	6	1	3

(a) F~ ~ School 1, find the correlation between the order of preference in interdepartment and in faculty-administration conflicts.

(b) Find the correlation of orders of preference in interdepartmental conflicts, between faculty at School 1 and faculty at School 2.

5. Fifty-one factory workers show a Spearman rank correlation of .50 between absenteeism and difficulty of family problems. For all workers, is there any significant correlation? (.05 level of significance.)

6. Here are data on trends in census reduction, monthly admissions, and monthly "seen but not admitted" at Worcester State Hospital for 1969–1977:**

Fiscal Year*	Average daily census	Average monthly admissions†	Average monthly "seen but not admitted"	Ratio admissions to "seen but not admitted"
1969	993	152	‡	‡
1970	932	125	35	3.9:1
1971	738	118	39	3.0:1
1972	708	110	37	3.0:1
1973	579	94	47	2.0:1
1974	587	95	38	2.5:1
1975	480	99	52	1.9:1
1976	452	104	57	1.8:1
1977	462	125	55	2.3:1

* Year ending June 30.
† Figures are based on total admissions. Due to the way hospital records are maintained, the numbers of voluntary as opposed to involuntary admissions during this time period could not be differentiated. This results in a conservative estimate of the actual nonadmission rate for voluntary applicants.
‡ Data unavailable.

* Weisbord, M. R., Lawrence, P. R., and Charns, M. P.: Three Dilemmas of Academical Medical Centers," *Journal of Applied Behavioral Science* 14 (1978), 284–304.
** Morissey, Tessler and Farrin, "Being Seen But Not Admitted. A Note on Some Neglected Aspects of State Hospital Deinstitutionalization," *Am Journal of Orthopsychiatry 49*, 1979, pp. 153–156.

(a) Calculate the rank correlation coefficient of each of the other variables with year.

(b) In each case, test the null hypothesis of no trend, at the 5% level of significance.

(c) Compare the rank correlation coefficients with the Pearson correlation coefficients computed in Exercise 8, page 352.

7. Attributes of electric blenders were ranked by 3 different methods,* with the following results:

Attribute	Method 1	Method 2	Method 3
Number of speeds	4	2	6
Safety of operation	1	3	1
Quietness of operation	3	7	4
Wattage	7	6	9
Brand name	10	4	8
Container material	9	10	7
Ice crushing option	5	9	10
Warranty	2	5	3
Price	6	1	2
Ease of cleaning	8	8	5

(a) Find the rank correlation between each pair of methods (3 tests).

(b) Test whether or not $\rho_S = 0$ for each of the 3 pairs of methods at a .05 level of significance.

8. Test H_0: $\rho_S = 0$, H_1: $\rho_S \neq 0$ for (i) $n = 10$, (ii) $n = 50$, (iii) $n = 100$ if (a) $r_s = .20$, (b) $r_s = .40$, (c) $r_s = .60$.

ANSWERS

1. d: 0, -3, 2, 1, 1, -1, -1, 2, -1, 1; $\Sigma d^2 = 23$, $r_S = \dfrac{6(23)}{10(99)} = .86$.

2. (a) $r = \dfrac{6(203) - 15(47)}{\sqrt{6(55) - 15^2}\sqrt{6(895) - 47^2}} = .89$.

(b) $d = 0$ for all pairs, so $r_S = 1$.

(c)

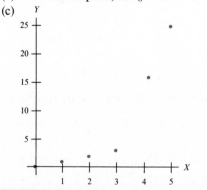

*Heeler, R., Okechuku, C., and Reid, S.: "Attribute Importance: Contrasting Measurements," *Journal of Marketing Research* 16 (1979) 60–63.

(d) $r_S = 1$ since Y increases whenever X increases. The relation is not linear, however, so $r \neq 1$.

3. $r_S = 1 - \dfrac{6(22)}{7(48)} = .61.$

4. (a) $\Sigma d^2 = 2^2 + 0^2 + 1^2 + 1^2 + 0^2 + 2^2 = 10, r_S = 1 - \dfrac{6(10)}{6(35)} = .71.$

    (b) $\Sigma d^2 = 8, r_S = 1 - \dfrac{6(8)}{6(35)} = .77.$

5. $H_0: \rho_S = 0, H_1: \rho_S \neq 0, t = \dfrac{.50\sqrt{49}}{\sqrt{1 - .50^2}} = 4.04; D = 4$, critical values of t are ± 1.96. Reject H_0; there is a significant correlation.

6. (a) (Remember to rank both variables from lowest to highest.) (i) $\Sigma d^2 = 8^2 + 6^2 + 4^2 + 2^2 + (-1)^2 + (-1)^2 + (-4)^2 + (-7)^2 + (-7)^2 = 236, r_S = 1 - \dfrac{6(236)}{9(80)} = -.97$; (ii) d:

    8, 5.5, 3, 1, -4, -4, -4, -4, $-1.5, r_S = 1 - \dfrac{6(170.5)}{9(80)} = -.42$; (iii) Renumber the years.

    $r_S = 1 - \dfrac{6(12)}{8(63)} = .86$; (iv) $r_S = 1 - \dfrac{6(150.5)}{8(63)} = -.79.$

    (b) (i) $t = \dfrac{-.97\sqrt{7}}{\sqrt{1 - (-.97)^2}} = -10.6$, critical $t = \pm 2.37$; reject H_0. (ii) $t = \dfrac{-.42\sqrt{7}}{\sqrt{1 - (-.42)^2}} = -1.22$, critical t again $= \pm 2.37$; fail to reject H_0; (iii) $t = \dfrac{.86\sqrt{6}}{\sqrt{1 - .86^2}} = 4.13; D = 6$, critical $t = \pm 2.45$; reject H_0. (iv) $t = \dfrac{-.79\sqrt{6}}{\sqrt{1 - (-.79)^2}} = -3.16$; critical $t = \pm 2.45$; reject H_0.

    (c)

	r_S	r
(i)	$-.97$	$-.96$
(ii)	$-.42$	$-.55$
(iii)	$.86$	$.88$
(iv)	$-.79$	$-.82$

7.

Methods 1, 2	Methods 1, 3	Methods 2, 3
(a) $r_S = 1 - \dfrac{6(112)}{10(99)} = .32$	$1 - \dfrac{6(68)}{10(99)} = .59$	$1 - \dfrac{6(78)}{10(99)} = .53$
(b) $t = \dfrac{.32\sqrt{8}}{\sqrt{1 - .32^2}} = .96$	$\dfrac{.59\sqrt{8}}{\sqrt{1 - .59^2}} = 2.07$	$\dfrac{.53\sqrt{8}}{\sqrt{1 - .53^2}} = 1.76$

    $D = 8$, 2-tail test, critical $t = 2.31$, so H_0 cannot be rejected for any pair.

8.

    | | Critical | (a) $r_s = .20$ | | (b) $r_s = .40$ | | (c) $r_s = .60$ | |
n	t	t	Conclusion	t	Conclusion	t	Conclusion
10	± 2.31	.58	Fail to reject	1.23	Fail to reject	2.12	Fail to reject
50	± 1.96	1.41	Fail to reject	3.02	Reject	5.20	Reject
100	± 1.96	2.02	Reject	4.32	Reject	7.42	Reject

16.7 VOCABULARY AND SYMBOLS

nonparametric test	Wilcoxon test
sign test	rank-sum test
Mann-Whitney test	Spearman rank correlation coefficient
U_X, U_Y, μ_U, σ_U	r_S, d, ρ_S

16.8 REVIEW EXERCISES

1. Eleven white children are chosen at random from seventh graders at Robbins Junior High School and tested on their prejudice towards blacks. They are retested after a series of eight very open discussions with some black students at the school. Six show less prejudice, 1 shows more, and 4 show no difference. Decide whether you have significant evidence that this type of discussion will lessen prejudice of all white seventh graders at the school. ($\alpha = .10$.)

2. Two mountain climbers had an argument about whether narrow or wide pitons were more popular, and enlisted the help of several climbing clubs in an experiment. Fifteen ropes (2 climbers each) climbed Class V pitches in Franconia Notch twice, once with narrow pitons and once with wide pitons. Of the 30 climbers, 20 preferred the narrow and 10 the wide pitons. At a .05 level of significance, is there a difference in preference?

3. Seventy-seven salesmen for the Coolidge Vacuum Cleaner Company are given a week's special training course. Forty of them show substantially better sales the following month, 28 show no change, and 9 don't sell as much as in the month before the course. At a .02 level of significance, have sales increased after the training course?

4. Twelve children in the same class at school are ranked for popularity (a) by group preference and (b) by rating on a social attitudes test. The results are as follows:

	Child											
	A	B	C	D	E	F	G	H	I	J	K	L
Group rating	6	2	7	3	4	11	1	5	10	8	9	12
Test rating	7	1	8	4	6	3	2	5	10	9	11	12

Use a sign test on differences in rank to see whether the two methods of ranking differ at a .05 level of significance.

5. Thirty-five children in a class are ranked for popularity (a) by group preference and (b) by score on a social attitudes test. Nineteen have higher rankings by group preference, 11 have lower, and 5 have the same ranking. Use a sign test on differences of rank to decide whether rankings by the two methods differ significantly. ($\alpha = .05$.)

6. The open discussions with black students discussed in Exercise 1 are carried out with 261 white children at Robbins Junior High School. In retesting after the discussions, 113 white children show less prejudice, 83 show more, and 65 show no change. Do you have significant evidence that this type of discussion will lessen prejudice of all white children at the school? ($\alpha = .05$.)

7. Does the drug CDX reduce anxiety? Fifteen patients were tested, and anxiety was measured by two different scales, the Hamilton Anxiety Rating Scale (HARS) and the Hopkins Symptoms Check List (HSCL). Differences in anxiety were as follows (a minus sign shows a decrease in anxiety):

Subject	1	2	3	4	5	6	7	8	9	10	11	12	13	14	15
HARS	−5	−2	−7	−1	−5	−5	3	3	−12	−12	−20	−7	−4	−18	−11
HSCL	−6	−9	−10	−7	0	0	9	0	−1	−4	−9	−9	−6	−9	−1

(a) Test the null hypothesis that CDX has no effect on a patient's anxiety level as measured by the HARS scale.

(b) Calculate the rank correlation coefficient between the change in HARS and the change in HSCL. Compare it with the correlation r calculated in Exercise 14, page 354.

(c) At a .05 level of significance, does the computed rank correlation result from just a chance effect?

8. Six courses, chosen at random, offered at Blackwood University were evaluated on a scale from 1 to 6, with 6 being "excellent," by the students in the courses for appropriateness of work expected for the number of credits given and for reasonableness of cost of textbooks. The mean scores of the six courses were as follows:

	A	B	C	D	E	F
Appropriateness of work expected	3.1	2.4	5.0	3.0	1.8	4.2
Reasonableness of textbook cost	4.0	5.7	5.4	3.4	1.2	5.0

(a) Use a t test to check, at a .05 level of significance, the hypothesis that all courses in the University have the same mean score on the two questions rated above.

(b) Use a Mann-Whitney test to see whether all courses in the University have the same distribution on these two questions on the evaluation form. ($\alpha = .05$.)

9. Carol Dahl in a cook-chill-and-reheat experiment measured the percent weight loss of 10 slices each of meat loaf (i) cooked initially to 45°C and (ii) not cooked before chilling:

Uncooked	19.1	24.3	21.8	19.1	19.1	17.4	21.8	26.1	27.8	17.4
45°C	36	39	38	38	38	36	38	40	38	38

(a) Calculate the rank correlation, test for significance at a .05 level, and interpret your finding.

(b) Why does it make no sense to do a Mann-Whitney test on the difference between meat cooked to 45°C and meat not cooked before chilling?

10. Refer to Exercise 8.

(a) For all courses in the University, is there a significant correlation between ratings on the two questions? ($\alpha = .05$; determine ρ_S.)

(b) Suppose that 100 courses, instead of 6, were evaluated, but that the computed Spearman correlation coefficient was unchanged. Is there a significant correlation between ratings on the two questions? ($\alpha = .05$.)

ANSWERS

1. Probability that 6 or more of 7 students will show less prejudice if $P = .5$ is $.055 + .008 = .063$, $< .10$ so H_0 is rejected. At a .10 level of significance, the series of discussions will cause a change in the attitudes of white seventh graders in the school. (But the binomial assumptions assume independence, and the 10 white children may influence each other.)

3. H_0: $P = .5$, H_1: $P > .5$, $n = 49$, $\mu = 24.5$, $\sigma = \sqrt{49(.5)(.5)} = 3.5$, $z = \dfrac{40 - .5 - 24.5}{3.5} =$ 4.29. $\alpha = .02$, so the critical value of z is 2.05. Reject H_0; sales have increased.

5. $n = 30$, $\mu = 15$, $\sigma = 2.7$, $z = \dfrac{19 - .5 - 15}{2.7} = 1.30$. $\alpha = .05$, the critical value of z is 1.64.

H_0 is not rejected; no difference in the two methods of ranking has been shown. But if you hope to show that the two methods have similar results, then you want a small β. How does that affect your choice of α?

7. (a) H_0: $P = .5$, H_1: $P < .5$. Pr(2 or less increases in 15 trials) $= .003$, $< .05$, so H_0 is rejected.

(b) HARS ranks: 7, 4, 9.5, 3, 7, 7, 1.5, 1.5, 12.5, 12.5, 15, 9.5, 5, 14, 11.
   HSCL ranks: 8.5, 12.5, 15, 10, 3, 3, 1, 3, 5.5, 7, 12.5, 12.5, 8.5, 12.5, 5.5.

$r_S = 1 - \dfrac{6(327.5)}{15(224)} = .42$, $< r = .46$.

(c) H_0: $\rho_S = 0$, H_1: $\rho_S > 0$. $t = \dfrac{.42\sqrt{13}}{\sqrt{1 - .42^2}} = 1.67$; $D = 13$, critical $t = 1.77$. H_0 cannot be rejected. (The sample is too small.)

9. Uncooked ranks: 4, 8, 6.5, 4, 4, 1.5, 6.5, 9, 10, 1.5.
      45°C ranks: 1.5, 9, 5.5, 5.5, 5.5, 1.5, 5.5, 10, 5.5, 5.5.

$r_S = 1 - \dfrac{6(51)}{10(99)} = .69$.

(b) Every score in one variable is less than every score in the other, so $U_Y = 0$ and H_0 (the medians are the same) is clearly rejected.

APPENDIX A

SYMBOLS

Numbers in parentheses indicate pages on which symbols are introduced.

a	Y-intercept of the line which best fits sample data (326)	d	difference in value between two paired scores (255)
A	an event; a subset of a sample space (93)	d	difference in rankings of scores in paired samples (389)
A	Y-intercept of the line which fits population data (357)	D, D_X, D_Y	number of degrees of freedom (244, 289)
$\bar{A}$	complement of the set A; the event "A does not occur" (101)	E	error (198)
		E	expected frequency (266)
$A(z)$	area under the normal curve between 0 and z (138)	f	frequency (31)
ANOVA	analysis of variance (296)	$F(D_X, D_Y)$	F score (289)
α (alpha)	probability of a Type I error; level of significance (213)	$F_\alpha(D_X, D_Y)$	critical value of F for a one-tail test, level of significance $= \alpha$ (290)
b	slope of the line which best fits sample data (326)	H_0	null hypothesis (208)
		H_1	alternative hypothesis (210)
B	slope of the line which fits population data (357)	k	number of samples (299)
β (beta)	probability of a Type II error (213)	Md	median (56)
		Mo	mode (55)
$1 - \beta$	power of a statistical test (385)	μ (mu)	population mean (57)

399

μ_d	mean of paired differences in the population (255)	Q	proportion of the time that an event does not occur in a population (178)
μ_{Y/X_0}	mean of population values for $X = X_0$ (362)	Q_1	first quartile (65)
		Q_3	third quartile (65)
n	sample size (57)		
$n!$	n factorial (125)	r	linear correlation coefficient for samples (338)
N	population size (57)		
O	observed frequency (266)	r^2	coefficient of determination (340)
p	proportion of the time that an event occurs in a sample (178)	r_S	Spearman rank correlation coefficient (389)
P	probability of success in 1 of n independent trials (124)	ρ (rho)	linear correlation coefficient for populations (369)
P	proportion of the time that an event occurs in a population (178)	s	standard deviation of a sample (70)
P_α	αth percentile (65)	s	number of successful outcomes among n equally likely outcomes (91)
$\Pr(a < X < b)$	probability that X has a value between a and b (137)	s^2	variance of a sample (69)
$\Pr(A)$	probability of the event A (89)	s_b	standard error of the slope of the line of best fit (359)
$\Pr(B/A)$	probability of event B, given that event A occurs (106)	s_d	standard deviation of differences in paired samples (255)
q	proportion of the time that an event does not occur in a sample (178)	s_e	standard error of estimate (338)

s_{Y_p}	standard deviation of the Y values predicted for $X = X_0$ (364)	Σ (capital sigma)	summation notation (3)
s_{Y/X_0}	standard error of the mean of Y values corresponding to X_0 (362)	t	Student's t distribution or t score (241)
S	sample space (91)	U	See Mann-Whitney test, page 383
S	See Scheffe's test, page 308	X	score (31)
σ (sigma)	standard deviation of a population (70)	$\bar{X}$	mean of a sample (57)
σ^2	variance of a population (69)	Y_c	value of Y computed from the line of best fit for samples (326)
$\hat{\sigma}^2\ between$	estimate of σ^2 based on variability between samples (298)	Y_p	value of Y predicted in a population for $X = X_0$ (364)
$\hat{\sigma}^2\ within$	estimate of σ^2 based on variability within samples (299)	z	score in standard deviation units (79)
σ_p	standard error of proportions (178)	Z	Fisher Z score (372)
$\sigma_{(p_X-p_Y)}$	standard error of differences of proportions (191)	$\neq$	does not equal
		$>$	greater than
		$<$	less than
$\sigma_{\bar{X}}$	standard error of the mean (156)	$\leq$	less than or equal to (101)
$\sigma_{(\bar{X}-\bar{Y})}$	standard error of differences of means (187)	$\approx$	approximately equals
		$|a|$	absolute value of a (69)

APPENDIX B

FORMULAS

Numbers in parentheses indicate pages.

Mean

of a sample
$$\bar{X} = \frac{\Sigma X}{n}$$
(57)

of a sample with repeated data
$$\bar{X} = \frac{\Sigma Xf}{n}$$
(59)

of a population
$$\mu = \frac{\Sigma X}{N}$$
(57)

of a probability distribution
$$\mu = \Sigma X \Pr(X)$$
(121)

of a binomial distribution
$$\mu = nP$$
(127)

Variance

of a sample
$$s^2 = \frac{\Sigma(X - \bar{X})^2}{n - 1}$$
(69)

$$s^2 = \frac{\Sigma X^2 - (\Sigma X)^2/n}{n - 1}$$
(71)

of a sample, repeated data
$$s^2 = \frac{\Sigma X^2 f - (\Sigma Xf)^2/n}{n - 1}$$
(74)

of a finite population
$$\sigma^2 = \frac{\Sigma(X - \mu)^2}{N}$$
(69)

$$\sigma^2 = \frac{\Sigma X^2 - (\Sigma X)^2/N}{N}$$
(71)

of a population, repeated data
$$\sigma^2 = \frac{\Sigma X^2 f - (\Sigma Xf)^2/N}{N}$$
(74)

of a binomial distribution
$$\sigma^2 = nP(1 - P)$$
(127)

Standard deviation

of a sample
$$s = \sqrt{s^2}$$
(71)

of a population
$$\sigma = \sqrt{\sigma^2}$$
(71)

Probability

if outcomes are equally likely
$$\Pr(A) = \frac{s}{n}$$
(91)

if A and B are mutually exclusive events	$\Pr(A \text{ or } B) = \Pr(A) + \Pr(B)$	(99)
if A and B are independent	$\Pr(A \text{ and } B) = \Pr(A)\Pr(B)$	(106)
if A and B are not independent	$\Pr(A \text{ and } B) = \Pr(A)\Pr(B/A)$	(106)

Slope of a straight line

$$\text{slope} = \frac{\text{change in } Y \text{ values}}{\text{change in } X \text{ values}} \tag{317}$$

Straight line

$$Y = a + bX \tag{315}$$

Regression equation

$$Y_c = a + bX \tag{326}$$

$$b = \frac{n\Sigma XY - (\Sigma X)(\Sigma Y)}{n\Sigma X^2 - (\Sigma X)^2} \tag{327}$$

$$a = \frac{\Sigma Y}{n} - b\frac{\Sigma X}{n} \tag{327}$$

$$b = r\left(\frac{s_Y}{s_X}\right) \tag{343}$$

Standard error of estimate

$$s_e = \sqrt{\frac{\Sigma(Y - Y_c)^2}{n - 2}} \tag{338}$$

$$\Sigma(Y - Y_c)^2 = \Sigma Y^2 - a\Sigma Y - b\Sigma XY \tag{359}$$

Error variation	$\Sigma(Y - Y_c)^2$	(338)
Explained variation	$\Sigma(Y_c - \bar{Y})^2$	(339)
Total variation	$\Sigma(Y - \bar{Y})^2$	(339)

Coefficient of determination

$$r^2 = \frac{\text{Explained variation}}{\text{Total variation}} \tag{340}$$

Coefficient of (linear) correlation

$$r = (\text{sign of } b)\ \sqrt{r^2} \tag{341}$$

$$r = \frac{n\Sigma XY - (\Sigma X)(\Sigma Y)}{\sqrt{n\Sigma X^2 - (\Sigma X)^2}\sqrt{n\Sigma Y^2 - (\Sigma Y)^2}} \tag{341}$$

$$r = b\frac{s_X}{s_Y} \tag{343}$$

Rough estimate

$$r = \pm\left(1 - \frac{W}{L}\right) \tag{336}$$

Coefficient of correlation between ranks (Spearman rank correlation coefficient)

$$r_S = 1 - \frac{6\Sigma d^2}{n(n^2 - 1)} \tag{389}$$

Formulas for Statistical Inference

Inference about	Formula	Confidence limits	Standard error	
Mean				
If σ is known and if population is normal or $n \geq 30$	$z = \dfrac{\bar{X} - \mu}{\sigma_{\bar{X}}}$	$\bar{X} \pm z\sigma_{\bar{X}}$	$\sigma_{\bar{X}} = \dfrac{\sigma}{\sqrt{n}}$	(159)
If σ is unknown and population is normal	$t = \dfrac{\bar{X} - \mu}{s/\sqrt{n}}, \; D = n - 1$	$\bar{X} \pm t\dfrac{s}{\sqrt{n}}$		(241)
Proportion				
If P is known	$z = \dfrac{p - P}{\sigma_p}$		$\sigma_p = \sqrt{\dfrac{PQ}{n}}$	(181)
If P is unknown	$z = \dfrac{p - P}{\sigma_p}$	$p \pm z\sigma_p$	$\sigma_p = \sqrt{\dfrac{pq}{n}}$	(182)
Differences of means, two independent samples				
If σ_X, σ_Y are known and both populations are normal or $n_X \geq 30$, $n_Y \geq 30$	$z = \dfrac{(\bar{X} - \bar{Y}) - (\mu_X - \mu_Y)}{\sigma_{(\bar{X}-\bar{Y})}}$	$(\bar{X} - \bar{Y}) \pm z\sigma_{(\bar{X}-\bar{Y})}$	$\sigma_{(\bar{X}-\bar{Y})} = \sqrt{\sigma_{\bar{X}}^2 + \sigma_{\bar{Y}}^2}$	(189)
If σ_X, σ_Y are unknown but equal and both populations are normal	$\begin{cases} t = \dfrac{(\bar{X} - \bar{Y}) - (\mu_X - \mu_Y)}{\sigma_{(\bar{X}-\bar{Y})}} \\ D = n_X + n_Y - 2 \end{cases}$	$(\bar{X} - \bar{Y}) \pm t\sigma_{(\bar{X}-\bar{Y})}$	$\sigma_{(\bar{X}-\bar{Y})} = \sqrt{\dfrac{(n_X - 1)s_X^2 + (n_Y - 1)s_Y^2}{n_X + n_Y - 2}\left(\dfrac{1}{n_X} + \dfrac{1}{n_Y}\right)}$	(250)

Inference about	Formula	Standard error	
Differences of means, two or more independent samples			
If all populations are normal and have the same variance	$F = \dfrac{\hat{\sigma}^2 \text{ between}}{\hat{\sigma}^2 \text{ within}}$		(299)
	$\hat{\sigma}^2 \text{ between} = \dfrac{A - B^2/n_T}{k - 1}$, $\hat{\sigma}^2 \text{ within} = \dfrac{C - A}{n_T - k}$ (See page 299 for A, B, and C.)		(308)
Which means are different? (Scheffe's test)	Compare $\lvert \bar{X}_i - \bar{X}_j \rvert$ and $S\sqrt{1/n_i + 1/n_j}$ $S = \sqrt{(k-1)F_a(k-1,\, n-k)(\hat{\sigma}^2 \text{ within})}$		(308)
Difference of means, paired samples			
If differences are normal	$t = \dfrac{d - \mu_d}{s_d}$, $D = n - 1$		(255)
Sign test (nonparametric)	$z = \dfrac{X \pm .5 - n/2}{\sqrt{n/2}}$		(382)
Difference of proportions	$z = \dfrac{(p_X - p_Y) - (P_X - P_Y)}{\sigma_{(p_X - p_Y)}}$	$\sigma_{(p_X - p_Y)} = \sqrt{\sigma_{p_X}^2 + \sigma_{p_Y}^2}$	(193)
Variances			
If both populations are normal	$F = \dfrac{s_X^2}{s_Y^2}$, $D_X = n_X - 1$, $D_Y = n_Y - 1$		(289)
Goodness of fit	$\chi_P^2 = \Sigma\left[\dfrac{(O - E)^2}{E}\right]$ $D = $ (number of E entries) $- 1$		(269)
Are two variables independent?	$\chi_P^2 = \Sigma\left[\dfrac{(O - E)^2}{E}\right]$ $D = (R - 1)(C - 1)$		(276)

Inference about	Formula	Standard error	
Regression line			
If, for each X, Y scores in a population are normal with the same variance:			
Slope B of regression line	$t = \dfrac{b - B}{s_b}, \quad D = n - 2$	$s_b = \sqrt{\dfrac{n(\Sigma Y^2 - a\Sigma Y - b\Sigma XY)}{(n-2)[n\Sigma X^2 - (\Sigma X)^2]}}$	(359)
Mean of Y scores corresponding to X_0	$t = \dfrac{Y_c - \mu_{Y/X_0}}{s_{Y/X_0}}$ $D = n - 2$	$s_{Y/X_0} = s_e \sqrt{\dfrac{1}{n} + \dfrac{n(X_0 - \bar{X})^2}{n\Sigma X^2 - (\Sigma X)^2}}$	(362)
		$s_e = \sqrt{\dfrac{\Sigma Y^2 - a\Sigma Y - b\Sigma XY}{n - 2}}$	
Prediction interval of Y for a given X_0 (not statistical inference)	$t = \dfrac{Y_c - Y_p}{s_{Y_p}}$ $D = n - 2$	$s_{Y_p} = s_e \sqrt{1 + \dfrac{1}{n} + \dfrac{n(X_0 - \bar{X})^2}{n\Sigma X^2 - (\Sigma X)^2}}$	(364)
Correlation			
If X, Y are normally distributed (have a bivariate normal distribution; see page 370)			
H_0: $\rho = 0$	$t = \dfrac{r\sqrt{n-2}}{\sqrt{1-r^2}}, \quad D = n - 2$		(370)
H_0: $\rho = c, c \neq 0$	$z = \dfrac{Z - \mu_Z}{\sigma_Z}$	$\sigma_Z = \dfrac{1}{\sqrt{n-3}}$	(373)
H_0: $\rho_S = 0$ (if $n > 10$)	$t = \dfrac{r_S\sqrt{n-2}}{\sqrt{1-r_S^2}}, \quad D = n - 2$		(391)
Are two populations the same?			
Mann-Whitney test: nonparametric (if two populations have the same shape and variance, do they have the same median?)	$\mu_U = \dfrac{n_X n_Y}{2}$	$s_U = \sqrt{\dfrac{n_X n_Y (n_X + n_Y + 1)}{12}}$	(384)

APPENDIX C

TABLES

TABLE 1   SQUARES AND SQUARE ROOTS

If n is between	SQUARE Multiply n^2 by	SQUARE ROOTS Use column headed	Multiply by
.000100 and .000999	.00000001	$\sqrt{n}$	.01
.00100 and .00999	.000001	$\sqrt{10n}$	.01
.0100 and .0999	.0001	$\sqrt{n}$	.1
.100 and .999	.01	$\sqrt{10n}$	.1
1.00 and 9.99	1	$\sqrt{n}$	1
10.0 and 99.9	100	$\sqrt{10n}$	1
100 and 999	10,000	$\sqrt{n}$	10
1,000 and 9,990	1,000,000	$\sqrt{10n}$	10
10,000 and 99,900	100,000,000	$\sqrt{n}$	100
100,000 and 999,000	10,000,000,000	$\sqrt{10n}$	100
1,100,000 and 9,990,000	10,000,000,000	$\sqrt{n}$	1,000

n	n^2	$\sqrt{n}$	$\sqrt{10n}$	n	n^2	$\sqrt{n}$	$\sqrt{10n}$	n	n^2	$\sqrt{n}$	$\sqrt{10n}$
1.00	1.0000	1.000	3.162	1.25	1.5625	1.118	3.536	1.50	2.2500	1.225	3.873
1.01	1.0201	1.005	3.178	1.26	1.5876	1.122	3.550	1.51	2.2801	1.229	3.886
1.02	1.0404	1.010	3.194	1.27	1.6129	1.127	3.564	1.52	2.3104	1.233	3.899
1.03	1.0609	1.015	3.209	1.28	1.6384	1.131	3.578	1.53	2.3409	1.237	3.912
1.04	1.0816	1.020	3.225	1.29	1.6641	1.136	3.592	1.54	2.3716	1.241	3.924
1.05	1.1025	1.025	3.240	1.30	1.6900	1.140	3.606	1.55	2.4025	1.245	3.937
1.06	1.1236	1.030	3.256	1.31	1.7161	1.145	3.619	1.56	2.4336	1.249	3.950
1.07	1.1449	1.034	3.271	1.32	1.7424	1.149	3.633	1.57	2.4649	1.253	3.962
1.08	1.1664	1.039	3.286	1.33	1.7689	1.153	3.647	1.58	2.4964	1.257	3.975
1.09	1.1881	1.044	3.302	1.34	1.7956	1.158	3.661	1.59	2.5281	1.261	3.987
1.10	1.2100	1.049	3.317	1.35	1.8225	1.162	3.674	1.60	2.5600	1.265	4.000
1.11	1.2321	1.054	3.332	1.36	1.8496	1.166	3.688	1.61	2.5921	1.269	4.012
1.12	1.2544	1.058	3.347	1.37	1.8769	1.170	3.701	1.62	2.6244	1.273	4.025
1.13	1.2769	1.063	3.362	1.38	1.9044	1.175	3.715	1.63	2.6569	1.277	4.037
1.14	1.2996	1.068	3.376	1.39	1.9321	1.179	3.728	1.64	2.6896	1.281	4.050
1.15	1.3225	1.072	3.391	1.40	1.9600	1.183	3.742	1.65	2.7225	1.285	4.062
1.16	1.3456	1.077	3.406	1.41	1.9881	1.187	3.755	1.66	2.7556	1.288	4.074
1.17	1.3689	1.082	3.421	1.42	2.0164	1.192	3.768	1.67	2.7889	1.292	4.087
1.18	1.3924	1.086	3.435	1.43	2.0449	1.196	3.782	1.68	2.8224	1.296	4.099
1.19	1.4161	1.091	3.450	1.44	2.0736	1.200	3.795	1.69	2.8561	1.300	4.111
1.20	1.4400	1.095	3.464	1.45	2.1025	1.204	3.808	1.70	2.8900	1.304	4.123
1.21	1.4641	1.100	3.479	1.46	2.1316	1.208	3.821	1.71	2.9241	1.308	4.135
1.22	1.4884	1.105	3.493	1.47	2.1609	1.212	3.834	1.72	2.9584	1.311	4.147
1.23	1.5129	1.109	3.507	1.48	2.1904	1.217	3.847	1.73	2.9929	1.315	4.159
1.24	1.5376	1.114	3.521	1.49	2.2201	1.221	3.860	1.74	3.0276	1.319	4.171

TABLE 1 SQUARES AND SQUARE ROOTS *Continued*

n	n^2	$\sqrt{n}$	$\sqrt{10n}$	n	n^2	$\sqrt{n}$	$\sqrt{10n}$	n	n^2	$\sqrt{n}$	$\sqrt{10n}$
1.75	3.0625	1.323	4.183	2.25	5.0625	1.500	4.743	2.75	7.5625	1.658	5.244
1.76	3.0976	1.327	4.195	2.26	5.1076	1.503	4.754	2.76	7.6176	1.661	5.254
1.77	3.1329	1.330	4.207	2.27	5.1529	1.507	4.764	2.77	7.6729	1.664	5.263
1.78	3.1684	1.334	4.219	2.28	5.1984	1.510	4.775	2.78	7.7284	1.667	5.273
1.79	3.2041	1.338	4.231	2.29	5.2441	1.513	4.785	2.79	7.7841	1.670	5.282
1.80	3.2400	1.342	4.243	2.30	5.2900	1.517	4.796	2.80	7.8400	1.673	5.292
1.81	3.2761	1.345	4.254	2.31	5.3361	1.520	4.806	2.81	7.8961	1.676	5.301
1.82	3.3124	1.349	4.266	2.32	5.3824	1.523	4.817	2.82	7.9524	1.679	5.310
1.83	3.3489	1.353	4.278	2.33	5.4289	1.526	4.827	2.83	8.0089	1.682	5.320
1.84	3.3856	1.356	4.290	2.34	5.4756	1.530	4.837	2.84	8.0656	1.685	5.329
1.85	3.4225	1.360	4.301	2.35	5.5225	1.533	4.848	2.85	8.1225	1.688	5.339
1.86	3.4596	1.364	4.313	2.36	5.5696	1.536	4.858	2.86	8.1796	1.691	5.348
1.87	3.4969	1.367	4.324	2.37	5.6169	1.539	4.868	2.87	8.2369	1.694	5.357
1.88	3.5344	1.371	4.336	2.38	5.6644	1.543	4.879	2.88	8.2944	1.697	5.367
1.89	3.5721	1.375	4.347	2.39	5.7121	1.546	4.889	2.89	8.3521	1.700	5.376
1.90	3.6100	1.378	4.359	2.40	5.7600	1.549	4.899	2.90	8.4100	1.703	5.385
1.91	3.6481	1.382	4.370	2.41	5.8081	1.552	4.909	2.91	8.4681	1.706	5.394
1.92	3.6864	1.386	4.382	2.42	5.8564	1.556	4.919	2.92	8.5264	1.709	5.404
1.93	3.7249	1.389	4.393	2.43	5.9049	1.559	4.930	2.93	8.5849	1.712	5.413
1.94	3.7636	1.393	4.405	2.44	5.9536	1.562	4.940	2.94	8.6436	1.715	5.422
1.95	3.8025	1.396	4.416	2.45	6.0025	1.565	4.950	2.95	8.7025	1.718	5.431
1.96	3.8416	1.400	4.427	2.46	6.0516	1.568	4.960	2.96	8.7616	1.720	5.441
1.97	3.8809	1.404	4.438	2.47	6.1009	1.572	4.970	2.97	8.8209	1.723	5.450
1.98	3.9204	1.407	4.450	2.48	6.1504	1.575	4.980	2.98	8.8804	1.726	5.459
1.99	3.9601	1.411	4.461	2.49	6.2001	1.578	4.990	2.99	8.9401	1.729	5.468
2.00	4.0000	1.414	4.472	2.50	6.2500	1.581	5.000	3.00	9.0000	1.732	5.477
2.01	4.0401	1.418	4.483	2.51	6.3001	1.584	5.010	3.01	9.0601	1.735	5.486
2.02	4.0804	1.421	4.494	2.52	6.3504	1.587	5.020	3.02	9.1204	1.738	5.495
2.03	4.1209	1.425	4.506	2.53	6.4009	1.591	5.030	3.03	9.1809	1.741	5.505
2.04	4.1616	1.428	4.517	2.54	6.4516	1.594	5.040	3.04	9.2416	1.744	5.514
2.05	4.2025	1.432	4.528	2.55	6.5025	1.597	5.050	3.05	9.3025	1.746	5.523
2.06	4.2436	1.435	4.539	2.56	6.5536	1.600	5.060	3.06	9.3636	1.749	5.532
2.07	4.2849	1.439	4.550	2.57	6.6049	1.603	5.070	3.07	9.4249	1.752	5.541
2.08	4.3264	1.442	4.561	2.58	6.6564	1.606	5.079	3.08	9.4864	1.755	5.550
2.09	4.3681	1.446	4.572	2.59	6.7081	1.609	5.089	3.09	9.5481	1.758	5.559
2.10	4.4100	1.449	4.583	2.60	6.7600	1.612	5.099	3.10	9.6100	1.761	5.568
2.11	4.4521	1.453	4.593	2.61	6.8121	1.616	5.109	3.11	9.6721	1.764	5.577
2.12	4.4944	1.456	4.604	2.62	6.8644	1.619	5.119	3.12	9.7344	1.766	5.586
2.13	4.5369	1.459	4.615	2.63	6.9169	1.622	5.128	3.13	9.7969	1.769	5.595
2.14	4.5796	1.463	4.626	2.64	6.9696	1.625	5.138	3.14	9.8596	1.772	5.604
2.15	4.6225	1.466	4.637	2.65	7.0225	1.628	5.148	3.15	9.9225	1.775	5.612
2.16	4.6656	1.470	4.648	2.66	7.0756	1.631	5.158	3.16	9.9856	1.778	5.621
2.17	4.7089	1.473	4.658	2.67	7.1289	1.634	5.167	3.17	10.0489	1.780	5.630
2.18	4.7524	1.476	4.669	2.68	7.1824	1.637	5.177	3.18	10.1124	1.783	5.639
2.19	4.7961	1.480	4.680	2.69	7.2361	1.640	5.187	3.19	10.1761	1.786	5.648
2.20	4.8400	1.483	4.690	2.70	7.2900	1.643	5.196	3.20	10.2400	1.789	5.657
2.21	4.8841	1.487	4.701	2.71	7.3441	1.646	5.206	3.21	10.3041	1.792	5.666
2.22	4.9284	1.490	4.712	2.72	7.3984	1.649	5.215	3.22	10.3684	1.794	5.675
2.23	4.9729	1.493	4.722	2.73	7.4529	1.652	5.225	3.23	10.4329	1.797	5.683
2.24	5.0176	1.497	4.733	2.74	7.5076	1.655	5.235	3.24	10.4976	1.800	5.692

TABLE 1 SQUARES AND SQUARE ROOTS *Continued*

n	n^2	$\sqrt{n}$	$\sqrt{10n}$	n	n^2	$\sqrt{n}$	$\sqrt{10n}$	n	n^2	$\sqrt{n}$	$\sqrt{10n}$
3.25	10.5625	1.803	5.701	3.75	14.0625	1.936	6.124	4.25	18.0625	2.062	6.519
3.26	10.6276	1.806	5.710	3.76	14.1376	1.939	6.132	4.26	18.1476	2.064	6.527
3.27	10.6929	1.808	5.718	3.77	14.2129	1.942	6.140	4.27	18.2329	2.066	6.535
3.28	10.7584	1.811	5.727	3.78	14.2884	1.944	6.148	4.28	18.3184	2.069	6.542
3.29	10.8241	1.814	5.736	3.79	14.3641	1.947	6.156	4.29	18.4041	2.071	6.550
3.30	10.8900	1.817	5.745	3.80	14.4400	1.949	6.164	4.30	18.4900	2.074	6.557
3.31	10.9561	1.819	5.753	3.81	14.5161	1.952	6.173	4.31	18.5761	2.076	6.565
3.32	11.0224	1.822	5.762	3.82	14.5924	1.954	6.181	4.32	18.6624	2.078	6.573
3.33	11.0889	1.825	5.771	3.83	14.6689	1.957	6.189	4.33	18.7489	2.081	6.580
3.34	11.1556	1.828	5.779	3.84	14.7456	1.960	6.197	4.34	18.8356	2.083	6.588
3.35	11.2225	1.830	5.788	3.85	14.8225	1.962	6.205	4.35	18.9225	2.086	6.595
3.36	11.2896	1.833	5.797	3.86	14.8996	1.965	6.213	4.36	19.0096	2.088	6.603
3.37	11.3569	1.836	5.805	3.87	14.9769	1.967	6.221	4.37	19.0969	2.090	6.611
3.38	11.4244	1.838	5.814	3.88	15.0544	1.970	6.229	4.38	19.1844	2.093	6.618
3.39	11.4921	1.841	5.822	3.89	15.1321	1.972	6.237	4.39	19.2721	2.095	6.626
3.40	11.5600	1.844	5.831	3.90	15.2100	1.975	6.245	4.40	19.3600	2.098	6.633
3.41	11.6281	1.847	5.840	3.91	15.2881	1.977	6.253	4.41	19.4481	2.100	6.641
3.42	11.6964	1.849	5.848	3.92	15.3664	1.980	6.261	4.42	19.5364	2.102	6.648
3.43	11.7649	1.852	5.857	3.93	15.4449	1.982	6.269	4.43	19.6249	2.105	6.656
3.44	11.8336	1.855	5.865	3.94	15.5236	1.985	6.277	4.44	19.7136	2.107	6.663
3.45	11.9025	1.857	5.874	3.95	15.6025	1.987	6.285	4.45	19.8025	2.110	6.671
3.46	11.9716	1.860	5.882	3.96	15.6816	1.990	6.293	4.46	19.8916	2.112	6.678
3.47	12.0409	1.863	5.891	3.97	15.7609	1.992	6.301	4.47	19.9809	2.114	6.686
3.48	12.1104	1.865	5.899	3.98	15.8404	1.995	6.309	4.48	20.0704	2.117	6.693
3.49	12.1801	1.868	5.908	3.99	15.9201	1.997	6.317	4.49	20.1601	2.119	6.701
3.50	12.2500	1.871	5.916	4.00	16.0000	2.000	6.325	4.50	20.2500	2.121	6.708
3.51	12.3201	1.874	5.925	4.01	16.0801	2.002	6.332	4.51	20.3401	2.124	6.716
3.52	12.3904	1.876	5.933	4.02	16.1604	2.005	6.340	4.52	20.4304	2.126	6.723
3.53	12.4609	1.879	5.941	4.03	16.2409	2.007	6.348	4.53	20.5209	2.128	6.731
3.54	12.5316	1.881	5.950	4.04	16.3216	2.010	6.356	4.54	20.6116	2.131	6.738
3.55	12.6025	1.884	5.958	4.05	16.4025	2.012	6.364	4.55	20.7025	2.133	6.745
3.56	12.6736	1.887	5.967	4.06	16.4836	2.015	6.372	4.56	20.7936	2.135	6.753
3.57	12.7449	1.889	5.975	4.07	16.5649	2.017	6.380	4.57	20.8849	2.138	6.760
3.58	12.8164	1.892	5.983	4.08	16.6464	2.020	6.387	4.58	20.9764	2.140	6.768
3.59	12.8881	1.895	5.992	4.09	16.7281	2.022	6.395	4.59	21.0681	2.142	6.775
3.60	12.9600	1.897	6.000	4.10	16.8100	2.025	6.403	4.60	21.1600	2.145	6.782
3.61	13.0321	1.900	6.008	4.11	16.8921	2.027	6.411	4.61	21.2521	2.147	6.790
3.62	13.1044	1.903	6.017	4.12	16.9744	2.030	6.419	4.62	21.3444	2.149	6.797
3.63	13.1769	1.905	6.025	4.13	17.0569	2.032	6.427	4.63	21.4369	2.152	6.804
3.64	13.2496	1.908	6.033	4.14	17.1396	2.035	6.434	4.64	21.5296	2.154	6.812
3.65	13.3225	1.910	6.042	4.15	17.2225	2.037	6.442	4.65	21.6225	2.156	6.819
3.66	13.3956	1.913	6.050	4.16	17.3056	2.040	6.450	4.66	21.7156	2.159	6.826
3.67	13.4689	1.916	6.058	4.17	17.3889	2.042	6.458	4.67	21.8089	2.161	6.834
3.68	13.5424	1.918	6.066	4.18	17.4724	2.045	6.465	4.68	21.9024	2.163	6.841
3.69	13.6161	1.921	6.075	4.19	17.5561	2.047	6.473	4.69	21.9961	2.166	6.848
3.70	13.6900	1.924	6.083	4.20	17.6400	2.049	6.481	4.70	22.0900	2.168	6.856
3.71	13.7641	1.926	6.091	4.21	17.7241	2.052	6.488	4.71	22.1841	2.170	6.863
3.72	13.8384	1.929	6.099	4.22	17.8084	2.054	6.496	4.72	22.2784	2.173	6.870
3.73	13.9129	1.931	6.107	4.23	17.8929	2.057	6.504	4.73	22.3729	2.175	6.878
3.74	13.9876	1.934	6.116	4.24	17.9776	2.059	6.512	4.74	22.4676	2.177	6.885

TABLE 1 SQUARES AND SQUARE ROOTS *Continued*

n	n^2	$\sqrt{n}$	$\sqrt{10n}$	n	n^2	$\sqrt{n}$	$\sqrt{10n}$	n	n^2	$\sqrt{n}$	$\sqrt{10n}$
4.75	22.5625	2.179	6.892	5.25	27.5625	2.291	7.246	5.75	33.0625	2.398	7.583
4.76	22.6576	2.182	6.899	5.26	27.6676	2.293	7.253	5.76	33.1776	2.400	7.589
4.77	22.7529	2.184	6.907	5.27	27.7729	2.296	7.259	5.77	33.2929	2.402	7.596
4.78	22.8484	2.186	6.914	5.28	27.8784	2.298	7.266	5.78	33.4084	2.404	7.603
4.79	22.9441	2.189	6.921	5.29	27.9841	2.300	7.273	5.79	33.5241	2.406	7.609
4.80	23.0400	2.191	6.928	5.30	28.0900	2.302	7.280	5.80	33.6400	2.408	7.616
4.81	23.1361	2.193	6.935	5.31	28.1961	2.304	7.287	5.81	33.7561	2.410	7.622
4.82	23.2324	2.195	6.943	5.32	28.3024	2.307	7.294	5.82	33.8724	2.412	7.629
4.83	23.3289	2.198	6.950	5.33	28.4089	2.309	7.301	5.83	33.9889	2.415	7.635
4.84	23.4256	2.200	6.957	5.34	28.5156	2.311	7.308	5.84	34.1056	2.417	7.642
4.85	23.5225	2.202	6.964	5.35	28.6225	2.313	7.314	5.85	34.2225	2.419	7.649
4.86	23.6196	2.205	6.971	5.36	28.7296	2.315	7.321	5.86	34.3396	2.421	7.655
4.87	23.7169	2.207	6.979	5.37	28.8369	2.317	7.328	5.87	34.4569	2.423	7.662
4.88	23.8144	2.209	6.986	5.38	28.9444	2.319	7.335	5.88	34.5744	2.425	7.668
4.89	23.9121	2.211	6.993	5.39	29.0521	2.322	7.342	5.89	34.6921	2.427	7.675
4.90	24.0100	2.214	7.000	5.40	29.1600	2.324	7.348	5.90	34.8100	2.429	7.681
4.91	24.1081	2.216	7.007	5.41	29.2681	2.326	7.355	5.91	34.9281	2.431	7.688
4.92	24.2064	2.218	7.014	5.42	29.3764	2.328	7.362	5.92	35.0464	2.433	7.694
4.93	24.3049	2.220	7.021	5.43	29.4849	2.330	7.369	5.93	35.1649	2.435	7.701
4.94	24.4036	2.223	7.029	5.44	29.5936	2.332	7.376	5.94	35.2836	2.437	7.707
4.95	24.5025	2.225	7.036	5.45	29.7025	2.335	7.382	5.95	35.4025	2.439	7.714
4.96	24.6016	2.227	7.043	5.46	29.8116	2.337	7.389	5.96	35.5216	2.441	7.720
4.97	24.7009	2.229	7.050	5.47	29.9209	2.339	7.396	5.97	35.6409	2.443	7.727
4.98	24.8004	2.232	7.057	5.48	30.0304	2.341	7.403	5.98	35.7604	2.445	7.733
4.99	24.9001	2.234	7.064	5.49	30.1401	2.343	7.409	5.99	35.8801	2.447	7.740
5.00	25.0000	2.236	7.071	5.50	30.2500	2.345	7.416	6.00	36.0000	2.449	7.746
5.01	25.1001	2.238	7.078	5.51	30.3601	2.347	7.423	6.01	36.1201	2.452	7.752
5.02	25.2004	2.241	7.085	5.52	30.4704	2.349	7.430	6.02	36.2404	2.454	7.759
5.03	25.3009	2.243	7.092	5.53	30.5809	2.352	7.436	6.03	36.3609	2.456	7.765
5.04	25.4016	2.245	7.099	5.54	30.6916	2.354	7.443	6.04	36.4816	2.458	7.772
5.05	25.5025	2.247	7.106	5.55	30.8025	2.356	7.450	6.05	36.6025	2.460	7.778
5.06	25.6036	2.249	7.113	5.56	30.9136	2.358	7.457	6.06	36.7236	2.462	7.785
5.07	25.7049	2.252	7.120	5.57	31.0249	2.360	7.463	6.07	36.8449	2.464	7.791
5.08	25.8064	2.254	7.127	5.58	31.1364	2.362	7.470	6.08	36.9664	2.466	7.797
5.09	25.9081	2.256	7.134	5.59	31.2481	2.364	7.477	6.09	37.0881	2.468	7.804
5.10	26.0100	2.258	7.141	5.60	31.3600	2.366	7.483	6.10	37.2100	2.470	7.810
5.11	26.1121	2.261	7.148	5.61	31.4721	2.369	7.490	6.11	37.3321	2.472	7.817
5.12	26.2144	2.263	7.155	5.62	31.5844	2.371	7.497	6.12	37.4544	2.474	7.823
5.13	26.3169	2.265	7.162	5.63	31.6969	2.373	7.503	6.13	37.5769	2.476	7.829
5.14	26.4196	2.267	7.169	5.64	31.8096	2.375	7.510	6.14	37.6996	2.478	7.836
5.15	26.5225	2.269	7.176	5.65	31.9225	2.377	7.517	6.15	37.8225	2.480	7.842
5.16	26.6256	2.272	7.183	5.66	32.0356	2.379	7.523	6.16	37.9456	2.482	7.849
5.17	26.7289	2.274	7.190	5.67	32.1489	2.381	7.530	6.17	38.0689	2.484	7.855
5.18	26.8324	2.276	7.197	5.68	32.2624	2.383	7.537	6.18	38.1924	2.486	7.861
5.19	26.9361	2.278	7.204	5.69	32.3761	2.385	7.543	6.19	38.3161	2.488	7.868
5.20	27.0400	2.280	7.211	5.70	32.4900	2.387	7.550	6.20	38.4400	2.490	7.874
5.21	27.1441	2.283	7.218	5.71	32.6041	2.390	7.556	6.21	38.5641	2.492	7.880
5.22	27.2484	2.285	7.225	5.72	32.7184	2.392	7.563	6.22	38.6884	2.494	7.887
5.23	27.3529	2.287	7.232	5.73	32.8329	2.394	7.570	6.23	38.8129	2.496	7.893
5.24	27.4576	2.289	7.239	5.74	32.9476	2.396	7.576	6.24	38.9376	2.498	7.899

TABLE 1 SQUARES AND SQUARE ROOTS *Continued*

n	n^2	$\sqrt{n}$	$\sqrt{10n}$	n	n^2	$\sqrt{n}$	$\sqrt{10n}$	n	n^2	$\sqrt{n}$	$\sqrt{10n}$
6.25	39.0625	2.500	7.906	6.75	45.5625	2.598	8.216	7.25	52.5625	2.693	8.515
6.26	39.1876	2.502	7.912	6.76	45.6976	2.600	8.222	7.26	52.7076	2.694	8.521
6.27	39.3129	2.504	7.918	6.77	45.8329	2.602	8.228	7.27	52.8529	2.696	8.526
6.28	39.4384	2.506	7.925	6.78	45.9684	2.604	8.234	7.28	52.9984	2.698	8.532
6.29	39.5641	2.508	7.931	6.79	46.1041	2.606	8.240	7.29	53.1441	2.700	8.538
6.30	39.6900	2.510	7.937	6.80	46.2400	2.608	8.246	7.30	53.2900	2.702	8.544
6.31	39.8161	2.512	7.944	6.81	46.3761	2.610	8.252	7.31	53.4361	2.704	8.550
6.32	39.9424	2.514	7.950	6.82	46.5124	2.612	8.258	7.32	53.5824	2.706	8.556
6.33	40.0689	2.516	7.956	6.83	46.6489	2.613	8.264	7.33	53.7289	2.707	8.562
6.34	40.1956	2.518	7.962	6.84	46.7856	2.615	8.270	7.34	53.8756	2.709	8.567
6.35	40.3225	2.520	7.969	6.85	46.9225	2.617	8.276	7.35	54.0225	2.711	8.573
6.36	40.4496	2.522	7.975	6.86	47.0596	2.619	8.283	7.36	54.1696	2.713	8.579
6.37	40.5769	2.524	7.981	6.87	47.1969	2.621	8.289	7.37	54.3169	2.715	8.585
6.38	40.7044	2.526	7.987	6.88	47.3344	2.623	8.295	7.38	54.4644	2.717	8.591
6.39	40.8321	2.528	7.994	6.89	47.4721	2.625	8.301	7.39	54.6121	2.718	8.597
6.40	40.9600	2.530	8.000	6.90	47.6100	2.627	8.307	7.40	54.7600	2.720	8.602
6.41	41.0881	2.532	8.006	6.91	47.7481	2.629	8.313	7.41	54.9081	2.722	8.608
6.42	41.2164	2.534	8.012	6.92	47.8864	2.631	8.319	7.42	55.0564	2.724	8.614
6.43	41.3449	2.536	8.019	6.93	48.0249	2.632	8.325	7.43	55.2049	2.726	8.620
6.44	41.4736	2.538	8.025	6.94	48.1636	2.634	8.331	7.44	55.3536	2.728	8.626
6.45	41.6025	2.540	8.031	6.95	48.3025	2.636	8.337	7.45	55.5025	2.729	8.631
6.46	41.7316	2.542	8.037	6.96	48.4416	2.638	8.343	7.46	55.6516	2.731	8.637
6.47	41.8609	2.544	8.044	6.97	48.5809	2.640	8.349	7.47	55.8009	2.733	8.643
6.48	41.9904	2.546	8.050	6.98	48.7204	2.642	8.355	7.48	55.9504	2.735	8.649
6.49	42.1201	2.548	8.056	6.99	48.8601	2.644	8.361	7.49	56.1001	2.737	8.654
6.50	42.2500	2.550	8.062	7.00	49.0000	2.646	8.367	7.50	56.2500	2.739	8.660
6.51	42.3801	2.551	8.068	7.01	49.1401	2.648	8.373	7.51	56.4001	2.740	8.666
6.52	42.5104	2.553	8.075	7.02	49.2804	2.650	8.379	7.52	56.5504	2.742	8.672
6.53	42.6409	2.555	8.081	7.03	49.4209	2.651	8.385	7.53	56.7009	2.744	8.678
6.54	42.7716	2.557	8.087	7.04	49.5616	2.653	8.390	7.54	56.8516	2.746	8.683
6.55	42.9025	2.559	8.093	7.05	49.7025	2.655	8.396	7.55	57.0025	2.748	8.689
6.56	43.0336	2.561	8.099	7.06	49.8436	2.657	8.402	7.56	57.1536	2.750	8.695
6.57	43.1649	2.563	8.106	7.07	49.9849	2.659	8.408	7.57	57.3049	2.751	8.701
6.58	43.2964	2.565	8.112	7.08	50.1264	2.661	8.414	7.58	57.4564	2.753	8.706
6.59	43.4281	2.567	8.118	7.09	50.2681	2.663	8.420	7.59	57.6081	2.755	8.712
6.60	43.5600	2.569	8.124	7.10	50.4100	2.665	8.426	7.60	57.7600	2.757	8.718
6.61	43.6921	2.571	8.130	7.11	50.5521	2.666	8.432	7.61	57.9121	2.759	8.724
6.62	43.8244	2.573	8.136	7.12	50.6944	2.668	8.438	7.62	58.0644	2.760	8.729
6.63	43.9569	2.575	8.142	7.13	50.8369	2.670	8.444	7.63	58.2169	2.762	8.735
6.64	44.0896	2.577	8.149	7.14	50.9796	2.672	8.450	7.64	58.3696	2.764	8.741
6.65	44.2225	2.579	8.155	7.15	51.1225	2.674	8.456	7.65	58.5225	2.766	8.746
6.66	44.3556	2.581	8.161	7.16	51.2656	2.676	8.462	7.66	58.6756	2.768	8.752
6.67	44.4889	2.583	8.167	7.17	51.4089	2.678	8.468	7.67	58.8289	2.769	8.758
6.68	44.6224	2.585	8.173	7.18	51.5524	2.680	8.473	7.68	58.9824	2.771	8.764
6.69	44.7561	2.587	8.179	7.19	51.6961	2.681	8.479	7.69	59.1361	2.773	8.769
6.70	44.8900	2.588	8.185	7.20	51.8400	2.683	8.485	7.70	59.2900	2.775	8.775
6.71	45.0241	2.590	8.191	7.21	51.9841	2.685	8.491	7.71	59.4441	2.777	8.781
6.72	45.1584	2.592	8.198	7.22	52.1284	2.687	8.497	7.72	59.5984	2.778	8.786
6.73	45.2929	2.594	8.204	7.23	52.2729	2.689	8.503	7.73	59.7529	2.780	8.792
6.74	45.4276	2.596	8.210	7.24	52.4176	2.691	8.509	7.74	59.9076	2.782	8.798

TABLE 1 SQUARES AND SQUARE ROOTS *Continued*

n	n^2	$\sqrt{n}$	$\sqrt{10n}$	n	n^2	$\sqrt{n}$	$\sqrt{10n}$	n	n^2	$\sqrt{n}$	$\sqrt{10n}$
7.75	60.0625	2.784	8.803	8.25	68.0625	2.872	9.083	8.75	76.5625	2.958	9.354
7.76	60.2176	2.786	8.809	8.26	68.2276	2.874	9.088	8.76	76.7376	2.960	9.359
7.77	60.3729	2.787	8.815	8.27	68.3929	2.876	9.094	8.77	76.9129	2.961	9.365
7.78	60.5284	2.789	8.820	8.28	68.5584	2.877	9.099	8.78	77.0884	2.963	9.370
7.79	60.6841	2.791	8.826	8.29	68.7241	2.879	9.105	8.79	77.2641	2.965	9.376
7.80	60.8400	2.793	8.832	8.30	68.8900	2.881	9.110	8.80	77.4400	2.966	9.381
7.81	60.9961	2.795	8.837	8.31	69.0561	2.883	9.116	8.81	77.6161	2.968	9.386
7.82	61.1524	2.796	8.843	8.32	69.2224	2.884	9.121	8.82	77.7924	2.970	9.391
7.83	61.3089	2.798	8.849	8.33	69.3889	2.886	9.127	8.83	77.9689	2.972	9.397
7.84	61.4656	2.800	8.854	8.34	69.5556	2.888	9.132	8.84	78.1456	2.973	9.402
7.85	61.6225	2.802	8.860	8.35	69.7225	2.890	9.138	8.85	78.3225	2.975	9.407
7.86	61.7796	2.804	8.866	8.36	69.8896	2.891	9.143	8.86	78.4996	2.977	9.413
7:87	61.9369	2.805	8.871	8.37	70.0569	2.893	9.149	8.87	78.6769	2.978	9.418
7.88	62.0944	2.807	8.877	8.38	70.2244	2.895	9.154	8.88	78.8544	2.980	9.423
7.89	62.2521	2.809	8.883	8.39	70.3921	2.897	9.160	8.89	79.0321	2.982	9.429
7.90	62.4100	2.811	8.888	8.40	70.5600	2.898	9.165	8.90	79.2100	2.983	9.434
7.91	62.5681	2.812	8.894	8.41	70.7281	2.900	9.171	8.91	79.3881	2.985	9.439
7.92	62.7264	2.814	8.899	8.42	70.8964	2.902	9.176	8.92	79.5664	2.987	9.445
7.93	62.8849	2.816	8.905	8.43	71.0649	2.903	9.182	8.93	79.7449	2.988	9.450
7.94	63.0436	2.818	8.911	8.44	71.2336	2.905	9.187	8.94	79.9236	2.990	9.455
7.95	63.2025	2.820	8.916	8.45	71.4025	2.907	9.192	8.95	80.1025	2.992	9.460
7.96	63.3616	2.821	8.922	8.46	71.5716	2.909	9.198	8.96	80.2816	2.993	9.466
7.97	63.5209	2.823	8.927	8.47	71.7409	2.910	9.203	8.97	80.4609	2.995	9.471
7.98	63.6804	2.825	8.933	8.48	71.9104	2.912	9.209	8.98	80.6404	2.997	9.476
7.99	63.8401	2.827	8.939	8.49	72.0801	2.914	9.214	8.99	80.8201	2.998	9.482
8.00	64.0000	2.828	8.944	8.50	72.2500	2.915	9.220	9.00	81.0000	3.000	9.487
8.01	64.1601	2.830	8.950	8.51	72.4201	2.917	9.225	9.01	81.1801	3.002	9.492
8.02	64.3204	2.832	8.955	8.52	72.5904	2.919	9.230	9.02	81.3604	3.003	9.497
8.03	64.4809	2.834	8.961	8.53	72.7609	2.921	9.236	9.03	81.5409	3.005	9.503
8.04	64.6416	2.835	8.967	8.54	72.9316	2.922	9.241	9.04	81.7216	3.007	9.508
8.05	64.8025	2.837	8.972	8.55	73.1025	2.924	9.247	9.05	81.9025	3.008	9.513
8.06	64.9636	2.839	8.978	8.56	73.2736	2.926	9.252	9.06	82.0836	3.010	9.518
8.07	65.1249	2.841	8.983	8.57	73.4449	2.927	9.257	9.07	82.2649	3.012	9.524
8.08	65.2864	2.843	8.989	8.58	73.6164	2.929	9.263	9.08	82.4464	3.013	9.529
8.09	65.4481	2.844	8.994	8.59	73.7881	2.931	9.268	9.09	82.6281	3.015	9.534
8.10	65.6100	2.846	9.000	8.60	73.9600	2.933	9.274	9.10	82.8100	3.017	9.539
8.11	65.7721	2.848	9.006	8.61	74.1321	2.934	9.279	9.11	82.9921	3.018	9.545
8.12	65.9344	2.850	9.011	8.62	74.3044	2.936	9.284	9.12	83.1744	3.020	9.550
8.13	66.0969	2.851	9.017	8.63	74.4769	2.938	9.290	9.13	83.3569	3.022	9.555
8.14	66.2596	2.853	9.022	8.64	74.6496	2.939	9.295	9.14	83.5396	3.023	9.560
8.15	66.4225	2.855	9.028	8.65	74.8225	2.941	9.301	9.15	83.7225	3.025	9.566
8.16	66.5856	2.857	9.033	8.66	74.9956	2.943	9.306	9.16	83.9056	3.027	9.571
8.17	66.7489	2.858	9.039	8.67	75.1689	2.944	9.311	9.17	84.0889	3.028	9.576
8.18	66.9124	2.860	9.044	8.68	75.3424	2.946	9.317	9.18	84.2724	3.030	9.581
8.19	67.0761	2.862	9.050	8.69	75.5161	2.948	9.322	9.19	84.4561	3.032	9.586
8.20	67.2400	2.864	9.055	8.70	75.6900	2.950	9.327	9.20	84.6400	3.033	9.592
8.21	67.4041	2.865	9.061	8.71	75.8641	2.951	9.333	9.21	84.8241	3.035	9.597
8.22	67.5684	2.867	9.066	8.72	76.0384	2.953	9.338	9.22	85.0084	3.036	9.602
8.23	67.7329	2.869	9.072	8.73	76.2129	2.955	9.343	9.23	85.1929	3.038	9.607
8.24	67.8976	2.871	9.077	8.74	76.3876	2.956	9.349	9.24	85.3776	3.040	9.612

TABLE 1   SQUARES AND SQUARE ROOTS *Continued*

n	n^2	$\sqrt{n}$	$\sqrt{10n}$	n	n^2	$\sqrt{n}$	$\sqrt{10n}$	n	n^2	$\sqrt{n}$	$\sqrt{10n}$
9.25	85.5625	3.041	9.618	9.50	90.2500	3.082	9.747	9.75	95.0625	3.122	9.874
9.26	85.7476	3.043	9.623	9.51	90.4401	3.084	9.752	9.76	95.2576	3.124	9.879
9.27	85.9329	3.045	9.628	9.52	90.6304	3.085	9.757	9.77	95.4529	3.126	9.884
9.28	86.1184	3.046	9.633	9.53	90.8209	3.087	9.762	9.78	95.6484	3.127	9.889
9.29	86.3041	3.048	9.638	9.54	91.0116	3.089	9.767	9.79	95.8441	3.129	9.894
9.30	86.4900	3.050	9.644	9.55	91.2025	3.090	9.772	9.80	96.0400	3.130	9.899
9.31	86.6761	3.051	9.649	9.56	91.3936	3.092	9.778	9.81	96.2361	3.132	9.905
9.32	86.8624	3.053	9.654	9.57	91.5849	3.094	9.783	9.82	96.4324	3.134	9.910
9.33	87.0489	3.055	9.659	9.58	91.7764	3.095	9.788	9.83	96.6289	3.135	9.915
9.34	87.2356	3.056	9.664	9.59	91.9681	3.097	9.793	9.84	96.8256	3.137	9.920
9.35	87.4225	3.058	9.670	9.60	92.1600	3.098	9.798	9.85	97.0225	3.138	9.925
9.36	87.6096	3.059	9.675	9.61	92.3521	3.100	9.803	9.86	97.2196	3.140	9.930
9.37	87.7969	3.061	9.680	9.62	92.5444	3.102	9.808	9.87	97.4169	3.142	9.935
9.38	87.9844	3.063	9.685	9.63	92.7369	3.103	9.813	9.88	97.6144	3.143	9.940
9.39	88.1721	3.064	9.690	9.64	92.9296	3.105	9.818	9.89	97.8121	3.145	9.945
9.40	88.3600	3.066	9.695	9.65	93.1225	3.106	9.823	9.90	98.0100	3.146	9.950
9.41	88.5481	3.068	9.701	9.66	93.3156	3.108	9.829	9.91	98.2081	3.148	9.955
9.42	88.7364	3.069	9.706	9.67	93.5089	3.110	9.834	9.92	98.4064	3.150	9.960
9.43	88.9249	3.071	9.711	9.68	93.7024	3.111	9.839	9.93	98.6049	3.151	9.965
9.44	89.1136	3.072	9.716	9.69	93.8961	3.113	9.844	9.94	98.8036	3.153	9.970
9.45	89.3025	3.074	9.721	9.70	94.0900	3.114	9.849	9.95	99.0025	3.154	9.975
9.46	89.4916	3.076	9.726	9.71	94.2841	3.116	9.854	9.96	99.2016	3.156	9.980
9.47	89.6809	3.077	9.731	9.72	94.4784	3.118	9.859	9.97	99.4009	3.157	9.985
9.48	89.8704	3.079	9.737	9.73	94.6729	3.119	9.864	9.98	99.6004	3.159	9.990
9.49	90.0601	3.081	9.742	9.74	94.8676	3.121	9.869	9.99	99.8001	3.161	9.995

TABLE 2 BINOMIAL PROBABILITIES

$$\frac{n!}{X!(n-X)!}P^X(1-P)^{n-X}$$

n	X	.01	.05	.10	.20	.30	.40	P .50	.60	.70	.80	.90	.95	.99
2	0	.980	.902	.810	.640	.490	.360	.250	.160	.090	.040	.010	.002	
	1	.020	.095	.180	.320	.420	.480	.500	.480	.420	.320	.180	.095	.020
	2			.002	.010	.040	.090	.160	.250	.360	.490	.640	.810	.902	.980
3	0	.970	.857	.729	.512	.343	.216	.125	.064	.027	.008	.001		
	1	.029	.135	.243	.384	.441	.432	.375	.288	.189	.096	.027	.007	
	2		.007	.027	.096	.189	.288	.375	.432	.441	.384	.243	.135	.029
	3			.001	.008	.027	.064	.125	.216	.343	.512	.729	.857	.970
4	0	.961	.815	.656	.410	.240	.130	.063	.026	.008	.002			
	1	.039	.171	.292	.410	.412	.346	.250	.154	.076	.026	.004		
	2	.001	.014	.049	.154	.265	.346	.375	.346	.265	.154	.049	.014	.001
	3			.004	.026	.076	.154	.250	.346	.412	.410	.292	.171	.039
	4				.002	.008	.026	.063	.130	.240	.410	.656	.815	.961
5	0	.951	.774	.590	.328	.168	.078	.031	.010	.002				
	1	.048	.204	.328	.410	.360	.259	.156	.077	.028	.006			
	2	.001	.021	.073	.205	.309	.346	.312	.230	.132	.051	.008	.001	
	3		.001	.008	.051	.132	.230	.312	.346	.309	.205	.073	.021	.001
	4				.006	.028	.077	.156	.259	.360	.410	.328	.204	.048
	5					.002	.010	.031	.078	.168	.328	.590	.774	.951
6	0	.941	.735	.531	.262	.118	.047	.016	.004	.001				
	1	.057	.232	.354	.393	.303	.187	.094	.037	.010	.002			
	2	.001	.031	.098	.246	.324	.311	.234	.138	.060	.015	.001		
	3		.002	.015	.082	.185	.276	.312	.276	.185	.082	.015	.002	
	4			.001	.015	.060	.138	.234	.311	.324	.246	.098	.031	.001
	5				.002	.010	.037	.094	.187	.303	.393	.354	.232	.057
	6					.001	.004	.016	.047	.118	.262	.531	.735	.941
7	0	.932	.698	.478	.210	.082	.028	.008	.002					
	1	.066	.257	.372	.367	.247	.131	.055	.017	.004				
	2	.002	.041	.124	.275	.318	.261	.164	.077	.025	.004			
	3		.004	.023	.115	.227	.290	.273	.194	.097	.029	.003		
	4			.003	.029	.097	.194	.273	.290	.227	.115	.023	.004	
	5				.004	.025	.077	.164	.261	.318	.275	.124	.041	.002
	6					.004	.017	.055	.131	.247	.367	.372	.257	.066
	7						.002	.008	.028	.082	.210	.478	.698	.932
8	0	.923	.663	.430	.168	.058	.017	.004	.001					
	1	.075	.279	.383	.336	.198	.090	.031	.008	.001				
	2	.003	.051	.149	.294	.296	.209	.109	.041	.010	.001			
	3		.005	.033	.147	.254	.279	.219	.124	.047	.009			
	4			.005	.046	.136	.232	.273	.232	.136	.046	.005		
	5				.009	.047	.124	.219	.279	.254	.147	.033	.005	
	6				.001	.010	.041	.109	.209	.296	.294	.149	.051	.003
	7					.001	.008	.031	.090	.198	.336	.383	.279	.075
	8						.001	.004	.017	.058	.168	.430	.663	.923
9	0	.914	.630	.387	.134	.040	.010	.002						
	1	.083	.299	.387	.302	.156	.060	.018	.004					
	2	.003	.063	.172	.302	.267	.161	.070	.021	.004				
	3		.008	.045	.176	.267	.251	.164	.074	.021	.003			
	4		.001	.007	.066	.172	.251	.246	.167	.074	.017	.001		
	5			.001	.017	.074	.167	.246	.251	.172	.066	.007	.001	

TABLE 2 BINOMIAL PROBABILITIES *Continued*

n	X	.01	.05	.10	.20	.30	.40	P .50	.60	.70	.80	.90	.95	.99
9	6				.003	.021	.074	.164	.251	.267	.176	.045	.008	
	7					.004	.021	.070	.161	.267	.302	.172	.063	.003
	8						.004	.018	.060	.156	.302	.387	.299	.083
	9							.002	.010	.040	.134	.387	.630	.914
10	0	.904	.599	.349	.107	.028	.006	.001						
	1	.091	.315	.387	.268	.121	.040	.010	.002					
	2	.004	.075	.194	.302	.233	.121	.044	.011	.001				
	3		.010	.057	.201	.267	.215	.117	.042	.009	.001			
	4		.001	.011	.088	.200	.251	.205	.111	.037	.006			
	5			.001	.026	.103	.201	.246	.201	.103	.026	.001		
	6				.006	.037	.111	.205	.251	.200	.088	.011	.001	
	7				.001	.009	.042	.117	.215	.267	.201	.057	.010	
	8					.001	.011	.044	.121	.233	.302	.194	.075	.004
	9						.002	.010	.040	.121	.268	.387	.315	.091
	10							.001	.006	.028	.107	.349	.599	.904
11	0	.895	.569	.314	.086	.020	.004							
	1	.099	.329	.384	.236	.093	.027	.005	.001					
	2	.005	.087	.213	.295	.200	.089	.027	.005	.001				
	3		.014	.071	.221	.257	.177	.081	.023	.004				
	4		.001	.016	.111	.220	.236	.161	.070	.017	.002			
	5			.002	.039	.132	.221	.226	.147	.057	.010			
	6				.010	.057	.147	.226	.221	.132	.039	.002		
	7				.002	.017	.070	.161	.236	.220	.111	.016	.001	
	8					.004	.023	.081	.177	.257	.221	.071	.014	
	9					.001	.005	.027	.089	.200	.295	.213	.087	.005
	10						.001	.005	.027	.093	.236	.384	.329	.099
	11							.004	.020	.086	.314	.569	.895	
12	0	.886	.540	.282	.069	.014	.002							
	1	.107	.341	.377	.206	.071	.017	.003						
	2	.006	.099	.230	.283	.168	.064	.016	.002					
	3		.017	.085	.236	.240	.142	.054	.012	.001				
	4		.002	.021	.133	.231	.213	.121	.042	.008	.001			
	5			.004	.053	.158	.227	.193	.101	.029	.003			
	6				.016	.079	.177	.226	.177	.079	.016			
	7				.003	.029	.101	.193	.227	.158	.053	.004		
	8				.001	.008	.042	.121	.213	.231	.133	.021	.002	
	9					.001	.012	.054	.142	.240	.236	.085	.017	
	10						.002	.016	.064	.168	.283	.230	.099	.006
	11							.003	.017	.071	.206	.377	.341	.107
	12								.002	.014	.069	.282	.540	.886
13	0	.878	.513	.254	.055	.010	.001							
	1	.115	.351	.367	.179	.054	.011	.002						
	2	.007	.111	.245	.268	.139	.045	.010	.001					
	3		.021	.100	.246	.218	.111	.035	.006	.001				
	4		.003	.028	.154	.234	.184	.087	.024	.003				
	5			.006	.069	.180	.221	.157	.066	.014	.001			
	6			.001	.023	.103	.197	.209	.131	.044	.006			
	7				.006	.044	.131	.209	.197	.103	.023	.001		
	8				.001	.014	.066	.157	.221	.180	.069	.006		
	9					.003	.024	.087	.184	.234	.154	.028	.003	
	10					.001	.006	.035	.111	.218	.246	.100	.021	
	11						.001	.010	.045	.139	.268	.245	.111	.007
	12							.002	.011	.054	.179	.367	.351	.115
	13								.001	.010	.055	.254	.513	.878

TABLE 2 BINOMIAL PROBABILITIES *Continued*

n	X	.01	.05	.10	.20	.30	.40	P .50	.60	.70	.80	.90	.95	.99
14	0	.869	.488	.229	.044	.007	.001							
	1	.123	.359	.356	.154	.041	.007	.001						
	2	.008	.123	.257	.250	.113	.032	.006	.001					
	3		.026	.114	.250	.194	.085	.022	.003					
	4		.004	.035	.172	.229	.155	.061	.014	.001				
	5			.008	.086	.196	.207	.122	.041	.007				
	6			.001	.032	.126	.207	.183	.092	.023	.002			
	7				.009	.062	.157	.209	.157	.062	.009			
	8				.002	.023	.092	.183	.207	.126	.032	.001		
	9					.007	.041	.122	.207	.196	.086	.008		
	10					.001	.014	.061	.155	.229	.172	.035	.004	
	11						.003	.022	.085	.194	.250	.114	.026	
	12						.001	.006	.032	.113	.250	.257	.123	.008
	13							.001	.007	.041	.154	.356	.359	.123
	14								.001	.007	.044	.229	.488	.869
15	0	.860	.463	.206	.035	.005								
	1	.130	.366	.343	.132	.031	.005							
	2	.009	.135	.267	.231	.092	.022	.003						
	3		.031	.129	.250	.170	.063	.014	.002					
	4		.005	.043	.188	.219	.127	.042	.007	.001				
	5		.001	.010	.103	.206	.186	.092	.024	.003				
	6			.002	.043	.147	.207	.153	.061	.012	.001			
	7				.014	.081	.177	.196	.118	.035	.003			
	8				.003	.035	.118	.196	.177	.081	.014			
	9				.001	.012	.061	.153	.207	.147	.043	.002		
	10					.003	.024	.092	.186	.206	.103	.010	.001	
	11					.001	.007	.042	.127	.219	.188	.043	.005	
	12						.002	.014	.063	.170	.250	.129	.031	
	13							.003	.022	.092	.231	.267	.135	.009
	14								.005	.031	.132	.343	.366	.130
	15									.005	.035	.206	.463	.860
16	0	.851	.440	.185	.028	.003								
	1	.138	.371	.329	.113	.023	.003							
	2	.010	.146	.274	.211	.073	.015	.002						
	3		.036	.142	.246	.146	.047	.008	.001					
	4		.006	.051	.200	.204	.101	.028	.004					
	5		.001	.014	.120	.210	.162	.067	.014	.001				
	6			.003	.055	.165	.198	.122	.039	.006				
	7				.020	.101	.189	.175	.084	.018	.001			
	8				.006	.049	.142	.196	.142	.049	.006			
	9				.001	.018	.084	.175	.189	.101	.020			
	10					.006	.039	.122	.198	.165	.055	.003		
	11					.001	.014	.067	.162	.210	.120	.014	.001	
	12						.004	.028	.101	.204	.200	.051	.006	
	13						.001	.008	.047	.146	.246	.142	.036	
	14							.002	.015	.073	.211	.274	.146	.010
	15								.003	.023	.113	.329	.371	.138
	16									.003	.028	.185	.440	.851
17	0	.843	.418	.167	.022	.002								
	1	.145	.374	.315	.096	.017	.002							
	2	.012	.158	.280	.191	.058	.010	.001						
	3	.001	.042	.156	.239	.124	.034	.005						
	4		.008	.060	.209	.187	.080	.018	.002					
	5		.001	.018	.136	.208	.138	.047	.008	.001				
	6			.004	.068	.178	.184	.094	.024	.003				
	7			.001	.027	.120	.193	.148	.057	.010				

TABLE 2 BINOMIAL PROBABILITIES *Continued*

n	X	.01	.05	.10	.20	.30	.40	P .50	.60	.70	.80	.90	.95	.99
17	8				.008	.064	.161	.186	.107	.028	.002			
	9				.002	.028	.107	.186	.161	.064	.008			
	10					.010	.057	.148	.193	.120	.027	.001		
	11					.003	.024	.094	.184	.178	.068	.004		
	12					.001	.008	.047	.138	.208	.136	.018	.001	
	13						.002	.018	.080	.187	.209	.060	.008	
	14							.005	.034	.124	.239	.156	.042	.001
	15							.001	.010	.058	.191	.280	.158	.012
	16								.002	.017	.096	.315	.374	.145
	17									.002	.022	.167	.418	.843
18	0	.835	.397	.150	.018	.002								
	1	.152	.376	.300	.081	.013	.001							
	2	.013	.168	.284	.172	.046	.007	.001						
	3	.001	.047	.168	.230	.105	.025	.003						
	4		.009	.070	.215	.168	.061	.012	.001					
	5		.001	.022	.151	.202	.115	.033	.004					
	6			.005	.082	.187	.166	.071	.014	.001				
	7			.001	.035	.138	.189	.121	.037	.005				
	8				.012	.081	.173	.167	.077	.015	.001			
	9				.003	.038	.128	.186	.128	.038	.003			
	10				.001	.015	.077	.167	.173	.081	.012			
	11					.005	.037	.121	.189	.138	.035	.001		
	12					.001	.014	.071	.166	.187	.082	.005		
	13						.004	.033	.115	.202	.151	.022	.001	
	14						.001	.012	.061	.168	.215	.070	.009	
	15							.003	.025	.105	.230	.168	.047	.001
	16							.001	.007	.046	.172	.284	.168	.013
	17								.001	.013	.081	.300	.376	.152
	18									.002	.018	.150	.397	.835
19	0	.826	.377	.135	.014	.001								
	1	.159	.377	.285	.069	.009	.001							
	2	.014	.179	.285	.154	.036	.005							
	3	.001	.053	.180	.218	.087	.018	.002						
	4		.011	.080	.218	.149	.047	.007						
	5		.002	.027	.164	.192	.093	.022	.002					
	6			.007	.096	.192	.145	.052	.008					
	7			.001	.044	.152	.180	.096	.024	.002				
	8				.017	.098	.180	.144	.053	.008				
	9				.005	.051	.146	.176	.098	.022	.001			
	10				.001	.022	.098	.176	.146	.051	.005			
	11					.008	.053	.144	.180	.098	.017			
	12					.002	.024	.096	.180	.152	.044	.001		
	13						.008	.052	.145	.192	.096	.007		
	14						.002	.022	.093	.192	.164	.027	.002	
	15							.007	.047	.149	.218	.080	.011	
	16							.002	.018	.087	.218	.180	.053	.001
	17								.005	.036	.154	.285	.179	.014
	18								.001	.009	.069	.285	.377	.159
	19									.001	.014	.135	.377	.826
20	0	.818	.358	.122	.012	.001								
	1	.165	.377	.270	.058	.007								
	2	.016	.189	.285	.137	.028	.003							
	3	.001	.060	.190	.205	.072	.012	.001						
	4		.013	.090	.218	.130	.035	.005						
	5		.002	.032	.175	.179	.075	.015	.001					
	6			.009	.109	.192	.124	.037	.005					

417

TABLE 2   BINOMIAL PROBABILITIES   *Continued*

n	X	.01	.05	.10	.20	.30	.40	P .50	.60	.70	.80	.90	.95	.99
20	7			.002	.054	.164	.166	.074	.015	.001				
	8				.022	.114	.180	.120	.036	.004				
	9				.007	.065	.160	.160	.071	.012				
	10				.002	.031	.117	.176	.117	.031	.002			
	11					.012	.071	.160	.160	.065	.007			
	12					.004	.036	.120	.180	.114	.022			
	13					.001	.015	.074	.166	.164	.054	.002		
	14						.005	.037	.124	.192	.109	.009		
	15						.001	.015	.075	.179	.175	.032	.002	
	16							.005	.035	.130	.218	.090	.013	
	17							.001	.012	.072	.205	.190	.060	.001
	18								.003	.028	.137	.285	.189	.016
	19									.007	.058	.270	.377	.165
	20									.001	.012	.122	.358	.818

TABLE 3   RANDOM NUMBERS

	1	2	3	4	5	6	7	8
1	04479	44211	81536	09686	26743	87001	62392	59946
2	87019	90503	16034	07862	19701	85949	85876	58188
3	54222	56179	09833	34227	43897	38517	11617	30338
4	17929	24021	50932	89349	08012	37925	59003	95503
5	56399	82269	69443	62020	03365	82164	01356	24871
6	79242	52682	36255	74168	28636	93043	65454	36152
7	88869	22489	50467	14964	93146	51852	32408	22545
8	83970	03473	42981	83127	98774	74392	12218	91841
9	52754	85751	92705	70949	24331	42672	04885	44251
10	27011	69215	90920	96218	81127	67792	08377	60773
11	67952	43155	65547	50055	97940	38833	08745	69207
12	93376	38289	89474	22350	84982	85224	29969	42745
13	13674	24899	60602	33203	91953	48635	43938	08285
14	19345	11394	09241	72723	09052	76987	89854	48849
15	91609	18375	16171	30692	37389	51879	29556	51315
16	68537	17630	70322	26128	15645	91691	81064	58083
17	10038	17181	93964	41122	13020	98243	46447	28675
18	57023	28928	73917	94774	62542	30536	14777	72360
19	70791	39030	11261	76783	31184	38669	95862	99067
20	88033	42447	17815	75351	40853	83513	88714	09887
21	71858	65129	60871	04586	90651	93207	85501	83600
22	40357	80097	82138	61279	70478	49731	94154	50436
23	03339	05350	61895	46420	81433	61995	16654	91274
24	43084	21898	98854	70139	31516	29990	40919	05125
25	77223	58612	93223	12495	12628	43715	88010	03080
26	87135	86620	56893	82220	33968	13380	38087	74056
27	82521	26025	67975	81512	85227	39786	82990	38936
28	63590	66694	35357	19452	67724	10912	58569	66929
29	38133	07569	71030	75769	89240	48888	27184	78014
30	70369	48709	65114	69725	42994	22584	18455	52022
31	16168	91235	17509	72148	34676	61011	03681	21135
32	38541	45056	27395	13139	57487	57389	10764	62267
33	35508	90052	94492	83678	11316	98396	20893	87494
34	27147	55333	29880	81775	05384	86224	43487	86643
35	95351	12900	12689	07330	29470	39802	79928	68896
36	48047	70852	63798	62452	83695	38200	17414	13151
37	86417	48099	72299	46033	88948	93459	89657	52339
38	10361	07412	48001	57271	13210	04328	23855	65719
39	29998	88220	63213	98976	78720	61138	90709	50003
40	13008	59213	55737	68130	74358	74687	79519	29409
41	92882	31482	05651	53952	00915	43967	62276	47818
42	57149	14046	02876	79221	76700	68078	67712	98230
43	87080	09985	68303	23068	73514	39328	56046	98785
44	33504	84019	91220	05463	19500	66509	87209	71293
45	44702	70429	73468	16316	87536	49921	00239	37743
46	40473	76124	12097	56736	84635	77172	47155	77306
47	31727	64165	28937	14805	22863	62154	87637	80982
48	98855	63471	83278	00131	90229	02976	49485	67541
49	56067	73922	05810	24125	09603	99539	04848	57223
50	87152	73758	86758	77787	47126	31822	72088	00927

0   z

TABLE 4   AREAS UNDER THE NORMAL CURVE

z	.00	.01	.02	.03	.04	.05	.06	.07	.08	.09
0.0	.0000	.0040	.0080	.0120	.0160	.0199	.0239	.0279	.0319	.0359
0.1	.0398	.0438	.0478	.0517	.0557	.0596	.0636	.0675	.0714	.0753
0.2	.0793	.0832	.0871	.0910	.0948	.0987	.1026	.1064	.1103	.1141
0.3	.1179	.1217	.1255	.1293	.1331	.1368	.1406	.1443	.1480	.1517
0.4	.1554	.1591	.1628	.1664	.1700	.1736	.1772	.1808	.1844	.1879
0.5	.1915	.1950	.1985	.2019	.2054	.2088	.2123	.2157	.2190	.2224
0.6	.2257	.2291	.2324	.2357	.2389	.2422	.2454	.2486	.2518	.2549
0.7	.2580	.2611	.2642	.2673	.2704	.2734	.2764	.2794	.2823	.2852
0.8	.2881	.2910	.2939	.2967	.2995	.3023	.3051	.3078	.3106	.3133
0.9	.3159	.3186	.3212	.3238	.3264	.3289	.3315	.3340	.3365	.3389
1.0	.3413	.3438	.3461	.3485	.3508	.3531	.3554	.3577	.3599	.3621
1.1	.3643	.3665	.3686	.3708	.3729	.3749	.3770	.3790	.3810	.3830
1.2	.3849	.3869	.3888	.3907	.3925	.3944	.3962	.3980	.3997	.4015
1.3	.4032	.4049	.4066	.4082	.4099	.4115	.4131	.4147	.4162	.4177
1.4	.4192	.4207	.4222	.4236	.4251	.4265	.4279	.4292	.4306	.4319
1.5	.4332	.4345	.4357	.4370	.4382	.4394	.4406	.4418	.4429	.4441
1.6	.4452	.4463	.4474	.4484	.4495	.4505	.4515	.4525	.4535	.4545
1.7	.4554	.4564	.4573	.4582	.4591	.4599	.4608	.4616	.4625	.4633
1.8	.4641	.4649	.4656	.4664	.4671	.4678	.4686	.4693	.4699	.4706
1.9	.4713	.4719	.4726	.4732	.4738	.4744	.4750	.4756	.4761	.4767
2.0	.4772	.4778	.4783	.4788	.4793	.4798	.4803	.4808	.4812	.4817
2.1	.4821	.4826	.4830	.4834	.4838	.4842	.4846	.4850	.4854	.4857
2.2	.4861	.4864	.4868	.4871	.4875	.4878	.4881	.4884	.4887	.4890
2.3	.4893	.4896	.4898	.4901	.4904	.4906	.4909	.4911	.4913	.4916
2.4	.4918	.4920	.4922	.4925	.4927	.4929	.4931	.4932	.4934	.4936
2.5	.4938	.4940	.4941	.4943	.4945	.4946	.4948	.4949	.4951	.4952
2.6	.4953	.4955	.4956	.4957	.4959	.4960	.4961	.4962	.4963	.4964
2.7	.4965	.4966	.4967	.4968	.4969	.4970	.4971	.4972	.4973	.4974
2.8	.4974	.4975	.4976	.4977	.4977	.4978	.4979	.4979	.4980	.4981
2.9	.4981	.4982	.4982	.4982	.4984	.4984	.4985	.4985	.4986	.4986
3.0	.4987	.4987	.4987	.4988	.4988	.4989	.4989	.4989	.4990	.4990
3.1	.4990	.4991	.4991	.4991	.4992	.4992	.4992	.4992	.4993	.4993
3.2	.4993	.4993	.4994	.4994	.4994	.4994	.4994	.4995	.4995	.4995
3.3	.4995									
3.4	.4997									
3.5	.4998									
3.6	.4998									
3.7	.4999									
3.8	.4999									
3.9	.5000									
4.0	.5000									

2.326

2.576

T-value close to 0 results in large p-val

T-val far from 0 results in small p-val

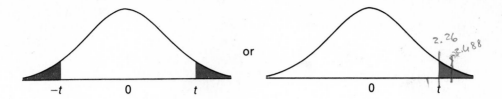

2.26

≈ 2.488

TABLE 5   STUDENT'S *t* DISTRIBUTIONS

	.50	.90	.95	.98	.99
One-tail:	.10	.05	.025	.01	.005
Two-tail:	.20	.10	.05	.02	.01
D	1.28	1.645	1.96	2.33	2.58	N(0,1)
1	3.08	6.31	12.71	31.82	63.66
2	1.89	2.92	4.30	6.97	9.92
3	1.64	2.35	3.18	4.54	5.84
4	1.53	2.13	2.78	3.75	4.60
5	1.48	2.02	2.57	3.37	4.03
6	1.44	1.94	2.45	3.14	3.71
7	1.42	1.90	2.37	3.00	3.50
8	1.40	1.86	2.31	2.90	3.36
9	1.38	1.83	2.26	2.82	3.25
10	1.37	1.81	2.23	2.76	3.17
11	1.36	1.80	2.20	2.72	3.11
12	1.36	1.78	2.18	2.68	3.06
13	1.35	1.77	2.16	2.65	3.01
14	1.35	1.76	2.15	2.62	2.98
15	1.34	1.75	2.13	2.60	2.95
16	1.34	1.75	2.12	2.58	2.92
17	1.33	1.74	2.11	2.57	2.90
18	1.33	1.73	2.10	2.55	2.88
19	1.33	1.73	2.09	2.54	2.86
20	1.33	1.73	2.09	2.53	2.85
21	1.32	1.72	2.08	2.52	2.83
22	1.32	1.72	2.07	2.51	2.82
23	1.32	1.71	2.07	2.50	2.81
24	1.32	1.71	2.06	2.49	2.80
25	1.32	1.71	2.06	2.49	2.79
26	1.32	1.71	2.06	2.48	2.78
27	1.31	1.70	2.05	2.47	2.77
28	1.31	1.70	2.05	2.47	2.76
29	1.31	1.70	2.05	2.46	2.76
30	1.31	1.70	2.04	2.46	2.75
40	1.30	1.68	2.02	2.42	2.70
∞	1.28	1.64	1.96	2.33	2.58

.02   .002   1.702

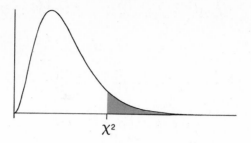

χ^2

TABLE 6 χ^2 DISTRIBUTIONS

D	.10	.05	.025	.01
1	2.71	3.84	5.02	6.63
2	4.61	5.99	7.38	9.21
3	6.25	7.81	9.35	11.3
4	7.78	9.49	11.1	13.3
5	9.24	11.1	12.8	15.1
6	10.6	12.6	14.4	16.8
7	12.0	14.1	16.0	18.5
8	13.4	15.5	17.5	20.1
9	14.7	16.9	19.0	21.7
10	16.0	18.3	20.5	23.2
11	17.3	19.7	21.9	24.7
12	18.5	21.0	23.3	26.2
13	19.8	22.4	24.7	27.7
14	21.1	23.7	26.1	29.1
15	22.3	25.0	27.5	30.6
16	23.5	26.3	28.8	32.0
17	24.8	27.6	30.2	33.4
18	26.0	28.9	31.5	34.8
19	27.2	30.1	32.9	36.2
20	28.4	31.4	34.2	37.6
21	29.6	32.7	35.5	38.9
22	30.8	33.9	36.8	40.3
23	32.0	35.2	38.1	41.6
24	33.2	36.4	39.4	43.0
25	34.4	37.7	40.6	44.3
26	35.6	38.9	41.9	45.6
27	36.7	40.1	43.2	47.0
28	37.9	41.3	44.5	48.3
29	39.1	42.6	45.7	49.6
30	40.3	43.8	47.0	50.9
40	51.8	55.8	59.3	63.7
50	63.2	67.5	71.4	76.2
60	74.4	79.1	83.3	88.4
70	85.5	90.5	95.0	100.4
80	96.6	101.9	106.6	112.3

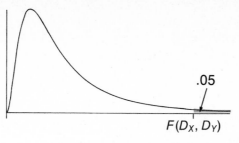

.05

$F(D_X, D_Y)$

TABLE 7a F DISTRIBUTIONS

$\alpha = .05$

always — momo variato
< x .05

D_X

D_Y	1	2	3	4	5	6	7	8	9
1	161	200	216	225	230	234	237	239	241
2	18.5	19.0	19.2	19.3	19.3	19.3	19.4	19.4	19.4
3	10.1	9.55	9.28	9.12	9.01	8.94	8.89	8.85	8.81
4	7.71	6.94	6.59	6.39	6.26	6.16	6.09	6.04	6.00
5	6.61	5.79	5.41	5.19	5.05	4.95	4.88	4.82	4.77
6	5.99	5.14	4.76	4.53	4.39	4.28	4.21	4.15	4.10
7	5.59	4.74	4.35	4.12	3.97	3.87	3.79	3.73	3.68
8	5.32	4.46	4.07	3.84	3.69	3.58	3.50	3.44	3.39
9	5.12	4.26	3.86	3.63	3.48	3.37	3.29	3.23	3.18
10	4.96	4.10	3.71	3.48	3.33	3.22	3.14	3.07	3.02
11	4.84	3.98	3.59	3.36	3.20	3.09	3.01	2.95	2.90
12	4.75	3.89	3.49	3.26	3.11	3.00	2.91	2.85	2.80
13	4.67	3.81	3.41	3.18	3.03	2.92	2.83	2.77	2.71
14	4.60	3.74	3.34	3.11	2.96	2.85	2.76	2.70	2.65
15	4.54	3.68	3.29	3.06	2.90	2.79	2.71	2.64	2.59
16	4.49	3.63	3.24	3.01	2.85	2.74	2.66	2.59	2.54
17	4.45	3.59	3.20	2.96	2.81	2.70	2.61	2.55	2.49
18	4.41	3.55	3.16	2.93	2.77	2.66	2.58	2.51	2.46
19	4.38	3.52	3.13	2.90	2.74	2.63	2.54	2.48	2.42
20	4.35	3.49	3.10	2.87	2.71	2.60	2.51	2.45	2.39
21	4.32	3.47	3.07	2.84	2.68	2.57	2.49	2.42	2.37
22	4.30	3.44	3.05	2.82	2.66	2.55	2.46	2.40	2.34
23	4.28	3.42	3.03	2.80	2.64	2.53	2.44	2.37	2.32
24	4.26	3.40	3.01	2.78	2.62	2.51	2.42	2.36	2.30
25	4.24	3.39	2.99	2.76	2.60	2.49	2.40	2.34	2.28
26	4.23	3.37	2.98	2.74	2.59	2.47	2.39	2.32	2.27
27	4.21	3.35	2.96	2.73	2.57	2.46	2.37	2.31	2.25
28	4.20	3.34	2.95	2.71	2.56	2.45	2.36	2.29	2.24
29	4.18	3.33	2.93	2.70	2.55	2.43	2.35	2.28	2.22
30	4.17	3.32	2.92	2.69	2.53	2.42	2.33	2.27	2.21
40	4.08	3.23	2.84	2.61	2.45	2.34	2.25	2.18	2.12
60	4.00	3.15	2.76	2.53	2.37	2.25	2.17	2.10	2.04
80	3.96	3.11	2.72	2.48	2.32	2.21	2.12	2.05	1.99
100	3.94	3.09	2.70	2.46	2.30	2.19	2.10	2.03	1.97
120	3.92	3.07	2.68	2.45	2.29	2.17	2.09	2.02	1.96
∞	3.84	3.00	2.60	2.37	2.21	2.10	2.01	1.94	1.88

TABLE 7a F DISTRIBUTIONS *Continued*

					D_X				
D_Y	12	14	16	20	24	30	40	60	100
1	244	245	246	248	249	250	251	252	253
2	19.4	19.4	19.4	19.5	19.5	19.5	19.5	19.5	19.5
3	8.74	8.71	8.69	8.66	8.64	8.62	8.59	8.57	8.56
4	5.91	5.87	5.84	5.80	5.77	5.75	5.72	5.69	5.66
5	4.68	4.64	4.60	4.56	4.53	4.50	4.46	4.43	4.40
6	4.00	3.96	3.92	3.87	3.84	3.81	3.77	3.74	3.71
7	3.57	3.52	3.49	3.44	3.41	3.38	3.34	3.30	3.28
8	3.28	3.23	3.20	3.15	3.12	3.08	3.04	3.01	2.98
9	3.07	3.02	2.98	2.94	2.90	2.86	2.83	2.79	2.76
10	2.91	2.87	2.82	2.77	2.74	2.70	2.66	2.62	2.59
11	2.79	2.74	2.70	2.65	2.61	2.57	2.53	2.49	2.45
12	2.69	2.64	2.60	2.54	2.51	2.47	2.43	2.38	2.35
13	2.60	2.55	2.51	2.46	2.42	2.38	2.34	2.30	2.26
14	2.53	2.48	2.44	2.39	2.35	2.31	2.27	2.22	2.19
15	2.48	2.43	2.39	2.33	2.29	2.25	2.20	2.16	2.12
16	2.42	2.37	2.33	2.28	2.24	2.19	2.15	2.11	2.07
17	2.38	2.33	2.29	2.23	2.19	2.15	2.10	2.06	2.02
18	2.34	2.29	2.25	2.19	2.15	2.11	2.06	2.02	1.98
19	2.31	2.26	2.21	2.16	2.11	2.07	2.03	1.98	1.94
20	2.28	2.23	2.18	2.12	2.08	2.04	1.99	1.95	1.90
21	2.25	2.20	2.15	2.10	2.05	2.01	1.96	1.92	1.87
22	2.23	2.18	2.13	2.07	2.03	1.98	1.94	1.89	1.84
23	2.20	2.14	2.10	2.05	2.01	1.96	1.91	1.86	1.82
24	2.18	2.13	2.09	2.03	1.98	1.94	1.89	1.84	1.80
25	2.16	2.11	2.06	2.01	1.96	1.92	1.87	1.82	1.77
26	2.15	2.10	2.05	1.99	1.95	1.90	1.85	1.80	1.76
27	2.13	2.08	2.03	1.97	1.93	1.88	1.84	1.79	1.74
28	2.12	2.06	2.02	1.96	1.91	1.87	1.82	1.77	1.72
29	2.10	2.05	2.00	1.94	1.90	1.85	1.81	1.75	1.71
30	2.09	2.04	1.99	1.93	1.89	1.84	1.79	1.74	1.69
40	2.00	1.95	1.90	1.84	1.79	1.74	1.69	1.64	1.59
60	1.92	1.86	1.81	1.75	1.70	1.65	1.59	1.53	1.48
80	1.88	1.82	1.77	1.70	1.65	1.60	1.54	1.47	1.42
100	1.85	1.79	1.75	1.68	1.63	1.57	1.51	1.44	1.39
120	1.83	1.77	1.73	1.66	1.61	1.55	1.50	1.43	1.37
∞	1.75	1.69	1.64	1.57	1.52	1.46	1.39	1.32	1.24

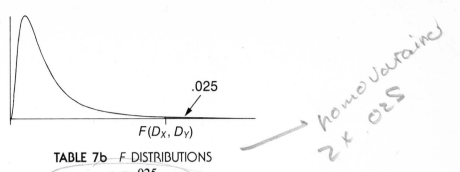

TABLE 7b _F_ DISTRIBUTIONS

$\alpha = .025$

D_X

D_Y	1	2	3	4	5	6	7	8	9
1	648	800	864	900	922	937	948	957	963
2	38.5	39.0	39.2	39.2	39.3	39.3	39.4	39.4	39.4
3	17.4	16.0	15.4	15.1	14.9	14.7	14.6	14.5	14.5
4	12.2	10.6	9.98	9.60	9.36	9.20	9.07	8.98	8.90
5	10.0	8.43	7.76	7.39	7.15	6.98	6.85	6.76	6.68
6	8.81	7.26	6.60	6.23	5.99	5.82	5.70	5.60	5.52
7	8.07	6.54	5.89	5.52	5.29	5.12	4.99	4.90	4.82
8	7.57	6.06	5.42	5.05	4.82	4.65	4.53	4.43	4.36
9	7.21	5.71	5.08	4.72	4.48	4.32	4.20	4.10	4.03
10	6.94	5.46	4.83	4.47	4.24	4.07	3.95	3.85	3.78
11	6.72	5.26	4.63	4.28	4.04	3.88	3.76	3.66	3.59
12	6.55	5.10	4.47	4.12	3.89	3.73	3.61	3.51	3.44
13	6.41	4.97	4.35	4.00	3.77	3.60	3.48	3.39	3.31
14	6.30	4.86	4.24	3.89	3.66	3.50	3.38	3.29	3.21
15	6.20	4.77	4.15	3.80	3.58	3.41	3.29	3.20	3.12
16	6.12	4.69	4.08	3.73	3.50	3.34	3.22	3.12	3.05
17	6.04	4.62	4.01	3.66	3.44	3.28	3.16	3.06	2.98
18	5.98	4.56	3.95	3.61	3.38	3.22	3.10	3.01	2.93
19	5.92	4.51	3.90	3.56	3.33	3.17	3.05	2.96	2.88
20	5.87	4.46	3.86	3.51	3.29	3.13	3.01	2.91	2.84
21	5.83	4.42	3.82	3.48	3.25	3.09	2.97	2.87	2.80
22	5.79	4.38	3.78	3.44	3.22	3.05	2.93	2.84	2.76
23	5.75	4.35	3.75	3.41	3.18	3.02	2.90	2.81	2.73
24	5.72	4.32	3.72	3.38	3.15	2.99	2.87	2.78	2.70
25	5.69	4.29	3.69	3.35	3.13	2.97	2.85	2.75	2.68
26	5.66	4.27	3.67	3.33	3.10	2.94	2.82	2.73	2.65
27	5.63	4.23	3.64	3.30	3.08	2.92	2.80	2.70	2.63
28	5.61	4.21	3.62	3.28	3.06	2.90	2.78	2.68	2.61
29	5.59	4.19	3.60	3.26	3.04	2.88	2.76	2.66	2.59
30	5.57	4.18	3.59	3.25	3.03	2.87	2.75	2.65	2.57
40	5.42	4.05	3.46	3.13	2.90	2.74	2.62	2.53	2.45
60	5.29	3.93	3.34	3.01	2.79	2.63	2.51	2.41	2.33
80	5.20	3.85	3.28	2.94	2.72	2.57	2.45	2.37	2.25
100	5.17	3.82	3.24	2.90	2.68	2.53	2.41	2.31	2.21
120	5.15	3.80	3.23	2.89	2.67	2.52	2.39	2.30	2.22
∞	5.02	3.69	3.12	2.79	2.57	2.41	2.29	2.19	2.11

TABLE 7b F DISTRIBUTIONS Continued

D_Y	12	14	16	20	24	30	40	60	100
					D_X				
1	977	984	989	993	997	1000	1006	1010	1013
2	39.4	39.4	39.4	39.4	39.5	39.5	39.5	39.5	39.5
3	14.3	14.3	14.3	14.2	14.1	14.1	14.0	14.0	13.9
4	8.75	8.69	8.64	8.56	8.51	8.46	8.41	8.36	8.32
5	6.52	6.45	6.41	6.33	6.28	6.23	6.18	6.12	6.08
6	5.37	5.28	5.25	5.17	5.12	5.07	5.01	4.96	4.92
7	4.67	4.59	4.55	4.47	4.42	4.36	4.31	4.25	4.21
8	4.20	4.13	4.08	4.00	3.95	3.89	3.84	3.78	3.74
9	3.87	3.80	3.75	3.67	3.62	3.56	3.51	3.45	3.40
10	3.62	3.55	3.50	3.42	3.37	3.31	3.26	3.20	3.15
11	3.43	3.36	3.31	3.23	3.17	3.12	3.06	3.00	2.95
12	3.28	3.20	3.15	3.07	3.02	2.96	2.91	2.85	2.80
13	3.15	3.08	3.03	2.95	2.89	2.84	2.78	2.72	2.67
14	3.05	2.98	2.92	2.84	2.79	2.73	2.67	2.61	2.56
15	2.96	2.89	2.84	2.76	2.70	2.64	2.59	2.52	2.47
16	2.89	2.82	2.76	2.68	2.63	2.57	2.51	2.45	2.40
17	2.82	2.75	2.70	2.62	2.56	2.50	2.44	2.38	2.33
18	2.77	2.70	2.64	2.56	2.50	2.44	2.38	2.32	2.27
19	2.72	2.65	2.59	2.51	2.45	2.39	2.33	2.27	2.22
20	2.68	2.60	2.54	2.46	2.41	2.35	2.29	2.22	2.17
21	2.64	2.57	2.50	2.42	2.37	2.31	2.25	2.18	2.13
22	2.60	2.53	2.47	2.39	2.33	2.27	2.21	2.14	2.09
23	2.57	2.50	2.44	2.36	2.30	2.24	2.18	2.11	2.06
24	2.54	2.47	2.41	2.33	2.27	2.21	2.15	2.08	2.03
25	2.51	2.44	2.38	2.30	2.24	2.18	2.12	2.05	2.00
26	2.48	2.41	2.35	2.27	2.21	2.15	2.09	2.02	1.97
27	2.46	2.39	2.33	2.25	2.19	2.13	2.07	2.00	1.95
28	2.44	2.37	2.31	2.23	2.17	2.11	2.05	1.98	1.93
29	2.42	2.35	2.29	2.21	2.15	2.09	2.03	1.96	1.91
30	2.41	2.34	2.28	2.20	2.14	2.07	2.01	1.94	1.89
40	2.29	2.21	2.15	2.07	2.01	1.94	1.88	1.80	1.76
60	2.17	2.09	2.03	1.94	1.88	1.82	1.74	1.67	1.61
80	2.10	2.03	1.96	1.87	1.81	1.75	1.67	1.59	1.52
100	2.07	1.99	1.93	1.84	1.78	1.71	1.63	1.55	1.48
120	2.05	1.98	1.92	1.82	1.76	1.69	1.61	1.53	1.46
∞	1.94	1.86	1.81	1.71	1.64	1.57	1.48	1.39	1.30

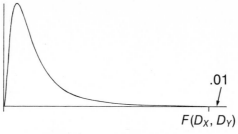

.01

$F(D_X, D_Y)$

homo-variance
.01 × 2

TABLE 7c _F_ DISTRIBUTIONS

$\alpha = .01$

D_X

D_Y	1	2	3	4	5	6	7	8	9
1	4052	4999	5403	5625	5764	5859	5928	5981	6022
2	98.5	99.0	99.2	99.3	99.3	99.3	99.3	99.4	99.4
3	34.1	30.8	29.5	28.7	28.2	27.9	27.7	27.5	27.3
4	21.2	18.0	16.7	16.0	15.5	15.2	15.0	14.8	14.7
5	16.3	13.3	12.1	11.4	11.0	10.7	10.5	10.3	10.2
6	13.7	10.9	9.78	9.15	8.75	8.47	8.26	8.10	7.98
7	12.3	9.55	8.45	7.85	7.46	7.19	7.00	6.84	6.71
8	11.3	8.65	7.59	7.01	6.63	6.37	6.19	6.03	5.91
9	10.6	8.02	6.99	6.42	6.06	5.80	5.62	5.47	5.35
10	10.0	7.56	6.55	5.99	5.64	5.39	5.21	5.06	4.95
11	9.65	7.21	6.22	5.67	5.32	5.07	4.88	4.74	4.63
12	9.33	6.93	5.95	5.41	5.06	4.82	4.65	4.50	4.39
13	9.07	6.70	5.74	5.20	4.86	4.62	4.44	4.30	4.19
14	8.86	6.51	5.56	5.03	4.69	4.46	4.28	4.14	4.03
15	8.68	6.36	5.42	4.89	4.56	4.32	4.14	4.00	3.89
16	8.53	6.23	5.29	4.77	4.44	4.20	4.03	3.89	3.78
17	8.40	6.11	5.18	4.67	4.34	4.10	3.93	3.79	3.68
18	8.28	6.01	5.09	4.58	4.25	4.01	3.85	3.71	3.60
19	8.18	5.93	5.01	4.50	4.17	3.94	3.77	3.63	3.52
20	8.10	5.85	4.94	4.43	4.10	3.87	3.71	3.56	3.45
21	8.02	5.78	4.87	4.37	4.04	3.81	3.65	3.51	3.40
22	7.94	5.72	4.82	4.31	3.99	3.76	3.59	3.45	3.35
23	7.88	5.66	4.76	4.26	3.94	3.71	3.54	3.40	3.30
24	7.82	5.61	4.72	4.22	3.90	3.67	3.50	3.36	3.25
25	7.77	5.57	4.68	4.18	3.86	3.63	3.46	3.32	3.21
26	7.72	5.53	4.64	4.14	3.82	3.59	3.42	3.29	3.17
27	7.68	5.49	4.60	4.11	3.79	3.56	3.39	3.26	3.14
28	7.64	5.45	4.57	4.08	3.76	3.53	3.36	3.23	3.11
29	7.60	5.42	4.54	4.05	3.73	3.50	3.33	3.20	3.09
30	7.56	5.39	4.51	4.02	3.70	3.47	3.30	3.17	3.06
40	7.31	5.18	4.31	3.83	3.51	3.29	3.12	2.99	2.88
60	7.08	4.98	4.13	3.65	3.34	3.12	2.95	2.82	2.72
80	6.96	4.88	4.04	3.56	3.25	3.04	2.87	2.74	2.64
100	6.90	4.82	3.98	3.51	3.20	2.99	2.82	2.69	2.59
120	6.85	4.79	3.95	3.48	3.17	2.96	2.79	2.66	2.56
∞	6.63	4.60	3.78	3.32	3.02	2.80	2.64	2.51	2.41

TABLE 7c *F* DISTRIBUTIONS *Continued*

D_X

D_Y	12	14	16	20	24	30	40	100
1	6106	6140	6170	6210	6235	6260	6287	6330
2	99.4	99.4	99.5	99.5	99.5	99.5	99.5	99.5
3	27.1	26.9	26.8	26.7	26.6	26.5	26.4	26.2
4	14.4	14.2	14.1	14.0	13.9	13.8	13.7	13.6
5	9.89	9.77	9.68	9.55	9.47	9.38	9.29	9.13
6	7.72	7.60	7.52	7.39	7.31	7.23	7.14	6.99
7	6.47	6.35	6.27	6.15	6.07	5.98	5.91	5.75
8	5.67	5.56	5.48	5.36	5.28	5.20	5.11	4.90
9	5.11	5.00	4.92	4.80	4.73	4.64	4.56	4.41
10	4.71	4.60	4.52	4.40	4.33	4.25	4.17	4.01
11	4.40	4.29	4.21	4.10	4.02	3.94	3.86	3.70
12	4.16	4.05	3.98	3.86	3.78	3.70	3.61	3.46
13	3.96	3.85	3.78	3.67	3.59	3.51	3.43	3.27
14	3.80	3.70	3.62	3.51	3.43	3.34	3.26	3.11
15	3.67	3.56	3.48	3.36	3.29	3.20	3.12	2.97
16	3.55	3.45	3.37	3.25	3.18	3.10	3.01	2.86
17	3.45	3.35	3.27	3.15	3.08	3.00	2.92	2.76
18	3.37	3.27	3.19	3.07	3.00	2.92	2.83	2.68
19	3.30	3.19	3.12	3.00	2.92	2.84	2.76	2.60
20	3.23	3.13	3.05	2.94	2.86	2.77	2.69	2.53
21	3.17	3.07	2.99	2.88	2.80	2.72	2.63	2.47
22	3.12	3.02	2.94	2.83	2.75	2.67	2.58	2.42
23	3.07	2.97	2.89	2.78	2.70	2.62	2.53	2.37
24	3.03	2.93	2.85	2.74	2.66	2.58	2.49	2.33
25	2.99	2.89	2.81	2.70	2.62	2.54	2.45	2.29
26	2.96	2.86	2.77	2.66	2.58	2.50	2.41	2.25
27	2.93	2.83	2.74	2.63	2.55	2.47	2.38	2.21
28	2.90	2.80	2.71	2.60	2.52	2.44	2.35	2.18
29	2.87	2.77	2.68	2.57	2.49	2.41	2.32	2.15
30	2.84	2.74	2.66	2.55	2.46	2.38	2.29	2.13
40	2.66	2.56	2.49	2.37	2.29	2.20	2.11	1.97
60	2.50	2.40	2.32	2.20	2.12	2.03	1.93	1.74
80	2.41	2.32	2.24	2.11	2.03	1.94	1.84	1.65
100	2.36	2.26	2.19	2.06	1.98	1.89	1.79	1.59
120	2.34	2.24	2.16	2.03	1.95	1.86	1.76	1.56
∞	2.18	2.07	1.99	1.87	1.79	1.70	1.59	

TABLE 8 TRANSFORMATION OF r TO Z AND ρ TO μ_Z VALUES

r	Z	r	Z	r	Z	r	Z
.00	.000	.25	.255	.50	.549	.75	.973
.01	.010	.26	.266	.51	.563	.76	.996
.02	.020	.27	.277	.52	.576	.77	1.020
.03	.030	.28	.288	.53	.590	.78	1.045
.04	.040	.29	.299	.54	.604	.79	1.071
.05	.050	.30	.310	.55	.618	.80	1.099
.06	.060	.31	.321	.56	.633	.81	1.127
.07	.070	.32	.332	.57	.648	.82	1.157
.08	.080	.33	.343	.58	.663	.83	1.188
.09	.090	.34	.354	.59	.678	.84	1.221
.10	.100	.35	.365	.60	.693	.85	1.256
.11	.110	.36	.377	.61	.709	.86	1.293
.12	.121	.37	.388	.62	.725	.87	1.333
.13	.131	.38	.400	.63	.741	.88	1.376
.14	.141	.39	.412	.64	.758	.89	1.422
.15	.151	.40	.424	.65	.775	.90	1.472
.16	.161	.41	.436	.66	.793	.91	1.528
.17	.172	.42	.448	.67	.811	.92	1.589
.18	.182	.43	.460	.68	.829	.93	1.658
.19	.192	.44	.472	.69	.848	.94	1.738
.20	.203	.45	.485	.70	.867	.95	1.832
.21	.213	.46	.497	.71	.887	.96	1.946
.22	.224	.47	.510	.72	.908	.97	2.092
.23	.234	.48	.523	.73	.929	.98	2.298
.24	.245	.49	.536	.74	.950	.99	2.647

TABLE 9   DISTRIBUTION OF U IN THE MANN-WHITNEY TEST
$\alpha = .05$

Any value of U which is equal to or smaller than that shown is significant at the .05 level for a two-tail test.

n of larger sample	2	3	4	5	6	7	8	9	10	11	12	13	14	15	16	17	18	19	20
4	-	-	0																
5	-	0	1	2															
6	-	1	2	3	5														
7	-	1	3	5	6	8													
8	0	2	4	6	8	10	13												
9	0	2	4	7	10	12	15	17											
10	0	3	5	8	11	14	17	20	23										
11	0	3	6	9	13	16	19	23	26	30									
12	1	4	7	11	14	18	22	26	29	33	37								
13	1	4	8	12	16	20	24	28	33	37	41	45							
14	1	5	9	13	17	22	26	31	36	40	45	50	55						
15	1	5	10	14	19	24	29	34	39	44	49	54	59	64					
16	1	6	11	15	21	26	31	37	42	47	53	59	64	70	75				
17	2	6	11	17	22	28	34	39	45	51	57	63	69	75	81	87			
18	2	7	12	18	24	30	36	42	48	55	61	67	74	80	86	93	99		
19	2	7	13	19	25	32	38	45	52	58	65	72	78	85	92	99	106	113	
20	2	8	14	20	27	34	41	48	55	62	69	76	83	90	98	105	112	119	127
21	3	8	15	22	29	36	43	50	58	65	73	80	88	96	103	111	119	126	134
22	3	9	16	23	30	38	45	53	61	69	77	85	93	101	109	117	125	133	141
23	3	9	17	24	32	40	48	56	64	73	81	89	98	106	115	123	132	140	149
24	3	10	17	25	33	42	50	59	67	76	85	94	102	111	120	129	138	147	156
25	3	10	18	27	35	44	53	62	71	80	89	98	107	117	126	135	145	154	163
26	4	11	19	28	37	46	55	64	74	83	93	102	112	122	132	141	151	161	171
27	4	11	20	29	38	48	57	67	77	87	97	107	117	127	137	147	158	168	178
28	4	12	21	30	40	50	60	70	80	90	101	111	122	132	143	154	164	175	186
29	4	13	22	32	42	52	62	73	83	94	105	116	127	138	149	160	171	182	193
30	5	13	23	33	43	54	65	76	87	98	109	120	131	143	154	166	177	189	200
31	5	14	24	34	45	56	67	78	90	101	113	125	136	148	160	172	184	196	208
32	5	14	24	35	46	58	69	81	93	105	117	129	141	153	166	178	190	203	215
33	5	15	25	37	48	60	72	84	96	108	121	133	146	159	171	184	197	210	222
34	5	15	26	38	50	62	74	87	99	112	125	138	151	164	177	190	203	217	230
35	6	16	27	39	51	64	77	90	103	116	129	142	156	169	183	196	210	224	237
36	6	16	28	40	53	66	79	92	106	119	133	147	161	174	188	202	216	231	245
37	6	17	29	41	55	68	81	95	109	123	137	151	165	180	194	209	223	238	252
38	6	17	30	43	56	70	84	98	112	127	141	156	170	185	200	215	230	245	259
39	7	18	31	44	58	72	86	101	115	130	145	160	175	190	206	221	236	252	267
40	7	18	31	45	59	74	89	103	119	134	149	165	180	196	211	227	243	258	274

INDEX